Partial Differential Equations

Methods, Applications and Theories

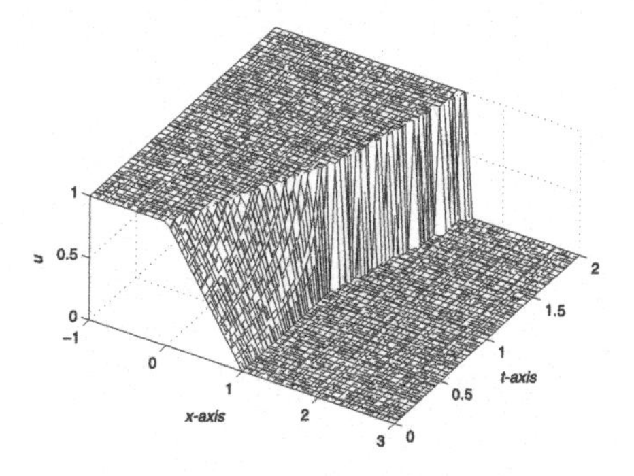

Partial Differential Equations

Methods, Applications and Theories

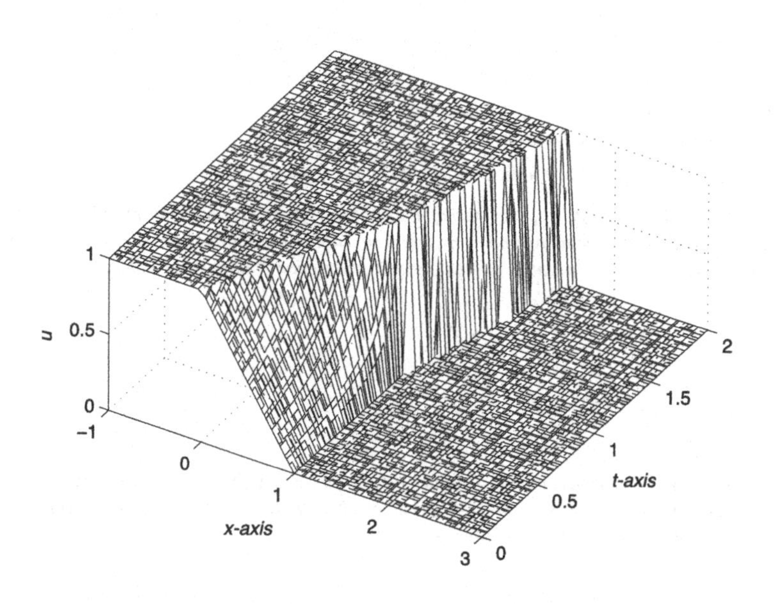

Harumi Hattori

West Virginia University, USA

NEW JERSEY • LONDON • SINGAPORE • BEIJING • SHANGHAI • HONG KONG • TAIPEI • CHENNAI

Published by

World Scientific Publishing Co. Pte. Ltd.
5 Toh Tuck Link, Singapore 596224
USA office: 27 Warren Street, Suite 401-402, Hackensack, NJ 07601
UK office: 57 Shelton Street, Covent Garden, London WC2H 9HE

British Library Cataloguing-in-Publication Data
A catalogue record for this book is available from the British Library.

PARTIAL DIFFERENTIAL EQUATIONS
Methods, Applications and Theories

ISBN 978-981-4407-56-4

Printed in Singapore by World Scientific Printers.

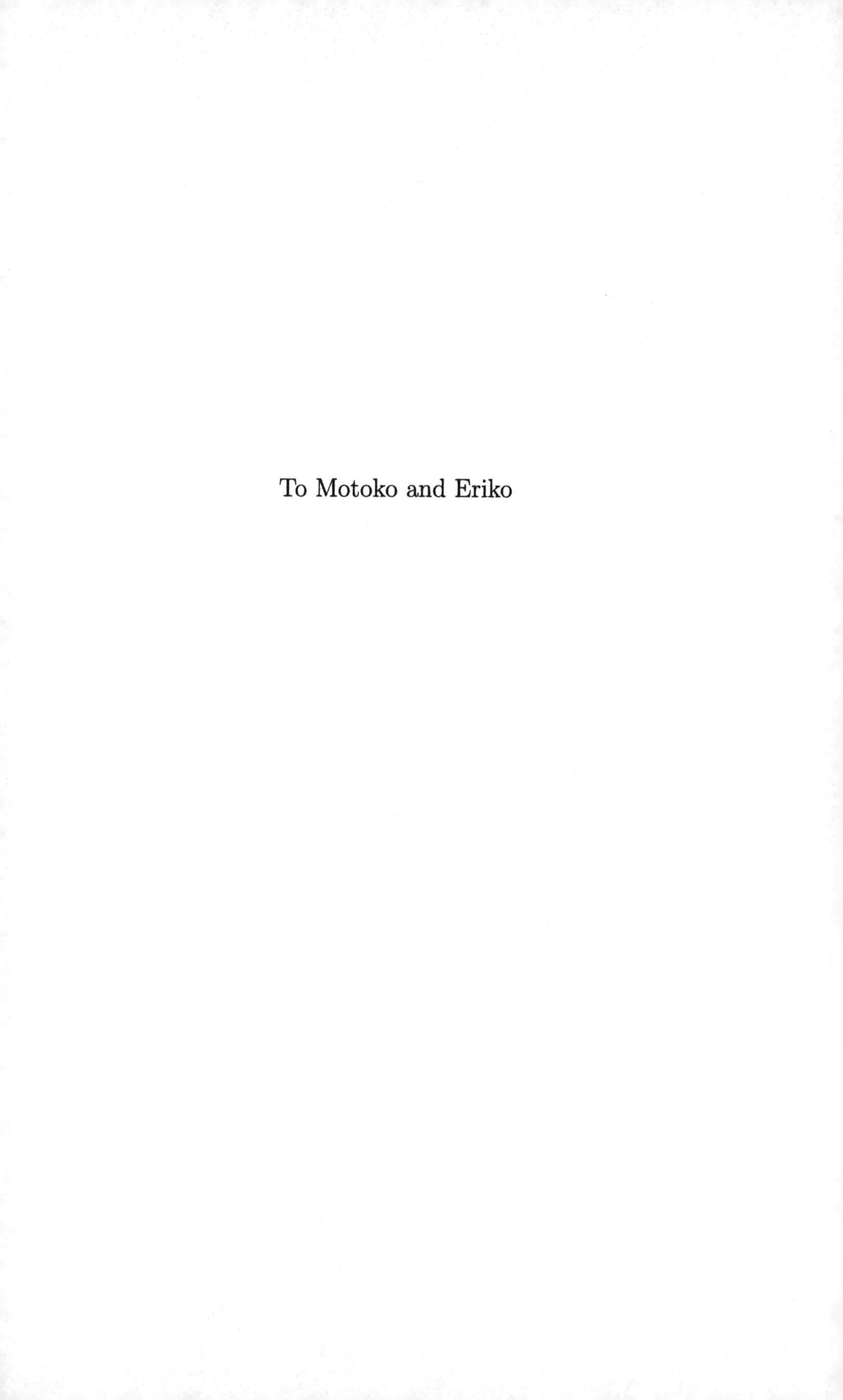

To Motoko and Eriko

Preface

Partial differential equations (PDE's) are equations containing partial derivatives of unknowns. This contrasts with ordinary differential equations (ODE's) where only ordinary derivatives of unknowns appear in the equations. PDE's are very diverse fields. There are many different PDE's as they arise in many areas in our effort to model physical to social phenomena. Different PDE's may require different methods of solutions. Also, the questions we deal with are diverse. They range from basic questions such as existence and uniqueness of solutions to practical questions such as the methods of solutions and approximations. Yet, there are a few important equations which appear often and the methods of solutions used recurrently. Also, there are properties or qualitative behaviors which characterize these important equations. In this book we learn basic equations, methods, and properties that are important and useful in studying more advanced topics or subjects related to PDE's in various disciplines.

There are three main goals in the PDE course. They are

1. To become familiar with the basic equations and problems.
2. To become familiar with the methods of solutions used.
3. To become familiar with the modeling aspects (or the derivation).

As far as the first goal is concerned, there are several equations and problems we need to be familiar with. They are

1. The first order equations
2. Three fundamental second order equations, heat equation, wave equation, and Laplace equation.
3. The initial value problems, the initial boundary value problems, and the boundary value problems.

The first five chapters are written with these goals in mind. The important equations and basic problems mentioned above are covered. The derivations of these equations are discussed as well. Also, the important methods are introduced. Chapter 6 and the later chapters are more focused on the methods of solutions.

Chapters and Sections can be arranged by the methods of solutions or by the types of equations. From Chapters 1 to 5, we organize chapters by equations and classify the equations by order. By the order of equation we mean the order of highest order derivatives in the equation. Chapter 1 is a preparatory chapter where we learn the basic definitions such as order, and linear and nonlinear equations. We study the first order linear equations and learn the method of characteristics. We also classify the second order linear equations using the characteristics and briefly discuss the concept of well-posedness. In Chapters 2 to 4 we learn three fundamental second order linear equations. They are the heat equation, the wave equation, and the Laplace equation. We learn the derivation of equations, how we pose problems, and we discuss basic methods such as the separation of variables and method of characteristics. We learn that the three fundamental second order equations are examples of parabolic, hyperbolic, and elliptic equations. Also at the end of Chapter 4 we give two examples related to one of the well-posedness issues. In Chapter 5 we revisit the first order equations and study nonlinear equations. We discuss two important applications, *i.e.*, the conservation laws and the Hamilton-Jacobi-Bellman (HJB) equations. First order equations are split into two chapters. This way students can learn the method of characteristics used in Chapter 3 an early stage and postpone harder nonlinear first order equations later in the course.

Chapters 6 to 12 are organized mainly by methods or properties. We study the methods discussed or mentioned in earlier chapters in more details. In Chapter 6 we study the Fourier series used in earlier chapters. We study convergence issue. We also study the theoretical aspects of the eigenvalues and eigenfunctions. In Chapter 7 we extend the separation of variables and the eigenvalue problems to higher dimensional cases in the rectangular coordinates. We also study the eigenfunction expansions to treat the non-homogeneous problems. In Chapter 8 we discuss how the separation of variables are carried out in more complex geometries such as cylinder and sphere. As a consequence of non-rectangular coordinates, we learn, among other things, the Bessel equations and the Legendre equations for the eigenvalue problems. In Chapter 9 we study the Fourier transform and distributions. We introduce the delta function in both classical and modern ways. We apply the method to the problems which involves infinite domain. The Laplace transform is discussed in Chapter 10. In Chapter 11 we study the higher dimensional problems which are not based on the separation of variables. We learn the method of spherical means and method of descent for the wave equation, and also discuss the Duhamel's principle for the heat and wave equations. In Chapter 12 we study the Green's functions for the time dependent problems. We (re)define the Green's functions using the delta function and find the explicit forms of the Green's functions for the second order equations.

This book can be used in a one-semester or in a one-year course. For a one-semester course most of Chapters 1 to 6 could be covered. A couple of possible organization of the course are listed as follows.

1. Chapter 1: 1.1-1.3, (1.4), Chapter 2: (2.1), 2.2-2.4, (2.5), 2.6, (2.7), Chapter 3: (3.1), 3.2-3.4, (3.5), Chapter 4: (4.1), 4.2-4.4, (4.5), Chapter 5: 5.1-5.4, Chapter 6: 6.1-6.3, (6.4), 6.5
2. Chapter 1: 1.1, 1.2, Chapter 2: (2.1), 2.2-2.4, (2.5), 2.6, (2.7), Chapter 3: (3.1), 3.2-3.4, (3.5), Chapter 4: (4.1), 4.2-4.4, (4.5), Chapter 1: (1.4), 1.3, Chapter 5: 5.1-5.4, Chapter 6: 6.1-6.3, (6.4), 6.5

For a one-year course some (or most) of Chapters 7 to 12 could be covered besides Chapters 1 to 6. In Chapter 12 we use the Duhamel's principle from Chapter 11, but it is not essential for Chapter 12. It is recommended that Chapters 7, 9, 12 are covered in this order. On the other hand, Chapters 8, 10, 11 are independent and could be covered in any order. Possible orderings of chapters are listed as follows.

1. Chapter 7: 7.1, (7.2), 7.3, Chapter 8, Chapter 9, Chapter 10, Chapter 11, Chapter 12
2. Chapter 7: 7.1, (7.2), 7.3, Chapter 9, Chapter 11, Chapter 12, Chapter 8, Chapter 10

Finally, I would like to express my appreciation to the people in World Scientific Publishing Co. Without their help this book would not have been completed.

Harumi Hattori

Contents

Chapter 1

First and Second Order Linear Equations - Preparation

This is a preparatory chapter where we learn basic notions used throughout the book, *i.e.*, the order of equations, the definition of solutions, and linear and nonlinear equations. The linearity plays an important role in PDE's, especially, the superposition principle. We study them in Section 1.2. In this book most PDE's we study are either the first or the second order. In this chapter we introduce basic aspects of these equations. For the first order linear equations we learn a method called the method of characteristics in Section 1.3. This method is used in Section 1.4 and later chapters. In this chapter we study linear equations only. Nonlinear equations are more difficult and studied in Chapter 5. For the second order equations, the classification of the second order linear equations are treated in Section 1.4. We also introduce the concept of well-posedness in Section 1.5.

1.1 Terminologies

The first order equations mean the PDE's where the highest order derivatives are the first order. For example, if u is a function of (x, t),

$$u_t + u_x = u$$

is a first order equation for u and we try to find u satisfying the equation. If such a u exists, it is called a solution. For example $u = e^t$ is a solution. We can verify it by computing $u_t = e^t$, $u_x = 0$ and substituting them in the equation.

Suppose that $x_1, \ldots, x_n$ are the independent variables, u is unknown, and $u_{x_i} (i = 1, 2, \ldots, n)$ are partial derivatives of u with respect to x_i. Then, the general form of the first order equations is written as

$$f(x_1, \ldots, x_n, u, u_{x_1}, u_{x_2}, \ldots, u_{x_n}) = 0, \tag{1.1.1}$$

where f is a scalar function of the arguments. A goal is to find u and to express u in terms of $x_1, \ldots, x_n$. We say that u is a classical solution to (1.1.1) if

1. u is continuously differentiable (denoted as C^1), *i.e*, all first derivatives u_{x_i}, $i = 1, \ldots, n$, exist and are continuous.
2. u satisfies (1.1.1).

Solutions satisfying the above conditions are also referred to as C^1-solutions.

If the highest order derivatives in an equation is the second order, it is called the second order equation and the general form of the second order equations is written as

$$f(x_1, \ldots, x_n, u, u_{x_1}, u_{x_2}, \ldots, u_{x_n}, u_{x_1 x_1}, u_{x_1 x_2}, \ldots, u_{x_n x_n}) = 0, \qquad (1.1.2)$$

where $u_{x_i x_j}$ $(i, j = 1, 2, \ldots, n)$ are the second order partial derivatives of u with respect to x_i and x_j. For example, the heat equation $u_t = u_{xx}$ we study in Chapter 2 is a second order equation. For (1.1.2) a classical solution is a C^2-solution satisfying

1. u is twice continuously differentiable (denoted as C^2), *i.e*, all first and second derivatives $u_{x_i x_j}$, $i, j = 1, \ldots, n$, exist and
2. u satisfies (1.1.2).

Exercises

1. Find the order of the following PDE's.

   (a) $u_{tt} - u_{xx} = -u_t$
   (b) $u_t - u_x = u_{xxx}$
   (c) $u_{xx} + u_{yy} = 0$
   (d) $u_{tt} + u_{xxxx} = 0$

2. (a) Verify that $u = e^t f(x - ct)$ is a solution to $u_t + cu_x = u$, where f is an arbitrary differentiable function and c is a constant.

   (b) Verify that $u = x^2/2 + a^2 t$ is a solution to the heat equation $u_t = a^2 u_{xx}$, where a is a positive constant.

1.2 Linearity

1.2.1 *Superposition Principle*

The linearity plays an important role in PDE's. Various usage will appear throughout the book. In this section we introduce the concept and give examples.

We say that $\mathcal{L}$ is a linear operator if it satisfies

$$\mathcal{L}(u_1 + u_2) = \mathcal{L}u_1 + \mathcal{L}u_2 \quad \text{and} \quad \mathcal{L}(cu) = c\mathcal{L}u, \tag{1.2.1}$$

where c is a constant. The above two relations are combined into one and we also say that $\mathcal{L}$ is a linear operator if it satisfies

$$\mathcal{L}(c_1u_1 + c_2u_2) = c_1\mathcal{L}u_1 + c_2\mathcal{L}u_2 \tag{1.2.2}$$

for any two functions u_1 and u_2, and any constants c_1 and c_2. For example, if we define $\mathcal{L}$ by

$$\mathcal{L} = a(x,t)\frac{\partial}{\partial t} + b(x,t)\frac{\partial}{\partial x} + d(x,t),$$

then, $\mathcal{L}u$ is given by

$$\mathcal{L}u = a(x,t)u_t + b(x,t)u_x + d(x,t)u.$$

Another example is the heat equation. If we define $\mathcal{L}$ by

$$\mathcal{L} = \frac{\partial}{\partial t} - a^2\frac{\partial^2}{\partial x^2},$$

$\mathcal{L}u$ is given by

$$\mathcal{L}u = u_t - a^2u_{xx}.$$

We can interpret that $\mathcal{L}$ operates on u to get $\mathcal{L}u$.

The question is how to show that the $\mathcal{L}$'s defined above are linear operators. This is explained in the following examples.

Example 1.2.1. Define $\mathcal{L} = a(x,t)\partial/\partial t + b(x,t)\partial/\partial x + d(x,t)$. Show that $\mathcal{L}$ is a linear operator.

Solution: We show that (1.2.2) holds.

$$\begin{aligned}
&\mathcal{L}(c_1u_1 + c_2u_2) \\
&= a(x,t)\frac{\partial}{\partial t}(c_1u_1 + c_2u_2) + b(x,t)\frac{\partial}{\partial x}(c_1u_1 + c_2u_2) + d(x,t)(c_1u_1 + c_2u_2) \\
&= c_1\{a(x,t)\frac{\partial u_1}{\partial t} + b(x,t)\frac{\partial u_1}{\partial x} + d(x,t)u_1\} \\
&\quad + c_2\{a(x,t)\frac{\partial u_2}{\partial t} + b(x,t)\frac{\partial u_2}{\partial x} + d(x,t)u_2\} \\
&= c_1\mathcal{L}u_1 + c_2\mathcal{L}u_2.
\end{aligned}$$

Example 1.2.2. Define $\mathcal{L} = \partial/\partial t - a^2\partial^2/\partial x^2$. Show that $\mathcal{L}$ is a linear operator.

Solution: We show that (1.2.2) holds.

$$\begin{aligned}\mathcal{L}(c_1u_1 + c_2u_2) &= \frac{\partial}{\partial t}(c_1u_1 + c_2u_2) - a^2\frac{\partial^2}{\partial x^2}(c_1u_1 + c_2u_2)\\ &= c_1(\frac{\partial}{\partial t}u_1 - a^2\frac{\partial^2}{\partial x^2}u_1) + c_2(\frac{\partial}{\partial t}u_2 - a^2\frac{\partial^2}{\partial x^2}u_2)\\ &= c_1(\frac{\partial}{\partial t} - a^2\frac{\partial^2}{\partial x^2})u_1 + c_2(\frac{\partial}{\partial t} - a^2\frac{\partial^2}{\partial x^2})u_2\\ &= c_1\mathcal{L}u_1 + c_2\mathcal{L}u_2.\end{aligned}$$

An equation of the form

$$\mathcal{L}u = g,$$

where $\mathcal{L}$ is a linear operator and g does not depend on u, is called a linear equation. If g is identically equal to zero, the equation is called homogeneous and otherwise it is called non-homogeneous. An important characteristic of linear equations is called the superposition principle and stated as follows.

Theorem 1.2.3. ***(Superposition Principle)*** *If u_1 and u_2 are solutions to a linear homogeneous equation, then the linear combination $c_1u_1 + c_2u_2$ is also a solution to the same linear homogeneous equation for arbitrary values of constants c_1 and c_2.*

Remark 1.2.4. Usually the superposition principle is referred to the equation only. However, if the boundary conditions are linear and homogeneous, the principle holds for a linear equation with the linear homogeneous boundary conditions.

A partial differential equation is linear if the corresponding homogeneous equation satisfies Theorem 1.2.3 or if we can define a linear operator from the equation. The following example shows how we can examine if a given equation is linear.

Example 1.2.5. Show that the following two dimensional heat equation is linear.

$$u_t - a^2(u_{xx} + u_{yy}) = g(x,t). \tag{1.2.3}$$

Solution: Note that the equation is non-homogeneous. The corresponding homogeneous equation is

$$u_t - a^2(u_{xx} + u_{yy}) = 0. \tag{1.2.4}$$

Therefore, we show that if u_1 and u_2 are solutions to (1.2.4), $u = c_1u_1+c_2u_2$ is also a solution to the same equation. For this purpose assume that u_1 and u_2 are solutions to (1.2.4), *i.e.*,

$$u_{1t} - a^2(u_{1xx} + u_{1yy}) = 0 \quad \text{and} \quad u_{2t} - a^2(u_{2xx} + u_{2yy}) = 0.$$

Then, $u = c_1u_1 + c_2u_2$ satisfies

$$\begin{aligned} &u_t - a^2(u_{xx} + u_{yy}) \\ &= (c_1u_1 + c_2u_2)_t - a^2[(c_1u_1 + c_2u_2)_{xx} + (c_1u_1 + c_2u_2)_{yy}] \\ &= c_1[u_{1t} - a^2(u_{1xx} + u_{1yy})] + c_2[u_{2t} - a^2(u_{2xx} + u_{2yy})] \\ &= 0. \end{aligned} \tag{1.2.5}$$

This shows that $c_1u_1 + c_2u_2$ is also a solution to the same heat equation (1.2.4). Therefore, (1.2.3) is linear.

Alternatively, we can define an operator $\mathcal{L} = \partial/\partial t - a^2(\partial^2/\partial x^2 + \partial^2/\partial y^2)$ and show that this satisfies (1.2.2). Since $\mathcal{L}u$ is given by

$$\mathcal{L}u = u_t - a^2(u_{xx} + u_{yy}),$$

we utilize the part of (1.2.5) to show that

$$\begin{aligned} &\mathcal{L}(c_1u_1 + c_2u_2) \\ &= (c_1u_1 + c_2u_2)_t - a^2[(c_1u_1 + c_2u_2)_{xx} + (c_1u_1 + c_2u_2)_{yy}]. \\ &= c_1[u_{1t} - a^2(u_{1xx} + u_{1yy})] + c_2[u_{2t} - a^2(u_{2xx} + u_{2yy})] \\ &= c_1\mathcal{L}u_1 + c_2\mathcal{L}u_2. \end{aligned}$$

Equations which are not linear are called nonlinear equations. An example of nonlinear equation is

$$u_t = a^2u_x + u^n,$$

where $n \neq 0, 1$. Note that if $n = 0$ or 1, the equation is linear. The term u^n is nonlinear if $n \neq 0, 1$. Another example is

$$u_t + uu_x = u_{xx}.$$

The term uu_x makes the equation nonlinear.

Example 1.2.6. Show that the following equation is nonlinear

$$u_t = a^2u_x + u^2. \tag{1.2.6}$$

Solution: We show that the principle of superposition does not hold. Assume that u_1 and u_2 are solutions to (1.2.6) and see if $u = c_1u_1 + c_2u_2$ is also a solution to the same equation. So, we assume that

$$u_{1t} - a^2u_{1x} - u_1^2 = 0, \quad u_{2t} - a^2u_{2x} - u_2^2 = 0.$$

Then,

$$\begin{aligned}
&u_t - a^2u_x - u^2 \\
&= (c_1u_1 + c_2u_2)_t - a^2(c_1u_1 + c_2u_2)_x - (c_1u_1 + c_2u_2)^2 \\
&= c_1(u_{1t} - a^2u_{1x} - u_1^2) + c_2(u_{2t} - a^2u_{2x} - u_2^2) - 2c_1c_2u_1u_2 \\
&= -2c_1c_2u_1u_2.
\end{aligned}$$

This is not zero in general. Therefore, the superposition principle does not hold.

1.2.2 *Linear Independence*

We learned the linear independence and dependence for vectors. We develop the same concept for a set of functions. For vectors their definitions are given as follows.

Definition 1.2.7. Suppose there are m vectors $\mathbf{v}_1,\ldots,\mathbf{v}_m$ in R^n. We say that they are linearly independent if the relation

$$c_1\mathbf{v}_1 + \cdots + c_m\mathbf{v}_m = \mathbf{0} \tag{1.2.7}$$

holds only when the coefficients $c_1, \ldots, c_m$ satisfy $c_1 = c_2 = \cdots = c_m = 0$. On the other hand, we say that they are linearly dependent if there exist the coefficients, not all zero, such that (1.2.7) hold.

If a set of vectors are linearly dependent, they are redundant. If $m > n$, they are linearly dependent. If $m \leq n$, the following example illustrate how to check the linear independence of the vectors.

Example 1.2.8. Show if the vectors $\mathbf{v}_1 = [1, 2, 3]$, $\mathbf{v}_2 = [1, 2, 1]$, and $\mathbf{v}_3 = [0, 2, 3]$ are linearly independent or dependent.

Solution: We check if there are nontrivial constants c_1, c_2, and c_3 such that (1.2.7) hold. This is equivalent to showing that whether

$$A\mathbf{c} = \begin{bmatrix} 1 & 1 & 0 \\ 2 & 2 & 2 \\ 3 & 1 & 3 \end{bmatrix} \begin{bmatrix} c_1 \\ c_2 \\ c_3 \end{bmatrix} = \mathbf{0}$$

has a nontrivial (nonzero) solution for $\mathbf{c}$. Since $\det A = 4$, by Appendix A.4.1 we see that there is no nontrivial solution for $\mathbf{c}$. Therefore, they are linearly independent.

We have the similar concepts for a set of functions.

Definition 1.2.9. Suppose there are m functions $f_1(\mathbf{x}),\ldots, f_m(\mathbf{x})$ defined on a domain $\Omega \subseteq R^n$. We say that they are linearly independent if the relation

$$c_1 f_1(\mathbf{x}) + \cdots + c_m f_m(\mathbf{x}) = 0 \tag{1.2.8}$$

holds for all $\mathbf{x} \in \Omega$ only when the coefficients $c_1, \ldots, c_m$ satisfy $c_1 = c_2 = \cdots = c_m = 0$. On the other hand, we say that they are linearly dependent if there exist the coefficients, not all zero, such that (1.2.8) hold for all $\mathbf{x} \in \Omega$.

An important difference between vectors and functions is that for the functions the relation (1.2.8) must hold for all $\mathbf{x} \in \Omega$.

Example 1.2.10. Show if the functions $f_1 = 1$, $f_2 = x$, and $f_3 = x^2$ are linearly independent or dependent on the interval $[0, 1]$.

Solution: They look different. So, it is likely that they are linearly independent. We need to show that there are no nontrivial solutions for

$$c_1 f_1(x) + c_2 f_2(x) + c_3 f_3(x) = 0 \tag{1.2.9}$$

for all $x \in [0, 1]$. We choose three different points and show that $c_1 = c_2 = c_3 = 0$. For example, if we choose $x = 0$, $1/2$, and 1, we have

$$A\mathbf{c} = \begin{bmatrix} f_1(0) & f_2(0) & f_3(0) \\ f_1(\frac{1}{2}) & f_2(\frac{1}{2}) & f_3(\frac{1}{2}) \\ f_1(1) & f_2(1) & f_3(1) \end{bmatrix} \begin{bmatrix} c_1 \\ c_2 \\ c_3 \end{bmatrix} = \begin{bmatrix} 1 & 0 & 0 \\ 1 & \frac{1}{2} & \frac{1}{4} \\ 1 & 1 & 1 \end{bmatrix} \begin{bmatrix} c_1 \\ c_2 \\ c_3 \end{bmatrix} = \mathbf{0}.$$

Since $\det A = 1/4$, $c_1 = c_2 = c_3 = 0$. Therefore, they are linearly independent.

Exercises

1. Show whether the following equations are linear or nonlinear. Here, a, b, and c are constants.
   (a) $u_t = a^2 u_{xx} + bu$
   (b) $u_x^2 + u_y^2 = 1$
   (c) $u_t + uu_x = u_{xxx}$
   (d) $u_t = a^2 u_x + bu + x$
   (e) $u_{tt} - c^2 u_{xx} + u^3 = 0$

2. Show that if u_1 and u_2 are solutions to (1.2.10) and (1.2.11), then $u = c_1 u_1 + c_2 u_2$ is also a solution to (1.2.10) and (1.2.11).

$$u_t - a^2 u_{xx} = 0, \quad 0 < x < L, \quad 0 < t, \tag{1.2.10}$$

$$\alpha_1 u_x(0,t) + \beta_1 u(0,t) = 0, \quad \alpha_2 u_x(L,t) + \beta_2 u(L,t) = 0. \tag{1.2.11}$$

3. The initial value problem for the wave equation in one dimension is given by

$$u_{tt} - c^2 u_{xx} = 0, \tag{1.2.12}$$

$$u(x,0) = f(x), \quad u_t(x,0) = g(x). \tag{1.2.13}$$

Suppose that u_p solves the initial value problem

$$u_{tt} - c^2 u_{xx} = 0, \tag{1.2.14}$$

$$u(x,0) = 0, \quad u_t(x,0) = p(x). \tag{1.2.15}$$

(a) Show that $v = \partial u_p / \partial t$ solves the initial value problem

$$v_{tt} - c^2 v_{xx} = 0,$$

$$v(x,0) = p(x), \quad v_t(x,0) = 0.$$

(b) Show that $u = \partial u_f / \partial t + u_g$ is a solution to (1.2.12) and (1.2.13).

4. Determine whether the set of functions are linearly independent or dependent. If linearly dependent, find the relation.

(a) $f_1(x) = e^{3x}, \quad f_2(x) = e^{3(x+1)}, \quad f_3(x) = e^{4x}$
(b) $f_1(x) = \cos 2x, \quad f_2(x) = \sin 2x, \quad f_3(x) = 1$
(c) $f_1(x) = \cos^2 x, \quad f_2(x) = \sin^2 x, \quad f_3(x) = 1$

1.3 First Order Linear Equations

1.3.1 *Initial Value Problems*

The most general form of the first order linear equation is given by

$$a(x,t)u_t + b(x,t)u_x + d(x,t)u = g(x,t). \tag{1.3.1}$$

In the initial value problems we find a function $u = u(x,t)$ to (1.3.1) passing through a given curve C_I in R^3. The curve C_I may be given by a parametric form

$$C_I : \; x = x_0(s), \; t = t_0(s), \; u = u_0(s) \tag{1.3.2}$$

or by a form of function

$$C_I : \; u = u_0(x,0).$$

In the latter case, the corresponding parametric form is

$$C_I : \ x_0 = s, \ t_0 = 0, \ u = u_0(s, 0).$$

In (1.3.2), it is more natural to interpret that the initial data $u = u_0(s)$ is specified along a curve $(x_0, t_0)(s)$ in the xt-plane. The curve C_I is called the initial curve but $(x_0, t_0)(s)$ in the xt-plane may be also called the initial curve. The confusion would be minimal.

The method of characteristics is a very important method and a nice application of the chain rule for multivariable calculus. The essence of the chain rule is the following. Suppose u is a function of (x, t). According to the chain rule, along a curve $x = x(\tau)$ and $t = t(\tau)$ in R^2, where τ is a parameter for the curve, $du/d\tau$ is given by

$$\frac{du}{d\tau} = u_x \frac{dx}{d\tau} + u_t \frac{dt}{d\tau}.$$

So, if we choose

$$\frac{dx}{d\tau} = a(x, t), \quad \frac{dt}{d\tau} = b(x, t), \tag{1.3.3}$$

by (1.3.1) we have

$$\frac{du}{d\tau} = a(x, t)u_x + b(x, t)u_t = -d(x, t)u + g(x, t). \tag{1.3.4}$$

The ODE's in (1.3.3) and (1.3.4) are often called the characteristic system. The solutions to (1.3.3) and (1.3.4) are called the characteristic curves or characteristics. The projections of the characteristics to the xt-plane, *i.e.*, the solutions to (1.3.3), are called the characteristic rays, but often they are also called the characteristics. The following is how the method of characteristics works for the linear problems.

1. First, we parametrize the initial curve C_I by s. To avoid using x or t for the initial curve, this step is important.
2. We solve the ODE's (1.3.3) and (1.3.4) with the initial conditions given along the curve C_I for each s. In the linear problems we can solve (1.3.3) first and express x and t in terms of τ and s. Next, we use these $x = x(s, \tau)$ and $t = t(s, \tau)$ to solve (1.3.4). Then, for each s we have
$$x = x(s, \tau), \quad t = t(s, \tau), \quad u = u(s, \tau).$$
3. We solve $x = x(s, \tau)$ and $t = t(s, \tau)$ for s and τ and use them to express u in terms of x and t.

The question is if the third step is possible. The following theorem gives a sufficient condition.

Theorem 1.3.1. *If the condition*

$$J = \begin{vmatrix} x_s(s,0) & x_\tau(s,0) \\ t_s(s,0) & t_\tau(s,0) \end{vmatrix} = \begin{vmatrix} x_0'(s) & a(x_0(s),t_0(s)) \\ t_0'(s) & b(x_0(s),t_0(s)) \end{vmatrix} \neq 0 \qquad (1.3.5)$$

is satisfied along the initial curve C_I, there is an interval of τ containing $\tau = 0$, in which we can solve $x = x(s,\tau)$ and $t = t(s,\tau)$ for s and τ.

Proof. The vector $[x_0'(s), t_0'(s)]^T$, where the superscript T is the transpose, is parallel to the x and t components of the initial curve. Similarly, the vector $[a(x_0(s),t_0(s)), b(x_0(s),t_0(s))]^T$ is parallel to the curve defined by (1.3.3) at $\tau = 0$. The condition $J \neq 0$ means that two curves $x = x(s,\tau)$ and $t = t(s,\tau)$ intersect along the initial curve C_I, *i.e.*, they are not parallel along C_I. If they are parallel, the solution may not be able to leave the plane $t = 0$. □

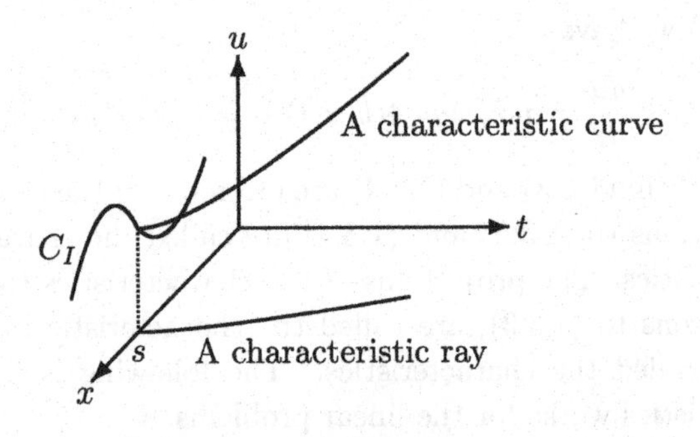

Fig. 1.1 The characteristics and its projection to the xt-plane.

This is basically the method of characteristics. In the following, examples illustrating the method are given. We start from a constant coefficient case.

Example 1.3.2. Find the solution to

$$u_t + cu_x = 0, \quad -\infty < x < \infty,\ 0 < t, \qquad (1.3.6)$$

$$u(x,0) = f(x), \qquad (1.3.7)$$

where c is a constant.

Solution: Step 1: We parametrize the initial curve C_I as follows. This is to avoid using x for the initial data and for the solution.

$$C_I: \ x_0 = s, \ t_0 = 0, \ u_0 = f(s).$$

Step 2: We derive the set of ODE's by the method of characteristics and solve them. For each s, following (1.3.3) and (1.3.4), we solve the system of ordinary differential equations

$$\frac{dx}{d\tau} = c, \ \frac{dt}{d\tau} = 1,$$

$$\frac{du}{d\tau} = u_x \frac{dx}{d\tau} + u_t \frac{dt}{d\tau} = u_t + cu_x = 0$$

with the initial conditions

$$x(0) = x_0 = s, \ t(0) = t_0 = 0, \ u(0) = u_0 = f(s).$$

These equations can be solved separately and we have

$$x(s, \tau) = c\tau + x_0 = c\tau + s, \tag{1.3.8}$$

$$t(s, \tau) = \tau + t_0 = \tau, \tag{1.3.9}$$

$$u(s, \tau) = u_0 = f(s). \tag{1.3.10}$$

Step 3: We solve (1.3.8) and (1.3.9) for (x, t) and express u in terms of (x, t). Substituting $\tau = t$ in $x = c\tau + s$, we have $s = x - ct$ as a characteristic ray. Therefore, substituting this in (1.3.10), we obtain

$$u = u(x, t) = f(x - ct). \tag{1.3.11}$$

If we identify t as time, we see that the initial data (or profile) $u(x, 0) = f(x)$ propagates at the velocity (speed) c. For this reason, Equation (1.3.6) is called the transport equation and c is called the wave speed.

The next example is a continuation of Example 1.3.2 with a specific $f(x)$.

Example 1.3.3. Find the solution.

$$u_t + cu_x = 0, \quad -\infty < x < \infty, \ 0 < t, \tag{1.3.12}$$

$$f(x) = \begin{cases} 1, & x \leq 0, \\ 1 - x, & 0 < x \leq 1, \\ 0, & 1 < x. \end{cases} \tag{1.3.13}$$

Solution: The solution to (1.3.12) and (1.3.13) is given by (1.3.11). The simplest way is to replace x in (1.3.13) by $x - ct$. Then,

$$u(x,t) = f(x-ct) = \begin{cases} 1, & x - ct \leq 0, \\ 1 - x + ct, & 0 < x - ct \leq 1, \\ 0, & 1 < x - ct. \end{cases}$$

Rewriting the intervals, we have

$$u(x,t) = \begin{cases} 1, & x \leq ct, \\ 1 - x + ct, & ct < x \leq 1 + ct, \\ 0, & 1 + ct < x. \end{cases}$$

For $c = 1/2$ the profile of the solution for $0 \leq t \leq 2$ is depicted in Figure 1.2.

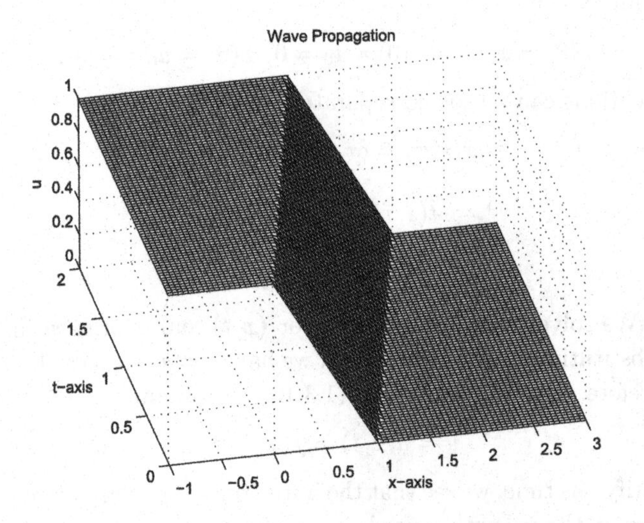

Fig. 1.2 Wave propagation with $c = 1/2$.

The following is the example where the condition (1.3.5) is not satisfied.

Example 1.3.4. Explain why we have a problem in solving the following initial value problem.

$$u_t + cu_x = 0, \quad -\infty < x < \infty, \ 0 < t,$$

$$u(cx, x) = f(x).$$

Solution: Step 1: First, we parametrize the initial curve C_I as follows.

$$C_I: \; x_0 = cs, \; t_0 = s, \; u_0 = f(s).$$

Step 2: Second, for each s, we solve the system of ODE's

$$\frac{dx}{d\tau} = c, \; \frac{dt}{d\tau} = 1, \; \frac{du}{d\tau} = 0$$

with the initial conditions

$$x(0) = x_0 = cs, \; t(0) = t_0 = s, \; u(0) = u_0 = f(s).$$

Solve them, we have

$$x(s,\tau) = c\tau + x_0 = c\tau + cs,$$

$$t(s,\tau) = \tau + t_0 = \tau + s,$$

$$u(s,\tau) = u_0 = f(s).$$

Step 3: Now, we see that we cannot express u using (x,t). The reason is that the direction of characteristics and the direction of initial curve are the same. It is easy to see that the condition (1.3.5) is violated.

$$\begin{vmatrix} x_0'(s) & a(x_0(s), t_0(s)) \\ t_0'(s) & b(x_0(s), t_0(s)) \end{vmatrix} = \begin{vmatrix} c & c \\ 1 & 1 \end{vmatrix} = 0.$$

Example 1.3.5. Solve the initial value problem

$$u_t + tu_x = x, \quad -\infty < x < \infty, \; 0 < t, \tag{1.3.14}$$

$$u(x,0) = x^2. \tag{1.3.15}$$

Solution: **Step 1**: First, the initial condition (1.3.15) is parametrized as follows.

$$C_I: \; x_0 = s, \; t_0 = 0, \; u_0 = s^2.$$

Step 2: Second, for each s we solve the system of ODE's

$$\frac{dx}{d\tau} = t, \; \frac{dt}{d\tau} = 1, \; \frac{du}{d\tau} = x$$

with the initial data

$$x(0) = x_0 = s, \; t(0) = t_0 = 0, \; u(0) = u_0 = s^2.$$

Examining the three equations, we notice that we need to solve $dt/d\tau = 1$ first. Then,

$$t(s,\tau) = \tau. \tag{1.3.16}$$

Now we can solve $dx/d\tau = t$ using (1.3.16). Then,

$$x(s,\tau) = \frac{1}{2}\tau^2 + x_0 = \frac{1}{2}\tau^2 + s. \tag{1.3.17}$$

Next, solving $du/d\tau = x$, we have

$$u(s,\tau) = \frac{1}{6}\tau^3 + s\tau + s^2. \tag{1.3.18}$$

Step 3: Since we want to express u in terms of x and t, we solve (1.3.16) and (1.3.17) for s and τ and substitute them in (1.3.18). Then, we obtain

$$u(x,t) = \frac{1}{6}t^3 + (x - \frac{1}{2}t^2)t + (x - \frac{1}{2}t^2)^2. \tag{1.3.19}$$

Let's check if this is the solution. Since at $t = 0$, $u(x,0) = x^2$, u in (1.3.19) satisfies the initial condition. Next, computing u_x and u_t, we see

$$u_x = t + 2(x - \frac{1}{2}t^2),\ u_t = \frac{1}{2}t^2 + x - \frac{3}{2}t^2 + 2(x - \frac{1}{2}t^2)(-t),$$

$$u_t + tu_x = t^2 + 2(x - \frac{1}{2}t^2)t + x - t^2 + 2(x - \frac{1}{2}t^2)(-t) = x.$$

So, it is the solution. The condition (1.3.5) is easily checked and

$$J = \begin{vmatrix} x_0'(s) & t_0'(s) \\ a(x_0(s),t_0(s)) & b(x_0(s),t_0(s)) \end{vmatrix} = \begin{vmatrix} 1 & 0 \\ 0 & 1 \end{vmatrix} \neq 0.$$

1.3.2 *General Solutions*

We started with the initial value problems for the first order linear equations. If we consider the solutions to PDE's without any further conditions, the solutions may be called the general solutions. As an example consider the equation

$$u_t + cu_x = 0.$$

The method of characteristics leads to

$$\frac{dt}{d\tau} = 1 \Rightarrow t = \tau + t_0, \tag{1.3.20}$$

$$\frac{dx}{d\tau} = x \Rightarrow x = x\tau + x_0, \tag{1.3.21}$$

$$\frac{du}{d\tau} = u_t + cu_x = 0 \Rightarrow u = u_0(x_0, t_0). \tag{1.3.22}$$

Eliminating τ from (1.3.20) and (1.3.21), we have the characteristics

$$(x - x_0) = c(t - t_0).$$

For simplicity assume $c \neq 0$. Then, this relation yields

$$u(x,t) = u_0(x_0, t_0) = u_0(x - c(t - t_0), t_0).$$

Since u_0 is an arbitrary function of the arguments, we may set

$$u(x,t) = f(x - ct).$$

We observe that the solutions are arbitrary function of characteristics. In ODE's the general solutions contain arbitrary constant(s). On the other hands in PDE's solutions are arbitrary functions of characteristics.

Example 1.3.6. Find the general solution.

$$\frac{1}{x}u_t + 2txu_x = 0. \tag{1.3.23}$$

Solution: If we use the method of characteristics, we have

$$\frac{dt}{d\tau} = \frac{1}{x}, \quad \frac{dx}{d\tau} = 2tx, \quad \frac{du}{d\tau} = \frac{1}{x}u_t + 2txu_x = 0.$$

This is difficult to solve. Either we multiply (1.3.23) by x or consider

$$\frac{dx}{dt} = \frac{\frac{dx}{d\tau}}{\frac{dt}{d\tau}} = 2tx^2. \tag{1.3.24}$$

This is a separable equation and the solution to (1.3.24) is given by

$$\frac{dx}{x^2} = 2tdt \Rightarrow -\frac{1}{x} + \frac{1}{x_0} = t^2 - t_0^2 \Rightarrow \frac{1}{x} + t^2 = \frac{1}{x_0} + t_0^2. \tag{1.3.25}$$

Equation (1.3.23) says that u is constant along the characteristics given by (1.3.25). Therefore,

$$u(x,t) = u_0(x_0, t_0). \tag{1.3.26}$$

Eliminating x_0 or t_0 from (1.3.26) using (1.3.25), we find that $u(x,t)$ is a function of $\frac{1}{x} + t^2$, *i.e.*,

$$u(x,t) = f(\frac{1}{x} + t^2), \tag{1.3.27}$$

where f is an arbitrary function.

Exercises

1. Solve the initial value problems in $-\infty < x < \infty$, $0 < t$.

   (a) $u_t - 2u_x = 0, \quad u(x,0) = \sin 2x$
   (b) $u_t + 2u_x = 0, \quad u(0,t) = \cos 3t$
   (c) $u_t + 2u_x = 0, \quad u(2t,t) = \cos 3t$

2. Verify that (1.3.27) satisfies (1.3.23).
3. Consider the initial value problem for the equation

$$u_t - u_x = 0, \quad -\infty < x < \infty,\ 0 < t,$$

   with the initial data

$$u(x,0) = f(x) = \begin{cases} 1, & x \leq 0, \\ 1 - x, & 0 < x \leq 1, \\ 0, & 1 < x. \end{cases}$$

   (a) Sketch the (projection of) characteristics on the xy-plane.
   (b) Find the solution.
   (c) Sketch $u(x,t)$ at $t = 1$, 2.

4. Solve the initial value problems in $-\infty < x < \infty$, $0 < t$.

   (a) $u_t + xu_x = 1, \quad u(x,0) = \cos 3x$
   (b) $u_t + tu_x = u, \quad u(x,0) = \cos 3x$
   (c) $2u_t + tu_x = u, \quad u(0,t) = t^2$

5. Consider

$$tu_t + xu_x = \alpha u, \quad -\infty < x < \infty,\ -\infty < t < \infty,$$

   where α is a nonzero constant.

   (a) Find the solution if the initial data is given by

$$u(x,1) = f(x).$$

   (b) Find the solution if the initial data is given by

$$u(x,\beta x) = f(x),$$

   where β is a nonzero constant.
   (c) Explain why we have a problem in solving (b) but not (a).

6. Solve the initial value problems.

$$tu_t + xu_x = u, \quad -\infty < x < \infty,\ 0 < t,$$

$$u(x,x^2) = 1.$$

7. In the following problems we verify the superposition principle.
   (a) Find the solution to
   $$u_t + u_x = -u + g(x,t), \quad -\infty < x < \infty,\ 0 < t,$$
   $$u(x,0) = 0.$$
   The solution may contain an integral.
   (b) Find the solution to
   $$u_t + u_x = -u, \quad -\infty < x < \infty,\ 0 < t,$$
   $$u(x,0) = f(x).$$
   (c) Using the results in (a) and (b) find the solution to
   $$u_t + u_x = -u + g(x,t), \quad -\infty < x < \infty,\ 0 < t,$$
   $$u(x,0) = f(x).$$
8. A large population of insects moves at a constant velocity c in the x-direction and decays at a constant rate r in time t. Assume that the movement is one dimensional and the initial population is $u(x,0) = f(x)$. Find the population $u(x,t)$.
9. Find the general solutions to
   (a) $(1+t)u_t + u_x = 0$
   (b) $u_t + xu_x = u$

1.4 Classification of Second Order Linear Equations

As we will learn in Chapters 2, 3, and 4, there are many equations in physics written as the second order equations. In this section, for the sake of preparation, we classify the second order linear equations in two independent variables (x,y). The most general form of the second order linear equation is given by

$$au_{xx} + bu_{xy} + cu_{yy} + du_x + eu_y + fu + g(x,y) = 0, \tag{1.4.1}$$

where the coefficients a to f depend only on (x,y). The highest order terms are called the principal parts. They determine the type of equations. We change the independent variables to obtain so called the canonical form to determine the types of equations. We introduce

$$\xi = \xi(x,y), \quad \eta = \eta(x,y)$$

to change the independent variables from (x,y) to (ξ,η). Using the chain rule, we see that

$$u_x = u_\xi \xi_x + u_\eta \eta_x, \quad u_y = u_\xi \xi_y + u_\eta \eta_y. \tag{1.4.2}$$

Similarly,

$$\begin{aligned} u_{xx} &= u_{\xi\xi}\xi_x^2 + 2u_{\xi\eta}\xi_x\eta_x + u_{\eta\eta}\eta_x^2 + u_\xi\xi_{xx} + u_\eta\eta_{xx}, \\ u_{xy} &= u_{\xi\xi}\xi_x\xi_y + u_{\xi\eta}\xi_x\eta_y + u_{\xi\eta}\xi_y\eta_x + u_{\eta\eta}\eta_x\eta_y + u_\xi\xi_{xy} + u_\eta\eta_{xy}, \\ u_{yy} &= u_{\xi\xi}\xi_y^2 + 2u_{\xi\eta}\xi_y\eta_y + u_{\eta\eta}\eta_y^2 + u_\xi\xi_{yy} + u_\eta\eta_{yy}. \end{aligned} \tag{1.4.3}$$

Substituting (1.4.2) and (1.4.3) in (1.4.1), for the principal parts we obtain

$$au_{xx} + bu_{xy} + cu_{yy} + \cdots = Au_{\xi\xi} + Bu_{\xi\eta} + Cu_{\eta\eta} + \cdots = 0,$$

where

$$\begin{aligned} A &= a\xi_x^2 + b\xi_x\xi_y + c\xi_y^2, \\ B &= 2a\xi_x\eta_x + b(\xi_x\eta_y + \xi_y\eta_x) + 2c\xi_y\eta_y, \\ C &= a\eta_x^2 + b\eta_x\eta_y + c\eta_y^2. \end{aligned} \tag{1.4.4}$$

Then, after a lengthy calculation we obtain

$$B^2 - 4AC = (b^2 - 4ac)(\xi_x\eta_y - \xi_y\eta_x)^2. \tag{1.4.5}$$

We define $\Delta = b^2 - 4ac$ as the discriminant. The above calculation shows that Δ does not change sign after the change of independent variables.

Definition 1.4.1. Equation (1.4.1) is classified as follows. It is hyperbolic if $\Delta > 0$, parabolic if $\Delta = 0$, and elliptic if $\Delta < 0$.

Example 1.4.2. Find the regions in the xy-plane where the Tricomi equation is hyperbolic, parabolic, or elliptic.

$$u_{yy} - yu_{xx} = 0.$$

Solution: If a, b, or c depends on x or y, the classification depends on the location in the xy-plane. It is hyperbolic if $y > 0$, parabolic if $y = 0$, and elliptic if $y < 0$.

In Chapters 2, 3, and 4 we study the heat, wave, and Laplace equations as examples of parabolic, hyperbolic, and elliptic equations, respectively. The importance of the above definition lies in Theorems 1.4.3, 1.4.6, and 1.4.8.

Theorem 1.4.3. *If Equation (1.4.1) is hyperbolic, (ξ, η) are chosen so that the principal part of u in the new coordinate system is expressed as*

$$u_{\xi\eta} + \cdots = 0. \tag{1.4.6}$$

Alternatively, there are $(\bar{\xi}, \bar{\eta})$ for which the principal part of u in the new coordinate system is expressed as

$$u_{\bar{\xi}\bar{\xi}} - u_{\bar{\eta}\bar{\eta}} + \cdots = 0. \tag{1.4.7}$$

Proof. We choose (ξ, η) so that $A = C = 0$. From (1.4.4), after dividing A by ξ_y, we see that $A = 0$ if

$$\frac{\xi_x}{\xi_y} = \frac{-b \pm \sqrt{b^2 - 4ac}}{2a}.$$

For example, if we choose

$$\xi_x + \frac{b + \sqrt{b^2 - 4ac}}{2a}\xi_y = 0 \tag{1.4.8}$$

for $A = 0$, we can choose

$$\eta_x + \frac{b - \sqrt{b^2 - 4ac}}{2a}\eta_y = 0 \tag{1.4.9}$$

for $C = 0$. Both (1.4.8) and (1.4.9) are the first order linear equations. Solving the equations, we obtain $\xi = \xi(x, y)$ and $\eta = \eta(x, y)$. Since $B = -\Delta\xi_y\eta_y/a$, B is not zero and we have (1.4.6). Once we find (ξ, η), it is not difficult to show that the transformation

$$\bar{\xi} = \xi + \eta, \quad \bar{\eta} = \xi - \eta$$

leads to (1.4.7). □

Remark 1.4.4. The curves $\xi(x, y) =$ constants and $\eta(x, y) =$ constants are called the characteristics or the characteristic curves. Therefore, for the hyperbolic equations there are two families of real characteristics, for the parabolic equations there is one family of real characteristics, and for the elliptic equations there are no real characteristics. The characteristics determine the way we pose the problems for the second order equations. This will be discussed in Section 4.6.

Example 1.4.5. Change the equation to the canonical form.

$$u_{xx} - 2u_{xy} - 3u_{yy} = 0.$$

Solution: Since the discriminant $b^2 - 4ac = 16 > 0$, there are two real characteristics and the equation is hyperbolic. To find them, replace the partial derivative with respect to x by ξ_x and the partial derivative with respect to y by ξ_y. In this example making the following replacements

$$u_{xx} \Rightarrow \xi_x^2,\ u_{xy} \Rightarrow \xi_x\xi_y,\ u_{yy} \Rightarrow \xi_y^2,$$

we have

$$\xi_x^2 - 2\xi_x\xi_y - 3\xi_y^2 = (\xi_x + \xi_y)(\xi_x - 3\xi_y) = 0.$$

So, ξ and η satisfy

$$\xi_x + \xi_y = 0,\ \eta_x - 3\eta_y = 0,$$

respectively. They are linear first order equations. Using the method of characteristics for ξ, we see that

$$\frac{dx}{d\tau} = 1,\ \frac{dy}{d\tau} = 1,\ \frac{d\xi}{d\tau} = 0.$$

Therefore, $x = x_0 + \tau$, $y = y_0 + \tau$, and $\xi = \xi_0$. There is a freedom in choosing the initial data. For example, if we take $x_0 = s$, $y_0 = 0$, and $\xi_0 = s$

$$\xi = x - y.$$

If the coefficients are constants, there is a shortcut. From the above experience we sense that $\xi = \alpha x + \beta y$, where α and β are constants, is sufficient and all we need to do is to find α and β satisfying $\xi_x + \xi_y = 0$. Similarly, for η we assume $\eta = \alpha x + \beta y$ and find α and β satisfying $\eta_x - 3\eta_y = 0$. Then, we obtain

$$\eta = 3x + y.$$

The characteristics are given by

$$x - y = \text{constants}, \quad 3x + y = \text{constants}.$$

If we perform the above change of variables, we obtain

$$\begin{aligned}
u_x &= u_\xi\xi_x + u_\eta\eta_x = u_\xi + 3u_\eta,\\
u_y &= u_\xi\xi_y + u_\eta\eta_y = -u_\xi + u_\eta,\\
u_{xx} &= u_{\xi\xi}\xi_x + u_{\xi\eta}\eta_x + 3u_{\eta\xi}\xi_x + 3u_{\eta\eta}\eta_x = u_{\xi\xi} + 4u_{\xi\eta} + 9u_{\eta\eta},\\
u_{xy} &= u_{\xi\xi}\xi_y + u_{\xi\eta}\eta_y + u_{\eta\xi}\xi_y + u_{\eta\eta}\eta_y = -u_{\xi\xi} + 3u_{\xi\eta} - u_{\eta\xi} + u_{\eta\eta},\\
u_{yy} &= -u_{\xi\xi}\xi_y - u_{\xi\eta}\eta_y + u_{\eta\xi}\xi_y + u_{\eta\eta}\eta_y = u_{\xi\xi} - u_{\xi\eta} - 3u_{\eta\xi} + u_{\eta\eta}.
\end{aligned}$$

Therefore, the canonical form (1.4.6) is

$$u_{xx} - 2u_{xy} - 3u_{yy} = 12u_{\xi\eta} = 0.$$

Furthermore, if we use

$$\bar{\xi} = \xi + \eta, \quad \bar{\eta} = \xi - \eta,$$

we obtain another canonical form (1.4.7)

$$\begin{aligned}
u_\xi &= u_{\bar{\xi}} + u_{\bar{\eta}},\ u_\eta = u_{\bar{\xi}} - u_{\bar{\eta}},\\
u_{\xi\eta} &= u_{\bar{\xi}\bar{\xi}} - u_{\bar{\xi}\bar{\eta}} + u_{\bar{\eta}\bar{\xi}} - u_{\bar{\eta}\bar{\eta}} = u_{\bar{\xi}\bar{\xi}} - u_{\bar{\eta}\bar{\eta}} = 0.
\end{aligned}$$

Theorem 1.4.6. *If Equation (1.4.1) is parabolic, (ξ, η) are chosen so that the principal part of u in the new coordinate system is expressed as*

$$u_{\xi\xi} + \cdots = 0. \tag{1.4.10}$$

Alternatively, there are (ξ, η) for which the principal part of u in the new coordinate system is expressed as

$$u_{\eta\eta} + \cdots = 0. \tag{1.4.11}$$

Proof. Since $\triangle = b^2 - 4ac = 0$, A is factored as

$$A = a(\xi_x + \frac{b}{2a}\xi_y)^2.$$

Therefore, we choose ξ so that

$$\frac{\xi_x}{\xi_y} = \frac{-b}{2a}. \tag{1.4.12}$$

Then, as we see below B is also zero.

$$\begin{aligned} B &= 2a\xi_x\eta_x + b(\xi_x\eta_y + \xi_y\eta_x) + 2c\xi_y\eta_y \\ &= -b\xi_y\eta_x - \frac{b^2}{2a}\xi_y\eta_y + b\xi_y\eta_x + 2c\xi_y\eta_y = 0. \end{aligned}$$

For η we choose the curve η so that

$$\frac{\eta_x}{\eta_y} \neq \frac{-b}{2a}. \tag{1.4.13}$$

Then $C \neq 0$ and we have the equation of the form given by (1.4.11). □

Example 1.4.7. Change the equation to the canonical form.

$$u_{xx} + 2u_{xy} + u_{yy} + au_x = 0.$$

Solution: Since $\triangle = 0$, the equation is parabolic. For this problem

$$A = \xi_x^2 + 2\xi_x\xi_y + \xi_y^2 = (\xi_x + \xi_y)^2 = 0.$$

Therefore,

$$\xi = x - y.$$

For η, we have freedom in choosing η as long as η satisfies (1.4.13). Taking advantage of this we choose the simplest η. So, we choose

$$\eta = y.$$

Then, the canonical form (1.4.11) is

$$\begin{aligned} u_{xx} + 2u_{xy} + u_{yy} + au_x &= u_{\xi\xi} + 2(-u_{\xi\xi} + u_{\xi\eta}) \\ &\quad +(u_{\xi\xi} - 2u_{\xi\eta} + u_{\eta\eta}) + a(-u_\xi) \\ &= u_{\eta\eta} - au_\xi = 0. \end{aligned}$$

Theorem 1.4.8. *If Equation (1.4.1) is elliptic, (ξ, η) are chosen so that the principal part of u in the new coordinate system is expressed as*

$$u_{\xi\xi} + u_{\eta\eta} + \cdots = 0.$$

Proof. In this case, for example, we choose (ξ, η) so that

$$\frac{\xi_x}{\xi_y} = \frac{-b + \sqrt{4ac - b^2}}{2a}, \quad \frac{\eta_x}{\eta_y} = \frac{-b - \sqrt{4ac - b^2}}{2a}. \tag{1.4.14}$$

Note that the inside of the square root is no longer $b^2 - 4ac$. We can show that

$$\begin{aligned} B &= 2a\xi_x\eta_x + b(\xi_x\eta_y + \xi_y\eta_x) + 2c\xi_y\eta_y \\ &= \frac{(b^2 - 4ac + b^2) - 2b^2 + 4ac}{2a}\xi_y\eta_y = 0. \end{aligned} \tag{1.4.15}$$

Also, the following are obtained

$$A = a\xi_x^2 + b\xi_x\xi_y + c\xi_y^2 = \frac{4ac - b^2}{2a}\xi_y^2, \tag{1.4.16}$$

$$C = a\eta_x^2 + b\eta_x\eta_y + c\eta_y^2 = \frac{4ac - b^2}{2a}\eta_y^2. \tag{1.4.17}$$

So, the principal parts are

$$Au_{\xi\xi} + Bu_{\xi\eta} + Cu_{\eta\eta} + \cdots = A[u_{\xi\xi} + (\frac{\eta_y}{\xi_y})^2 u_{\eta\eta} + \cdots] = 0.$$

Now, set

$$\bar{\xi} = \xi, \quad \bar{\eta} = \int \frac{\xi_y}{\eta_y} d\eta.$$

Then,

$$u_\eta = u_{\bar{\eta}}\frac{\partial\bar{\eta}}{\partial\eta} + u_{\bar{\xi}}\frac{\partial\bar{\xi}}{\partial\eta} = u_{\bar{\eta}}\frac{\partial\bar{\eta}}{\partial\eta},$$

$$u_{\eta\eta} = u_{\bar{\eta}\bar{\eta}}(\frac{\partial\bar{\eta}}{\partial\eta})^2 + u_{\bar{\eta}}\frac{\partial^2\bar{\eta}}{\partial\eta^2}.$$

From this we have

$$u_{\bar{\xi}\bar{\xi}} + u_{\bar{\eta}\bar{\eta}} + \cdots = 0.$$

□

Example 1.4.9. Change the equation to the canonical form.

$$u_{xx} + 2u_{xy} + 2u_{yy} = 0.$$

Solution: Since $\triangle = -4 < 0$, the equation is elliptic. We set

$$\frac{\xi_x}{\xi_y} = \frac{-b + \sqrt{4ac - b^2}}{2a} = \frac{-2+2}{2} = 0,$$

$$\frac{\eta_x}{\eta_y} = \frac{-b - \sqrt{4ac - b^2}}{2a} = \frac{-2-2}{2} = -2.$$

Then,

$$\xi = y, \ \eta = 2x - y.$$

Therefore, the canonical form is

$$\begin{aligned} u_{xx} + 2u_{xy} + 2u_{yy} &= 4u_{\eta\eta} + 2(2u_{\eta\xi} - 2u_{\eta\eta}) + 2(u_{\xi\xi} - 2u_{\xi\eta} + u_{\eta\eta}) \\ &= 2u_{\xi\xi} + 2u_{\eta\eta} = 0. \end{aligned}$$

Exercises

1. State the type of each equation and change the equations to the canonical form.

   (a) $u_{xx} + 2u_{xy} + 5u_{yy} = 0$
   (b) $9u_{xx} + 6u_{xy} + u_{yy} + u_x = 0$
   (c) $u_{xx} - 3u_{xy} + 2u_{yy} = 0$

2. Find the regions in xy-plane where the equation is hyperbolic, parabolic, or elliptic. Sketch the regions.

   (a) $x^2u_{xx} - 2xyu_{xy} - (1 + y)u_{yy} = 0$
   (b) $yu_{xx} - yu_{xy} + x^2u_{yy} = 0$

3. Substitute (1.4.2) and (1.4.3) in (1.4.1) to obtain (1.4.4).
4. Show that (1.4.5) holds.
5. In Theorem 1.4.8 show that $B = 0$, (1.4.16), and (1.4.17) hold.

1.5 Well-posedness

In Examples 1.3.3 and 1.3.4 we had the same equation but totally different outcomes, *i.e.*, a unique solution in Example 1.3.3 and no solution in Example 1.3.4. The difference is the initial condition. In PDE not only the equations but also the initial conditions or the boundary conditions will affect the solutions. (We must say that we have not seen the boundary conditions yet.) This observation leads to the concept of well-posedness of the problems. We say that a PDE problem is well-posed if it satisfies the three conditions.

1. A solution of the problem exists.
2. The solution is unique.
3. The solution depends on the initial or boundary conditions of the problem continuously.

We already encountered the existence issue in the examples mentioned above. In what follows we briefly explain what we mean by these questions and why we discuss such questions.

A classical solution to a partial differential equation means a function satisfying the given equation and having derivatives up to the order of equation. For example, in the first order equation, a classical solution is a C^1-function satisfying the equation and in the second order equation, a classical solution is a C^2-function satisfying the equation. Here, a function f is C^1 means that the first derivatives of f exist and they are continuous. For most equations we studied so far we know the explicit solutions and in this book we mostly discuss the case where explicit solutions are available. So, we do not pay attention to this issue as much as we should. However, these are exceptional cases. Once we leave the constant coefficient cases and the coefficients depend on (x, y), (x, t), or the unknown u and its derivatives, the situation become more complicated because of the fact that it is very hard to find the explicit solutions if not impossible. For those problems where the explicit solutions are not easy to find, the existence of solutions becomes an issue.

The uniqueness of solutions is also an important issue. Physically we expect that the solutions are unique. For example, if we start from the same initial data, we expect that we get the same outcome. However, there are cases where we have more than one solution satisfying the same initial data. We may need to specify more conditions so that we have a unique solution. We encounter such a situation in Section 5.2.

Physically it is rather difficult to specify the exact initial conditions or exact boundary conditions due to measurement errors. So, it is important to ask how the small errors affect the solutions. We expect that if the initial data are changed a little bit, then the solution will change accordingly. We do not expect a drastic change in the solution when the change of the initial condition is small. This is roughly speaking the continuous dependence of solutions to the initial or boundary conditions. Examples concerning this issue will be discussed in Section 4.6.

Chapter 2

Heat Equation

The heat equation we study in this chapter explains the heat conduction in a material. It also describes the way substances diffuse in media. For this reason it is also called the diffusion equation. Our main focus is one-dimensional problems. Multidimensional problems except for the derivation are discussed in Chapter 7 and later chapters.

In Section 2.1 we derive the heat equation. The basic principle we use is called the conservation law. This law states that the rate of change of the heat energy in a region is equal to the heat flux across the boundary of the region. The heat flux is the heat energy crossing the boundary per unit area per unit time. From Sections 2.2 to 2.5 we study the initial boundary value problems. A goal is to learn the solution method called the separation of variables. This technique is applied for various partial differential equations including the wave equation and the Laplace equation. It is important to become familiar with the method. Another goal is to learn how the boundary conditions affect the solutions. We discuss the several well-known boundary conditions.

The initial value problems are discussed in Section 2.6. In the one dimensional case, they are related to the heat conduction in an infinitely long bar.

Qualitative behavior of solutions is also important. The maximum principle is studied in Subsection 2.7. This basically says that the maximum temperature of the rod is either on the boundary or on the initial data. We also study the energy methods. Both will be useful to prove the uniqueness of the solutions.

2.1 Derivation of the Heat Equation

2.1.1 *One-dimensional Case*

To derive the heat equation we use a basic principle in physics called the conservation law. This law says that the rate of change of the amount of heat in a region is equal to the heat flux from the boundary. In what follows, we apply this law to derive the one-dimensional heat equation. The derivation in the multi-dimensional case will be discussed after the divergence theorem is introduced.

A typical example concerning the derivation of a one-dimensional heat equation is a thin rod. We consider a rod of length L with uniform cross section A as depicted in Figure 2.1. The one end of rod is located at $x = 0$ and the other end is at $x = L$. We assume that thermal properties of rod are uniform across each cross section so that we can treat the rod as a one-dimensional object. We also assume that the cylindrical surface is perfectly insulated so that there is no heat gain or loss through it. We derive an equation for heat conduction based on the conservation law of energy. This is stated as follows.

$$\begin{array}{c} \text{Rate of change} \\ \text{of heat energy} \\ \text{in a region} \end{array} = \begin{array}{c} \text{Rate at which} \\ \text{the heat flows} \\ \text{into the region} \end{array} + \begin{array}{c} \text{Rate at which the} \\ \text{heat is generated} \\ \text{in the region.} \end{array} \qquad (2.1.1)$$

The first and second terms on the right hand side are called the flux and source terms, respectively. For the sake of simplicity, we assume that the source term is zero and derive the heat equation. The case where there is a source term is discussed in Exercise 3 at the end of the section.

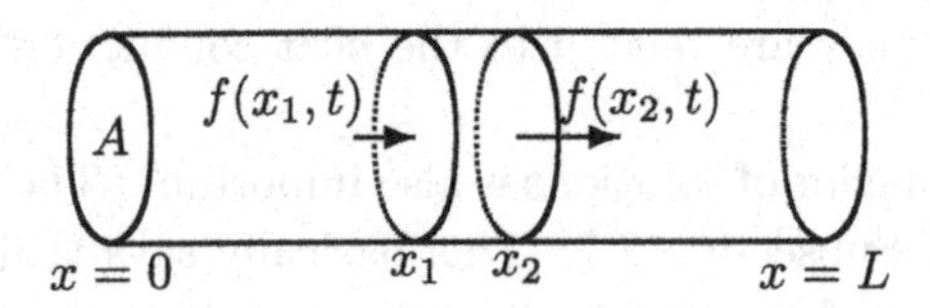

Fig. 2.1 One-dimensional heat conducting solid bar.

Our task is to express each term in (2.1.1) in mathematical terms starting from the left hand side. Perhaps this is the most important aspect of derivation (modeling). Let c be the specific heat of the material. It is defined as the amount of heat required to raise the temperature of the material by one degree Celsius per unit mass. Let ρ be the density of the material and defined as the mass per unit volume. We denote $u(x,t)$ to be the temperature of the rod at cross section x at time t. We take the region to be a segment $[x_1, x_2]$, where x_1 and x_2 are the locations satisfying $0 < x_1 < x_2 < L$. Then, the amount of heat in the segment is given by

$$A \int_{x_1}^{x_2} c\rho u dx.$$

Therefore, the left hand side is given by

$$A \frac{d}{dt} \int_{x_1}^{x_2} c\rho u dx = A \int_{x_1}^{x_2} \frac{\partial}{\partial t}(c\rho u) dx. \tag{2.1.2}$$

It is actually difficult to see when we can change the order of integration and differentiation. In Appendix A.1 we discuss the conditions under which we can exchange the order of integration and differentiation.

Next, we express the first term on the right hand side in (2.1.1) in mathematical terms. To do so let $f = f(x,t)$ be the rate at which the heat passes through x from left to right at t. If f is negative, the heat is moving from right to left. This rate is commonly called the heat flux and it is measured by the amount of heat passing the cross section x per unit time per unit area. Then, this should be equal to the rate at which the heat comes in at x_1 minus the rate at which the heat leaves at x_2. (If the heat leaves from x_1 at t, $f(x_1,t)$ is negative.) Therefore, this term is equal to

$$A(f(x_1,t) - f(x_2,t)) = -A \int_{x_1}^{x_2} \frac{\partial f}{\partial x}(x,t) dx. \tag{2.1.3}$$

We now derive a partial differential equation relating the heat and the flux. Equating (2.1.2) and (2.1.3), we have

$$A \int_{x_1}^{x_2} \frac{\partial}{\partial t}(c\rho u)(x,t) dx = -A \int_{x_1}^{x_2} \frac{\partial f}{\partial x}(x,t) dx. \tag{2.1.4}$$

Therefore,

$$\int_{x_1}^{x_2} [\frac{\partial}{\partial t}(c\rho u)(x,t) + \frac{\partial f}{\partial x}(x,t)] dx = 0.$$

Since x_1, x_2 are arbitrary, we obtain

$$(c\rho u)_t + f_x = 0. \tag{2.1.5}$$

This is the equation relating the heat and the flux.

Next, we express f in terms of u. For the heat flux we use the empirical law called the Fourier law which says that the amount of heat flux $f(x,t)$ is proportional to the temperature gradient. In one dimensional case, the temperature gradient is the slope of temperature and the heat flux is given by

$$f(x,t) = -ku_x, \tag{2.1.6}$$

where $k > 0$ is called the thermal conductivity whose units are $J/(C^o MT)$, where J = Joule, C^o = Celsius, M = length, and T = time. The negative sign shows that the heat flows from higher to lower temperature. Substituting the above f in (2.1.5), we obtain

$$(c\rho u)_t = (ku_x)_x. \tag{2.1.7}$$

In what follows for the sake of simplicity we assume that c, ρ, and k are constants. Then, (2.1.7) reduces to

$$u_t = a^2 u_{xx}, \tag{2.1.8}$$

where $a = \sqrt{k/(c\rho)}$ is called the thermal diffusivity. The equation is called the heat (conduction) equation or the diffusion equation.

2.1.2 *Divergence Theorem*

We explain the divergence theorem in two dimensional case. Consider a connected region Ω with the boundary $\partial\Omega$ as in Figure 2.2. The divergence of vector function $\mathbf{u} = [u_1, u_2]$ is given by

$$\text{div}\mathbf{u} = \nabla \cdot \mathbf{u} = u_{1x} + u_{2y}, \tag{2.1.9}$$

where $\nabla = [\partial/\partial x, \partial/\partial y]$ is a gradient operator. If it is operated on a scalar function u, we get a gradient vector of u

$$\nabla u = [\frac{\partial u}{\partial x}, \frac{\partial u}{\partial y}].$$

On the other hand, the dot product between the gradient operator ∇ and the vector function $\mathbf{u}$ produces a scalar as shown in (2.1.9).

The divergence theorem relates the divergence of a vector function in a region Ω with its flux along the boundary $\partial\Omega$ and is stated as follows.

Theorem 2.1.1. (Divergence Theorem) We have

$$\iint_\Omega \nabla \cdot \mathbf{u}\, d\mathbf{x} = \int_{\partial\Omega} \mathbf{u} \cdot \mathbf{n} ds, \tag{2.1.10}$$

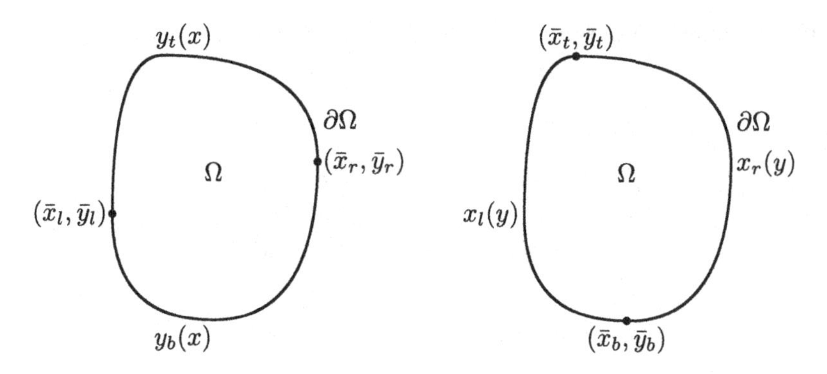

Fig. 2.2 The boundary curves and their end points.

where the left hand side is an area integral over the domain Ω, the integral on the right is a line integral on the boundary $\partial\Omega$, and $\mathbf{n}$ is the unit outward normal vector to the boundary $\partial\Omega$. The line integral is evaluated counter clockwise.

Proof. Consider the integral

$$\iint_{\Omega} \nabla \cdot \mathbf{u} d\mathbf{x} = \iint_{\Omega} (u_{1x} + u_{2y}) dy dx.$$

We integrate the above integral separately. Then,

$$\iint_{\Omega} u_{1x} dx dy = \int_{\bar{y}_b}^{\bar{y}_t} [u_1]_{x_l(y)}^{x_r(y)} dy = \int_{\bar{y}_b}^{\bar{y}_t} [u_1(x_r(y), y) - u_1(x_l(y), y)] dy,$$

$$\iint_{\Omega} u_{2y} dy dx = \int_{\bar{x}_l}^{\bar{x}_r} [u_2]_{y_b(x)}^{y_t(x)} dx = \int_{\bar{x}_l}^{\bar{x}_r} [u_2(x, y_t(x)) - u_2(x, y_b(x))] dx,$$

where $x_l(y)$, $x_r(y)$, $y_b(x)$, and $y_t(x)$ are the left, right, bottom, and top halves of the boundary, respectively. See Figure 2.2. The above integrals are line integrals parametrized by y and x, respectively. Note that the each portion of the boundary is parametrized by x and y. We rearrange the terms as follows and evaluate the integrals pairwise.

$$\begin{aligned}
\iint\limits_{\Omega} \nabla \cdot \mathbf{u} d\mathbf{x} &= \iint_{\Omega} (u_{1x} + u_{2y}) dy dx \\
&= \int_{\bar{y}_b}^{\bar{y}_t} [u_1(x_r(y), y) - u_1(x_l(y), y)] dy \\
&\quad + \int_{\bar{x}_l}^{\bar{x}_r} [u_2(x, y_t(x)) - u_2(x, y_b(x))] dx \\
&= \int_{\bar{y}_b}^{\bar{y}_r} u_1(x_r(y), y) dy - \int_{\bar{x}_b}^{\bar{x}_r} u_2(x, y_b(x)) dx \\
&\quad + \int_{\bar{y}_r}^{\bar{y}_t} u_1(x_r(y), y) dy - \int_{\bar{x}_r}^{\bar{x}_t} u_2(x, y_t(x)) dx \\
&\quad - \int_{\bar{y}_l}^{\bar{y}_t} u_1(x_l(y), y) dy + \int_{\bar{x}_l}^{\bar{x}_t} u_2(x, y_t(x)) dx \\
&\quad - \int_{\bar{y}_b}^{\bar{y}_l} u_1(x_l(y), y) dy + \int_{\bar{x}_b}^{\bar{x}_l} u_2(x, y_b(x)) dx. \qquad (2.1.11)
\end{aligned}$$

We change the parameter to the arc length s. As illustrated in Figure 2.3 θ is the angle between the arc length element ds and the positive direction of x. Therefore, the relations $dx = ds \cos\theta$ and $dy = ds \sin\theta$ hold. The outward normal vector $\mathbf{n}$ is perpendicular to ds and given by $\mathbf{n} = (\sin\theta, -\cos\theta)$. The first pair in (2.1.11) is the integral over the portion between $(\bar{x}_b, \bar{y}_b)$ and $(\bar{x}_r, \bar{y}_r)$ of $\partial\Omega$ and evaluated as follows.

$$\begin{aligned}
&\int_{\bar{y}_b}^{\bar{y}_r} u_1(x_r(y), y) dy - \int_{\bar{x}_b}^{\bar{x}_r} u_2(x, y_b(x)) dx \\
&= \int_0^{s_r} (u_1 \sin\theta - u_2 \cos\theta) ds = \int_0^{s_r} \mathbf{u} \cdot \mathbf{n} ds. \qquad (2.1.12)
\end{aligned}$$

Similarly, the second pair in (2.1.11) is the integral over the portion between $(\bar{x}_r, \bar{y}_r)$ and $(\bar{x}_t, \bar{y}_t)$ of $\partial\Omega$ and given by

$$\begin{aligned}
&\int_{\bar{y}_r}^{\bar{y}_t} u_1(x_r(y), y) dy - \int_{\bar{x}_r}^{\bar{x}_t} u_2(x, y_t(x)) dx \\
&= \int_{s_r}^{s_t} (u_1 \sin\theta - u_2 \cos\theta) ds = \int_{s_r}^{s_t} \mathbf{u} \cdot \mathbf{n} ds. \qquad (2.1.13)
\end{aligned}$$

Performing the similar parameter changes to the last two pairs in (2.1.11), we obtain

$$\iint\limits_{\Omega} \nabla \cdot \mathbf{u} d\mathbf{x} = \int\limits_{\partial\Omega} \mathbf{u} \cdot \mathbf{n} ds.$$

□

Remark 2.1.2. (a) If Ω is more complicated, we can decompose Ω to a union of smaller regions where the above result is applicable.

(b) The divergence theorem can be extended to the three dimensional case. The area integral and the line integral should be changed to the volume integral and the surface integral, respectively. Therefore, in three dimensional case the divergence theorem is expressed as

$$\iiint_{\Omega} \nabla \cdot \mathbf{u} d\mathbf{x} = \iint_{\partial\Omega} \mathbf{u} \cdot \mathbf{n} dS,$$

where the left hand side is a volume integral and the right hand side is a surface integral.

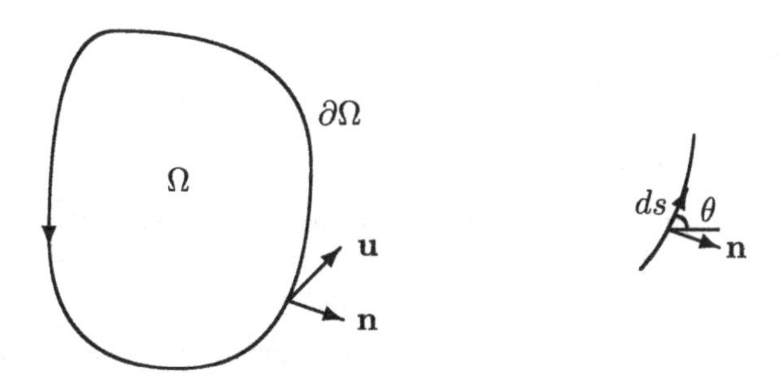

Fig. 2.3 Divergence theorem.

2.1.3 *Multi-dimensional Case*

The heat equations in two and three dimensions are given, respectively, by

$$u_t = a^2(u_{xx} + u_{yy}) = a^2\nabla \cdot \nabla u = a^2\triangle u \tag{2.1.14}$$

and

$$u_t = a^2(u_{xx} + u_{yy} + u_{zz}) = a^2\nabla \cdot \nabla u = a^2\triangle u.$$

We derive the heat equation in two dimension. Consider a fixed domain Ω. Using the divergence theorem, we see that the heat flux from the boundary $\partial\Omega$ is given by

$$\int_{\partial\Omega} \mathbf{f} \cdot \mathbf{n} ds = \iint_{\Omega} \text{div}\mathbf{f} d\mathbf{x},$$

where **f** is the heat flux and **n** is the outward normal to the boundary. An extension of the Fourier law to the multi-dimensional case is to assume that the heat diffuses in the direction of the steepest descent of temperature, *i.e.*,

$$\mathbf{f} = -k\nabla u.$$

The conservation law leads to

$$\frac{d}{dt}\iint_{\Omega} c\rho u d\mathbf{x} = \iint_{\Omega} (c\rho u)_t d\mathbf{x} = -\int_{\partial\Omega} \mathbf{f}\cdot\mathbf{n}ds = \iint_{\Omega} \mathrm{div}(k\nabla u)d\mathbf{x}.$$

Since this relation holds for every Ω, we conclude that if c, ρ, and k are constants, the heat equation in the multi-dimensional case is given by

$$u_t = a^2\Delta u.$$

Exercises

1. Find the units of a^2.
2. Answer the following.
   (a) Verify that $u(x,t) = \exp[-x^2/(4a^2t)]/(2a\sqrt{\pi t})$ is a solution to (2.1.8), where $\exp(a) = e^a$.
   (b) Find the heat flux of u at (x,t).
   (c) Verify that $v(x,y,t) = \exp[-(x^2+y^2)/(4a^2t)]/(4a^2\pi t)$ is a solution to (2.1.14).
   (d) Find the heat flux of v at (x,y,t).
3. In the one-dimensional heat equation, assume that there is a heat source in the rod so that it generates heat in the rod. Suppose $g(x,t)$ is the heat generated in the rod per unit volume per unit time at x and t. The corresponding heat equation is given by
$$u_t = a^2u_{xx} + g(x,t). \tag{2.1.15}$$
   (a) Show that the last term in (2.1.1) is given by
$$A\int_{x_1}^{x_2} g dx.$$
   (b) Modify (2.1.4) in the derivation so that we obtain (2.1.15).
   (c) If g is negative, what does it mean?
4. As is done in (2.1.12) and (2.1.10), show the following relations.
   (a) $-\int_{\bar{y}_l}^{\bar{y}_t} u_1(x_l(y),y)dy + \int_{\bar{x}_l}^{\bar{x}_t} u_2(x,y_t(x))dx = \int_{s_t}^{s_l} \mathbf{u}\cdot\mathbf{n}ds$
   (b) $-\int_{\bar{y}_b}^{\bar{y}_l} u_1(x_l(y),y)dy + \int_{\bar{x}_b}^{\bar{x}_l} u_2(x,y_b(x))dx = \int_{s_l}^{s_b} \mathbf{u}\cdot\mathbf{n}ds$

2.2 Initial Boundary Value Problems

We wish to find the solution $u(x,t)$, the temperature at a cross section x and at time t, to the heat equation (2.1.8). To do so we specify the conditions at the ends of the rod. We call them the boundary conditions. We also specify the temperature distribution in the rod at $t = 0$. This is called the initial condition.

There are several typical boundary conditions. For example,

$$u(0,t) = b_l(t)$$

means that the temperature of rod at the end $x = 0$ is given by $b_l(t)$. In particular if $u(0,t) = 0$, the temperature at $x = 0$ is kept at 0. The boundary conditions where the temperature is specified on the boundary are called the Dirichlet boundary conditions.

Another typical boundary condition is $u_x(0,t) = 0$, which means that the end $x = 0$ is insulated, *i.e.*, there is no heat flux across $x = 0$. More generally, if

$$-ku_x(0,t) = b_l(t),$$

the heat flux at $x = 0$ is given by $b_l(t)$. The boundary conditions where the heat flux is specified on the boundary are called the Neumann boundary conditions.

The third typical boundary condition is

$$-ku_x(0,t) = -d[u(0,t) - b_l(t)], \tag{2.2.1}$$

where d is called the heat transfer coefficient. This boundary condition is referred as the Robin boundary conditions. It says that the heat flux on the boundary $x = 0$ is proportional to the difference between the temperature at the end of rod and the ambient temperature $b_l(t)$. This relation is called the Newton's law of cooling.

There are cases where the boundary conditions are different at the left and right ends. For example if $u(0,t) = 0$ and $u_x(L,t) = 0$, the temperature at the end $x = 0$ is fixed at 0 and the end $x = L$ is thermally insulated. This type of boundary conditions is generally called the mixed boundary conditions.

The initial condition is given by

$$u(x,0) = f(x), \quad 0 \le x \le L$$

and this expresses the initial temperature distribution in the rod or the temperature distribution at $t = 0$.

The general initial boundary value problems are stated as

$$u_t - a^2 u_{xx} = h(x,t), \quad 0 < x < L, \quad t > 0, \tag{2.2.2}$$

$$\alpha_1 u_x(0,t) + \beta_1 u(0,t) = b_l(t), \quad \alpha_2 u_x(L,t) + \beta_2 u(L,t) = b_r(t), \quad t > 0, \tag{2.2.3}$$

$$u(x,0) = f(x), \quad 0 \le x \le L. \tag{2.2.4}$$

We find the solution u to the heat equation (2.6.25) satisfying the boundary conditions (2.2.3) and the initial condition (2.2.4). The function h is often called a source term. If h is identically equal to zero ($h \equiv 0$), the equation is called homogeneous and otherwise it is called non-homogeneous. Similarly, if both $b_l(t) \equiv 0$ and $b_r(t) \equiv 0$, the boundary conditions are called homogeneous and otherwise they are called non-homogeneous. In this chapter we study the case where $h \equiv 0$ or $h = h(x)$. The case where $h = h(x,t)$ will be discussed in Sections 7.3 and 11.2.

To see the effects of the various boundary conditions, in Section 2.3 we consider the following homogeneous boundary conditions.

$$u(0,t) = 0, \quad u(L,t) = 0, \tag{2.2.5}$$

$$u_x(0,t) = 0, \quad u_x(L,t) = 0, \tag{2.2.6}$$

$$u(0,t) = 0, \quad u_x(L,t) = 0, \tag{2.2.7}$$

$$u_x(0,t) = 0, \quad u(L,t) = 0. \tag{2.2.8}$$

For example, (2.2.8) says that the temperature at the end $x = 0$ is insulated and $x = L$ is fixed at zero. In Section 2.4 the non-homogeneous boundary conditions such as

$$u(0,t) = T_1, \quad u(L,t) = T_2 \tag{2.2.9}$$

will be discussed. They state that the temperature at $x = 0$ is fixed at T_1 and at $x = L$ is fixed at T_2. The Robin boundary conditions are more difficult than the other boundary conditions and treated in Section 2.5.

Exercises

1. Explain the boundary conditions (2.2.8).

   (a) $u_x(0,t) = T_1, \quad u(L,t) = T_2$
   (b) $u(0,t) = T_1, \quad u_x(L,t) = 0$

2. This problem is about the Robin boundary conditions.

   (a) At $x = 0$ if the temperature at $x = 0$ is lower than the ambient temperature, the heat crosses $x = 0$ from left to right. Based on this observation, determine the sign of d in (2.2.1).
   (b) Suppose that the following boundary condition is given at $x = L$

   $$-ku_x(L,t) = d[u(L,t) - b_r(t)].$$

   Determine the sign of d based on the similar argument to (a).

2.3 Homogeneous Boundary Conditions

In this section we study the initial boundary value problems with homogeneous boundary conditions. The main method for solutions is called the separation of variables. This is a very well-known method and useful for the linear PDE's. The main goal of this section is to learn how the separation of variables works. Another goal is to see how the boundary conditions affect the solutions.

2.3.1 *Temperature is Fixed at Zero at Both Ends*

We consider the boundary conditions (2.2.5) to illustrate how the separation of variables works. The initial boundary value problem is stated as

$$u_t - a^2 u_{xx} = 0, \quad 0 < x < L, \ 0 < t, \tag{2.3.1}$$

$$u(0,t) = 0, \quad u(L,t) = 0, \quad 0 < t, \tag{2.3.2}$$

$$u(x,0) = f(x). \tag{2.3.3}$$

What this says is that we find the solution $u(x,t)$ to the heat equation $u_t - a^2 u_{xx} = 0$ for $t > 0$ on the interval $0 < x < L$ satisfying the boundary conditions (2.3.2) and the initial condition (2.3.3). The method is decomposed into several steps.

Step 1 (Separating Variables): We assume

$$u(x,t) = X(x)T(t)$$

and derive the set of ordinary differential equations (ODE's) for X and T. This is the reason why the method is called the separation of variables. Substituting this form of solution in (2.3.1), we obtain

$$XT' - a^2X''T = 0. \tag{2.3.4}$$

Dividing (2.3.4) by a^2XT leads to

$$\frac{T'(t)}{a^2T(t)} = \frac{X''(x)}{X(x)} = -\lambda. \tag{2.3.5}$$

Note that the left hand side of (2.3.5) is a function of t only and the middle term is a function of x only. Suppose we fix x and change t. Since the middle term does not change, the left hand side does not change even if t is changed. This shows that the left hand side is actually a constant and it implies that both the middle term and λ are also constants. Therefore, we obtain two ODE's from (2.3.5)

$$X'' + \lambda X = 0, \tag{2.3.6}$$

$$T' + \lambda a^2 T = 0. \tag{2.3.7}$$

Our goal is to find the solutions to the above ODE's and to determine the values of λ.

Step 2 (Solving ODE's): We solve the set of ODE's (2.3.6) and (2.3.7) obtained in Step 1 starting from (2.3.6). The boundary conditions for X are obtained from (2.3.2) in the following way.

$$u(0,t) = 0 \Rightarrow X(0)T(t) = 0 \Rightarrow X(0) = 0, \tag{2.3.8}$$

$$u(L,t) = 0 \Rightarrow X(L)T(t) = 0 \Rightarrow X(L) = 0. \tag{2.3.9}$$

If we choose $T(t) = 0$, we end up with $u(x,t) = XT = 0$ and this may not satisfy the initial condition $u(x,0) = f(x)$. This is why we choose $X(0) = X(L) = 0$. As we see $X(x) = 0$ is a solution to (2.3.6) satisfying the boundary conditions (2.3.8) and (2.3.9), we call $X(x) = 0$ a trivial solution. If we choose $X = 0$ as a solution for X, $u(x,t) = XT = 0$ and again this may not satisfy the initial condition. Therefore, we look for non-trivial solutions for (2.3.6), (2.3.8), and (2.3.9). It turns out that there are values of λ for which nontrivial solutions exist. Such values of λ are called eigenvalues and the corresponding nontrivial solutions are called the eigenfunctions. It is possible to show that the eigenvalues are real. We take this for granted now and find them. Theoretical aspects of eigenvalues and eigenfunctions including the proof concerning the eigenvalues being real are discussed in Section 6.5. The general solutions of (2.3.6) are given by

$$\lambda > 0: \quad X = c_1 \cos\sqrt{\lambda}x + c_2 \sin\sqrt{\lambda}x, \tag{2.3.10}$$

$$\lambda = 0: \quad X = c_1 + c_2 x, \tag{2.3.11}$$

$$\lambda < 0: \quad X = c_1 e^{\sqrt{-\lambda}x} + c_2 e^{-\sqrt{-\lambda}x}. \tag{2.3.12}$$

We use the boundary conditions to figure out what values of λ are eigenvalues. For $\lambda > 0$, (2.3.10) with the boundary conditions (2.3.8) and (2.3.9) yields

$$X(0) = c_1 = 0,$$

$$X(L) = c_2 \sin\sqrt{\lambda}L = 0.$$

To have a nontrivial solution, $c_2 \neq 0$. This forces us to choose λ such that $\sin\sqrt{\lambda}L = 0$ holds. Therefore,

$$\sqrt{\lambda}L = n\pi \Rightarrow \quad \lambda = (\frac{n\pi}{L})^2, \quad n = 1, 2, \ldots$$

are the eigenvalues and the corresponding eigenfunctions are

$$X_n = \sin\frac{n\pi x}{L}, \quad n = 1, 2, \ldots.$$

Actually, $X_n = a_n \sin(n\pi x/L)$, where a_n is a nonzero constant, can be an eigenfunction. However, it is customary to take $a_n = 1$ for eigenfunctions.

For $\lambda = 0$, from (2.3.11) and the boundary conditions (2.3.8) and (2.3.9), we obtain

$$X(0) = c_1 = 0,$$

$$X(L) = c_2 L = 0.$$

Therefore, $c_1 = c_2 = 0$. So, $\lambda = 0$ is not an eigenvalue.

For $\lambda < 0$, (2.3.12) with the boundary conditions (2.3.8) and (2.3.9) implies

$$X(0) = c_1 + c_2 = 0, \tag{2.3.13}$$

$$X(L) = c_1 e^{\sqrt{-\lambda}L} + c_2 e^{-\sqrt{-\lambda}L} = 0. \tag{2.3.14}$$

The easiest way to solve the equations (2.3.13) and (2.3.14) for c_1 and c_2 is to substitute $c_2 = -c_1$ to (2.3.14). Then,

$$X(L) = c_1 e^{\sqrt{-\lambda}L} - c_1 e^{-\sqrt{-\lambda}L} = 2c_1 \sinh e^{\sqrt{-\lambda}L} = 0.$$

Since $\lambda < 0$, this happens only when $c_1 = 0$ and this implies that $c_1 = c_2 = 0$ is the only solution. So, there is no nontrivial solution.

In summary, the eigenvalues and the corresponding eigenfunctions are

$$\lambda_n = (\frac{n\pi}{L})^2, \ X_n = \sin\frac{n\pi x}{L}, \quad n = 1, 2, \ldots \tag{2.3.15}$$

Now, we solve (2.3.7) for the eigenvalues of λ in (2.3.15). As explained in Appendix A.3.1 this is a separable or linear first order ODE. Substituting $\lambda = \lambda_n = (n\pi/L)^2$ in (2.3.7),

$$\frac{dT_n}{T_n} = -(\frac{an\pi}{L})^2 dt \Rightarrow \ln|T_n| = -(\frac{an\pi}{L})^2 t + c_n \Rightarrow T_n = e^{c_n} e^{-(\frac{an\pi}{L})^2 t}.$$

Setting $b_n = e^{c_n}$, we have

$$T_n = b_n e^{-(\frac{an\pi}{L})^2 t}.$$

The constants b_n are determined in the next step by the initial condition. (We could set $b_n = 1$ and reintroduce b_n in (2.3.16) below as coefficients when we apply the superposition principle.)

Step 3 (Finding the Solution): Since we assume that $u = XT$, the solutions are

$$u_n = X_n T_n = b_n e^{-(\frac{an\pi}{L})^2 t} \sin\frac{n\pi x}{L}, \quad n = 1, 2, \ldots.$$

Using the superposition principle, we take

$$u(x,t) = \textstyle\sum_{n=1}^{\infty} u_n = \displaystyle\sum_{n=1}^{\infty} b_n e^{-(\frac{an\pi}{L})^2 t} \sin\frac{n\pi x}{L} \tag{2.3.16}$$

as the solution to the problem. Our task now is to find b_n from the initial condition. At $t = 0$,we have

$$u(x,0) = \sum_{n=1}^{\infty} b_n X_n = \sum_{n=1}^{\infty} b_n \sin\frac{n\pi x}{L} = f(x). \tag{2.3.17}$$

The above series is called the Fourier sine series and a special case of the Fourier series. A brief discussion is given in the next subsection. The series is also called the eigenfunction expansion. To find b_m we multiply (2.3.17) by $X_m = \sin(m\pi x/L)$ and integrate over $0 < x < L$. Using the relation called the orthogonality relation,

$$\int_0^L X_m(x) X_n(x) dx = \int_0^L \sin\frac{m\pi x}{L} \sin\frac{n\pi x}{L} dx$$

$$= \begin{cases} \int_0^L \frac{1}{2}[\cos\frac{(m-n)}{L}\pi x - \cos\frac{(m+n)}{L}\pi x] dx = 0 & m \neq n \\ \int_0^L \frac{1}{2}[1 - \cos\frac{(m+n)}{L}\pi x] dx = \frac{L}{2} & m = n \end{cases}, \tag{2.3.18}$$

we obtain

$$b_m = \frac{\int_0^L f(x)X_m(x)dx}{\int_0^L X_m^2(x))dx} = \frac{2}{L}\int_0^L f(x)\sin\frac{m\pi x}{L}dx. \tag{2.3.19}$$

The coefficients b_m are called the Fourier coefficients and the integral $\int_0^L X_m^2(x))dx$ is called the normalizing integral.

We are interested in the behavior of solutions as time goes to infinity. In (2.3.16) n starts from one. This shows that u_n exponentially decays to zero for all $n \geq 1$ as t approaches infinity. Therefore, we expect that

$$\lim_{t\to\infty} u(x,t) = 0$$

for rather large class of initial data $f(x)$. Rigorously speaking, the conclusion is not obvious since we are adding infinitely many terms. If we need to be rigorous, we may have to use the Weierstrass M-test described in Appendix A.2 to show the convergence. However, we do not pursue such rigor here.

Remark 2.3.1. The following trigonometric identities are useful. If we remember the two formulas

$$\cos(\alpha+\beta) = \cos\alpha\cos\beta - \sin\alpha\sin\beta, \tag{2.3.20}$$

$$\sin(\alpha+\beta) = \sin\alpha\cos\beta + \cos\alpha\sin\beta, \tag{2.3.21}$$

we can derive most of the trigonometric identities. Changing the sign of b and using $\cos(-b) = \cos b$ and $\sin(-b) = -\sin b$, we obtain

$$\cos(\alpha-\beta) = \cos\alpha\cos\beta + \sin\alpha\sin\beta, \tag{2.3.22}$$

$$\sin(\alpha-\beta) = \sin\alpha\cos\beta - \cos\alpha\sin\beta. \tag{2.3.23}$$

Also, subtracting (2.3.20) from (2.3.22), we have the trigonometric identity used in (2.3.18)

$$\sin\alpha\sin\beta = \frac{1}{2}[\cos(\alpha-\beta) - \cos(\alpha+\beta)].$$

Example 2.3.2. (a) Solve

$$u_t - 4u_{xx} = 0, \quad 0 < x < L,\ 0 < t,$$

$$u(0,t) = 0, \quad u(L,t) = 0, \quad 0 < t,$$

$$u(x,0) = x.$$

(b) Find $\lim_{t\to\infty} u(x,t)$.

Solution: (a) In this problem the thermal diffusivity is $a = 2$ and $f(x) = x$. From (2.3.19)

$$\begin{aligned} b_n &= \frac{2}{L}\int_0^L f(x)\sin\frac{n\pi x}{L}dx = \frac{2}{L}\int_0^L x\sin\frac{n\pi x}{L}dx \\ &= \frac{2}{L}[-x\frac{L}{n\pi}\cos\frac{n\pi x}{L}]_0^L + \frac{2}{L}\int_0^L \frac{L}{n\pi}\cos\frac{n\pi x}{L}dx \\ &= \frac{2L}{n\pi}(-1)^{n-1}. \end{aligned}$$

Therefore, the solution (2.3.16) is

$$u(x,t) = \sum_{n=1}^{\infty}\frac{2L}{n\pi}(-1)^{n-1}e^{-(\frac{2n\pi}{L})^2 t}\sin\frac{n\pi x}{L}. \tag{2.3.24}$$

(b) Since all terms of u decays to zero, $\lim_{t\to\infty} u(x,t)=0$.

2.3.2 *Brief Discussion of the Fourier Series*

The series in (2.3.17) is a special case of the Fourier series given by

$$f(x) = \frac{a_0}{2} + \sum_{n=1}^{\infty}[a_n\cos\frac{n\pi x}{L} + b_n\sin\frac{n\pi x}{L}].$$

There are several questions concerning the Fourier series.

1. Can we express f in terms of trigonometric series? Also, one might want to ask what kind of f can be expressed in terms of trigonometric series.
2. Does the series converge to f?
3. Are the eigenvalues real?
4. Are the eigenfunctions mutually orthogonal?

These questions are affirmatively answered in Chapter 6. In this section we define a few more terminologies and visually observe how the Fourier series behaves (or converges).

Concerning the orthogonality relations used in (2.3.18), the integral $\int_a^b f(x)g(x)dx$ between two functions f and g is called the inner product. For a set of functions $\{y_1(x), y_2(x), \ldots\}$ defined on (a,b) if their inner products satisfy the orthogonality relations

$$\int_a^b y_i(x)y_j(x)dx = \begin{cases} 0 & i \neq j \\ c_i > 0 & i = j, \end{cases}$$

we say that the functions in the set are mutually orthogonal. The set $\{\sin(n\pi x/L)\}_{n=1}^{\infty}$ satisfies this property with $a = 0$ and $b = L$. If (2.3.18) did not hold, it would be impossible to find the Fourier coefficients.

Each term of the Fourier series is called the (Fourier) mode. Since we often use the first few Fourier modes to approximate the solutions, we use the solution (2.3.24) in Example 2.3.2 to illustrate how the partial sum

$$u_N(x,t) = \sum_{n=1}^{N} \frac{2}{n\pi}(-1)^{n-1} e^{-(\frac{2n\pi}{L})^2 t} \sin \frac{n\pi x}{L}$$

of the Fourier series solution approximates the solution as we increase the value of N. For simplicity we choose $t = 0$ so that we know that $u(x,0) = x$. In this case $u_N(x,0)$ is given by

$$u_N(x,0) = \sum_{n=1}^{N} \frac{2}{n\pi}(-1)^{n-1} \sin \frac{n\pi x}{L}.$$

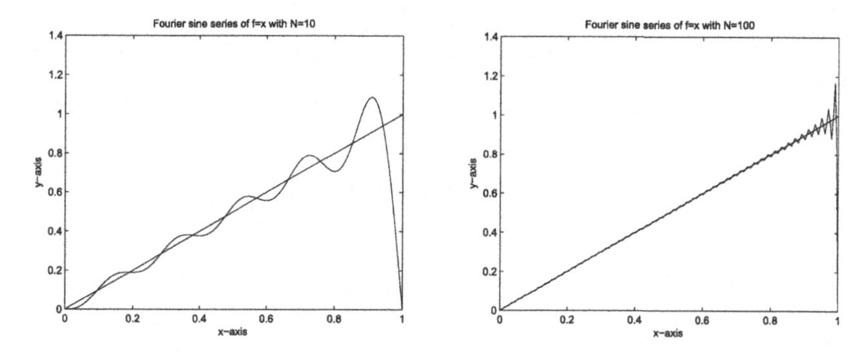

Fig. 2.4 Partial sums of the Fourier series of $f = \begin{cases} x, & 0 \le x < 1 \\ 0, & x = 1 \end{cases}$ with $N = 10$ (Left) and $N = 100$ (Right).

In Figure 2.4 the graphs of u_N for $N = 10, 100$ with $L = 1$ are plotted. We can observe a few interesting properties of approximation. Since we approximate $f = x$ as a partial sum of sine functions which are zero at $x = 1$ $(= L)$, there is a discontinuity at $x = 1$. We observe that the approximation is good away from $x = 1$ but not good near $x = 1$ at which there is a discontinuity. The approximation oscillates very badly around $f = x$ near the discontinuity. This phenomenon is called the Gibbs phenomenon and is typical behavior for the Fourier series. More details of Fourier series including convergence is discussed in Chapter 6.

The graphs of solutions at $t = 0.05$ and 0.1 with $N = 3$ and $N = 100$ are plotted in Figure 2.5. We see that the solution is smoothed out for $t > 0$ and $N = 3$ gives adequate results. Also, the graph of solution with $N = 10$ for $0 \le t \le 0.1$ is given in Figure 2.6. It is interesting to see how the solution decays to zero and how the Fourier series solution of the initial data is smooth out.

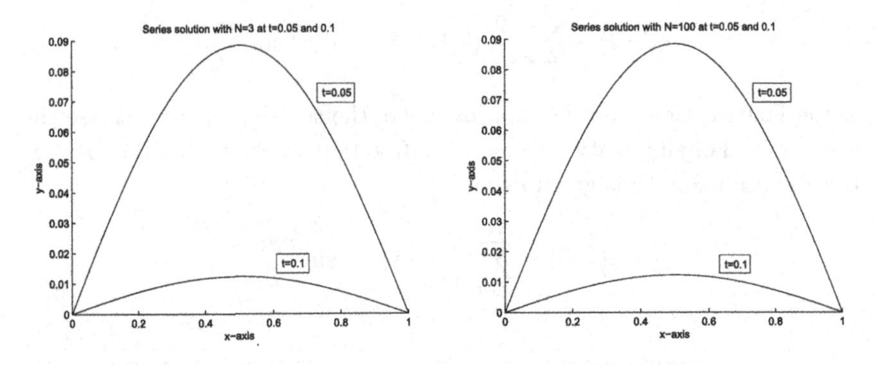

Fig. 2.5 Fourier series solutions at $t = 0.05$ and 0.1 with $N = 3$ (Left) and $N = 100$ (Right).

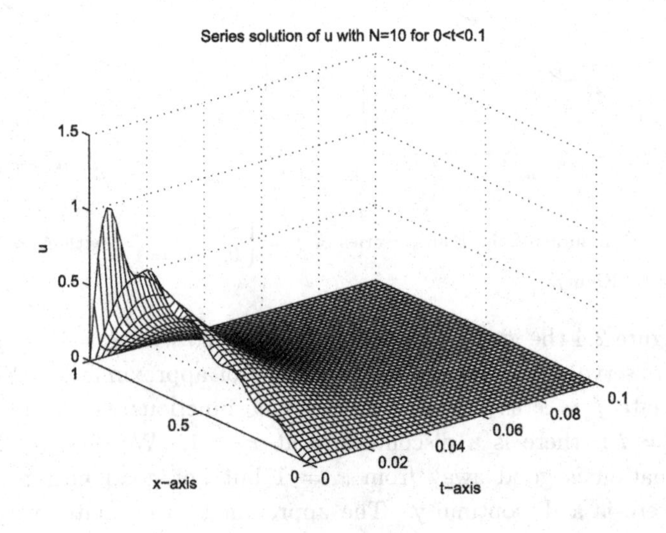

Fig. 2.6 Fourier series solution with $N = 10$ for $0 \le t \le 0.1$.

2.3.3 *Both Ends are Insulated*

As a second example of the separation of variables we consider the initial boundary value problem where the both ends of the bar are insulated. The initial boundary value problem is stated as follows.

$$u_t - a^2 u_{xx} = 0, \quad 0 < x < L,\ 0 < t,$$

$$u_x(0,t) = 0, \quad u_x(L,t) = 0, \quad 0 < t,$$

$$u(x,0) = f(x).$$

The goals are to see how the difference in the boundary conditions affects the solution and to get used to the procedure of the separation of variables. We will see the effects of the insulated boundary conditions to the solution and will see that the procedure is basically the same as Subsection 2.3.1.

Step 1: This step is precisely the same as Step 1 in Section 2.3.1. We assume

$$u(x,t) = X(x)T(t)$$

and derive ODE's for X and T

$$X'' + \lambda X = 0, \tag{2.3.25}$$

$$T' + \lambda a^2 T = 0. \tag{2.3.26}$$

Step 2: We solve the eigenvalue problem (2.3.25) for the spacial variable X first and then solve the initial value problem for T. The boundary conditions for X are

$$u_x(0,t) = 0 \Rightarrow \quad X'(0)T(t) = 0 \Rightarrow \quad X'(0) = 0, \tag{2.3.27}$$

$$u_x(L,t) = 0 \Rightarrow \quad X'(L)T(t) = 0 \Rightarrow \quad X'(L) = 0. \tag{2.3.28}$$

Notice that they are different from (2.3.8) and (2.3.9). The general solutions of (2.3.25) are given by

$$\lambda > 0 : \quad X = c_1 \cos\sqrt{\lambda}x + c_2 \sin\sqrt{\lambda}x, \tag{2.3.29}$$

$$\lambda = 0 : \quad X = c_1 + c_2 x, \tag{2.3.30}$$

$$\lambda < 0 : \quad X = c_1 e^{\sqrt{-\lambda}x} + c_2 e^{-\sqrt{-\lambda}x}. \tag{2.3.31}$$

We use the boundary conditions to find the eigenvalues. If $\lambda > 0$, differentiating (2.3.29), we have $X' = -c_1\sqrt{\lambda}\sin\sqrt{\lambda}x + c_2\sqrt{\lambda}\cos\sqrt{\lambda}x$ and from the boundary conditions (2.3.27) and (2.3.28),

$$X'(0) = c_2\sqrt{\lambda} = 0 \Rightarrow \quad c_2 = 0,$$

$$X'(L) = -c_1\sqrt{\lambda}\sin\sqrt{\lambda}L = 0.$$

To have a nontrivial solution, $c_1 \neq 0$ and therefore, $\sin\sqrt{\lambda}L = 0$. We conclude that

$$\sqrt{\lambda}L = n\pi \Rightarrow \lambda = (\frac{n\pi}{L})^2, \quad n = 1, 2, \ldots$$

are the eigenvalues and the corresponding eigenfunctions are

$$X_n = \cos\frac{n\pi x}{L}, \quad n = 1, 2, \ldots.$$

If $\lambda = 0$, from (2.3.30) $X' = c_2$ and from the boundary conditions (2.3.27) and (2.3.28),

$$X'(0) = c_2 = 0,$$

$$X'(L) = c_2 = 0.$$

Therefore, c_1 is arbitrary. So, $\lambda = 0$ is an eigenvalue and as a corresponding eigenfunction we take

$$X_0 = 1.$$

If $\lambda < 0$, from (2.3.31) $X' = c_1\sqrt{-\lambda}e^{\sqrt{-\lambda}x} - c_2\sqrt{-\lambda}e^{-\sqrt{-\lambda}x}$ and from the boundary conditions (2.3.27) and (2.3.28),

$$X'(0) = \sqrt{-\lambda}(c_1 - c_2) = 0,$$

$$X'(L) = c_1\sqrt{-\lambda}e^{\sqrt{-\lambda}L} - c_2\sqrt{-\lambda}e^{-\sqrt{-\lambda}L} = 0.$$

Substituting $c_1 = c_2$ in the second equation yields

$$X'(L) = 2c_1\sqrt{-\lambda}\sinh\sqrt{-\lambda}L = 0.$$

Since $\lambda < 0$, we see that $c_1 = c_2 = 0$ is the only solution. So, there is no non-trivial solution. In summary, the eigenvalues are

$$\lambda_n = (\frac{n\pi}{L})^2, \quad n = 0, 1, 2, \ldots$$

and the corresponding eigenfunctions are

$$X_n = \cos\frac{n\pi x}{L}, \quad n = 0, 1, 2, \ldots.$$

Now, we solve (2.3.7) for the eigenvalues $\lambda_n = (n\pi/L)^2$. Then,

$$T_n = a_n e^{-(\frac{an\pi}{L})^2 t}.$$

Step 3: Since we assume that $u = XT$, the solutions are

$$u_n = X_n T_n = a_n e^{-(\frac{an\pi}{L})^2 t}\cos\frac{n\pi x}{L}, \quad n = 0, 1, 2, \ldots.$$

Using the superposition principle, we take

$$u(x,t) = \frac{u_0}{2} + \sum_{n=1}^{\infty} u_n$$
$$= \frac{a_0}{2} + \sum_{n=1}^{\infty} a_n e^{-(\frac{an\pi}{L})^2 t} \cos\frac{n\pi x}{L} \tag{2.3.32}$$

as the solution to the problem. It looks strange that a_0 (or u_0) is divided by 2. The reason is that if we do so, the case $m = 0$ is included in the formula (2.3.34) below. Our task now is to find a_n from the initial condition. At $t = 0$,

$$u(x,0) = \frac{a_0}{2} + \sum_{n=1}^{\infty} a_n \cos\frac{n\pi x}{L} = f(x).$$

To find a_m we multiply by $\cos(m\pi x/L)$ and integrate over $0 < x < L$. Using the orthogonality relations

$$\int_0^L X_m(x)X_n(x)dx = \int_0^L \cos\frac{m\pi x}{L}\cos\frac{n\pi x}{L}dx = \begin{cases} 0 & m \neq n \\ \frac{L}{2} & m = n \end{cases}, \tag{2.3.33}$$

we obtain

$$a_m = \frac{\int_0^L f(x)X_m(x)dx}{\int_0^L X_m^2(x)dx} = \frac{2}{L}\int_0^L f(x)\cos\frac{m\pi x}{L}dx, \quad m = 0,1,2,\ldots. \tag{2.3.34}$$

Since in (2.3.32)

$$\frac{a_0}{2} = \frac{1}{L}\int_0^L f(x)dx$$

and the rest of terms decays to zero, u approaches the average temperature of the initial data as $t \to \infty$, *i.e.*,

$$\lim_{t\to\infty} u(x,t) = \frac{1}{L}\int_0^L f(x)dx.$$

As both ends are insulated, the heat in the rod does not escape. It is rather natural to see that the solution approaches the average temperature of the initial data as $t \to \infty$,

Example 2.3.3. (a) Solve

$$u_t - a^2 u_{xx} = 0, \quad 0 < x < L,\ 0 < t,$$

$$u_x(0,t) = 0, \quad u_x(L,t) = 0,$$

$$u(x,0) = x.$$

(b) Find $\lim_{t\to\infty} u(x,t)$.
Solution: (a) The initial condition is given by $f = x$. Then, from (2.3.34)

$$a_0 = \frac{2}{L}\int_0^L f(x)\cos\frac{0\pi x}{L}dx = \frac{2}{L}\int_0^L xdx = L,$$

$$\begin{aligned} a_n &= \frac{2}{L}\int_0^L f(x)\cos\frac{n\pi x}{L}dx = \frac{2}{L}\int_0^L x\cos\frac{n\pi x}{L}dx \\ &= \frac{2}{L}[x\frac{L}{n\pi}\sin\frac{n\pi x}{L}]_0^L - \frac{2}{L}\int_0^L \frac{L}{n\pi}\sin\frac{n\pi x}{L}dx \\ &= \frac{2}{L}(\frac{L}{n\pi})^2[\cos\frac{n\pi x}{L}]_0^L = \frac{2L}{(n\pi)^2}[(-1)^n - 1], \quad n = 1, 2, \ldots. \end{aligned}$$

Therefore, the solution (2.3.32) is

$$u(x,t) = \frac{L}{2} + \sum_{n=1}^{\infty} \frac{2L}{(n\pi)^2}[(-1)^n - 1]e^{-(\frac{an\pi}{L})^2 t}\cos\frac{n\pi x}{L}.$$

(b) $\lim_{t\to\infty} u(x,t) = \frac{L}{2}$.

2.3.4 *Temperature of One End is Zero and the Other End is Insulated*

We consider the following problem.

$$u_t - a^2 u_{xx} = 0, \quad 0 < x < L,\ 0 < t,$$

$$u(0,t) = 0, \quad u_x(L,t) = 0,$$

$$u(x,0) = f(x).$$

In this problem the temperature is fixed at zero at the left end and insulated at the right end. We will see how the above boundary conditions affect the solution. As in the previous subsections, we use the separation of variables.

Step 1: This step is precisely the same as Step 1 in Section 2.3.1. We assume

$$u(x,t) = X(x)T(t)$$

and derive ODE's for X and T

$$X'' + \lambda X = 0, \tag{2.3.35}$$

$$T' + \lambda a^2 T = 0. \tag{2.3.36}$$

Step 2: We solve the eigenvalue problem (2.3.35) first. We look for nontrivial solutions. The boundary conditions are

$$u(0,t) = 0 \Rightarrow X(0)T(t) = 0 \Rightarrow X(0) = 0, \tag{2.3.37}$$

$$u_x(L,t) = 0 \Rightarrow X'(L)T(t) = 0 \Rightarrow X'(L) = 0. \tag{2.3.38}$$

Notice the difference from the previous subsections. The general solution of (2.3.35) is given by

$$\lambda > 0 : \quad X = c_1 \cos\sqrt{\lambda}x + c_2 \sin\sqrt{\lambda}x, \tag{2.3.39}$$

$$\lambda = 0 : \quad X = c_1 + c_2 x, \tag{2.3.40}$$

$$\lambda < 0 : \quad X = c_1 e^{\sqrt{-\lambda}x} + c_2 e^{-\sqrt{-\lambda}x}. \tag{2.3.41}$$

We use the boundary conditions to figure out what values of λ are eigenvalues. If $\lambda > 0$, from (2.3.39) $X' = -c_1\sqrt{\lambda}\sin\sqrt{\lambda}x + c_2\sqrt{\lambda}\cos\sqrt{\lambda}x$ and from the boundary conditions (2.3.37) and (2.3.38),

$$X(0) = c_1 = 0,$$

$$X'(L) = c_2\sqrt{\lambda}\cos\sqrt{\lambda}L = 0.$$

To have a nontrivial solution, $c_2 \neq 0$. Therefore, from $\cos\sqrt{\lambda}L = 0$

$$\sqrt{\lambda}L = (n - \frac{1}{2})\pi \Rightarrow \lambda = (\frac{(n-\frac{1}{2})\pi}{L})^2, \quad n = 1, 2, \ldots$$

are the eigenvalues and the corresponding eigenfunctions are

$$X_n = \sin\frac{(n-\frac{1}{2})\pi x}{L}, \quad n = 1, 2, \ldots.$$

If $\lambda = 0$, from (2.3.40) $X' = c_2$ and from the boundary conditions (2.3.37) and (2.3.38),

$$X(0) = c_1 = 0,$$

$$X'(L) = c_2 = 0.$$

Therefore, $\lambda = 0$ is not an eigenvalue.

If $\lambda < 0$, from (2.3.41) $X' = c_1\sqrt{-\lambda}e^{\sqrt{-\lambda}x} - c_2\sqrt{-\lambda}e^{-\sqrt{-\lambda}x}$ and from the boundary conditions (2.3.37) and (2.3.38),

$$X(0) = c_1 + c_2 = 0,$$

$$X'(L) = c_1\sqrt{-\lambda}e^{\sqrt{-\lambda}L} - c_2\sqrt{-\lambda}e^{-\sqrt{-\lambda}L} = 0.$$

Substituting $c_2 = -c_1$ to the second equation, we have

$$X'(L) = c_1\sqrt{-\lambda}e^{\sqrt{-\lambda}L} + c_1\sqrt{-\lambda}e^{-\sqrt{-\lambda}L} = 2c_1\sqrt{-\lambda}\cosh e^{\sqrt{-\lambda}L} = 0.$$

Since $\lambda < 0$, we see that $c_1 = c_2 = 0$ is the only solution. So, there is no non-trivial solution. In summary, the eigenvalues are

$$\lambda_n = (\frac{(n-\frac{1}{2})\pi}{L})^2, \quad n = 1, 2, \ldots$$

and the corresponding eigenfunctions are

$$X_n = \sin\frac{(n-\frac{1}{2})\pi x}{L}, \quad n = 1, 2, \ldots.$$

Now, solving (2.3.36) for the eigenvalues $\lambda_n = (\frac{(n-\frac{1}{2})\pi}{L})^2$, we obtain

$$T_n = a_n e^{-(\frac{a(n-\frac{1}{2})\pi}{L})^2 t}.$$

Step 3: Since we assume that $u = XT$, the solutions are

$$u_n = X_n T_n = a_n e^{-(\frac{a(n-\frac{1}{2})\pi}{L})^2 t}\sin\frac{(n-\frac{1}{2})\pi x}{L}, \quad n = 1, 2, \ldots.$$

Using the superposition principle, we take

$$\begin{aligned} u(x,t) &= \sum_{n=1}^{\infty} u_n \\ &= \sum_{n=1}^{\infty} a_n e^{-(\frac{a(n-\frac{1}{2})\pi}{L})^2 t}\sin\frac{(n-\frac{1}{2})\pi x}{L} \end{aligned} \tag{2.3.42}$$

as the solution to the problem. We use the initial condition to find a_n. At $t = 0$,

$$u(x,0)\sum_{n=1}^{\infty} a_n X_n(x) = \sum_{n=1}^{\infty} a_n \sin\frac{(n-\frac{1}{2})\pi x}{L} = f(x). \tag{2.3.43}$$

To find a_m we multiply by $\sin[(m-1/2)\pi x/L]$ and integrate over $0 < x < L$. Using the orthogonality relation

$$\int_0^L X_m(x)X_n(x)dx = \int_0^L \sin\frac{(m-\frac{1}{2})\pi x}{L}\sin\frac{(n-\frac{1}{2})\pi x}{L}dx = \begin{cases} \frac{L}{2} & m = n \\ 0 & m \neq n \end{cases}, \tag{2.3.44}$$

we obtain

$$a_m = \frac{\int_0^L f(x)X_m(x)dx}{\int_0^L X_m^2(x)dx} = \frac{2}{L}\int_0^L f(x)\sin\frac{(m-\frac{1}{2})\pi x}{L}dx, \quad m = 1, 2, \ldots. \tag{2.3.45}$$

Example 2.3.4. (a) Solve

$$u_t - 9u_{xx} = 0, \quad 0 < x < L,\ 0 < t,$$

$$u(0,t) = 0, \quad u_x(L,t) = 0,$$

$$u(x,0) = x.$$

(b) Find $\lim_{t\to\infty} u(x,t)$.

Solution: (a) From (2.3.45)

$$\begin{aligned} a_n &= \frac{2}{L}\int_0^L f(x)\sin\frac{(n-\frac{1}{2})\pi x}{L}dx = \frac{2}{L}\int_0^L x\sin\frac{(n-\frac{1}{2})\pi x}{L}dx \\ &= \frac{2}{L}[-x\frac{L}{(n-\frac{1}{2})\pi}\cos\frac{(n-\frac{1}{2})\pi x}{L}]_0^L + \frac{2}{L}\int_0^L \frac{L}{(n-\frac{1}{2})\pi}\cos\frac{(n-\frac{1}{2})\pi x}{L}dx \\ &= \frac{2L}{[(n-\frac{1}{2})\pi]^2}[\sin\frac{(n-\frac{1}{2})\pi x}{L}]_0^L = \frac{2L}{[(n-\frac{1}{2})\pi]^2}(-1)^{n-1}, \quad n = 1, 2, \ldots. \end{aligned}$$

Therefore, the solution is

$$u(x,t) = \sum_{n=1}^{\infty} \frac{2L}{[(n-\frac{1}{2})\pi]^2}(-1)^{n-1}e^{-(\frac{3(n-\frac{1}{2})\pi}{L})^2 t}\sin\frac{(n-\frac{1}{2})\pi x}{L}.$$

(b) $\lim_{t\to\infty} u(x,t) = 0$.

Exercises

1. Determine whether the method of separation of variables can be used to replace the given partial differential equations by a set of ordinary differential equations. If so, find the equations.
   (a) $u_t = ku_{xx} - v_0 u_x$
   (b) $u_t + u_{xt} + u_{yy} = 0$
   (c) $u_{tt} + (x-t)u_{xx} = 0$
   (d) $\frac{\partial u}{\partial t} = \frac{t}{r}\frac{\partial}{\partial r}(r\frac{\partial u}{\partial r})$
2. Verify (2.3.33) and (2.3.44).
3. Find the solutions to the heat equation

   $$u_t - a^2 u_{xx} = 0, \quad 0 < x < 2, \quad 0 < t,$$

   with the initial condition

   $$u(x,0) = 1,$$

   for the following boundary conditions.

(a) $u(0,t) = 0, \quad u(2,t) = 0$
(b) $u_x(0,t) = 0, \quad u_x(2,t) = 0$
(c) $u(0,t) = 0, \quad u_x(2,t) = 0$
(d) $u_x(0,t) = 0, \quad u(2,t) = 0$

4. Consider the heat equation

$$u_t = a^2 u_{xx}$$

with the boundary conditions

$$u(0,t) = 0 \quad \text{and} \quad u(L,t) = 0.$$

Solve the initial boundary value problem if the initial temperature is give by

$$u(x,0) = \begin{cases} 1 & 0 < x < \frac{L}{2} \\ 2 & \frac{L}{2} < x < L. \end{cases}$$

5. Consider

$$\begin{aligned} u_t &= a^2 u_{xx}, \ 0 < x < 1, \ t > 0, \\ u_x(0,t) &= 0, \ u_x(1,t) = 0, \\ u(x,0) &= 1 - x. \end{aligned}$$

(a) Describe the physical meaning of the boundary conditions.
(b) Find the solution.
(c) What does the solution in (b) approach as t goes to infinity?

6. Consider

$$u_t = a^2 u_{xx}, \quad 0 < x < L, \quad 0 < t,$$

$$u(x,0) = 0, \quad u(x,L) = 0,$$

$$u(x,0) = f(x).$$

(a) What is the total heat energy in the rod as a function of time?
(b) What is the flux of heat energy at $x = 0$ and $x = L$?
(c) What relationship should exist between (a) and (b)?

7. Find the eigenvalues and the eigenfunctions of (2.3.46) for the boundary conditions (a) and (b). You do not have to show the details. If you have developed your shortcuts, use them. If you know the solutions, you can just write them down.

$$\frac{d^2 X}{dx^2} + \lambda X = 0, \ 0 < x < L. \tag{2.3.46}$$

(a) $X(0) = 0, \quad X(L) = 0$
(b) $X(0) = 0, \quad \frac{dX}{dx}(L) = 0$

8. Consider the heat equation where there is heat loss from the cylindrical side. If the ambient temperature is zero, the equation for the temperature is given by

$$u_t = a^2 u_{xx} - bu, \quad 0 < x < L, \ 0 < t,$$

where b is a positive constant. Suppose the initial condition is

$$u(x, 0) = L - x,$$

find the solutions for the following boundary conditions.

(a) $u(0,t) = 0, \quad u(L,t) = 0$
(b) $u_x(0,t) = 0, \quad u_x(L,t) = 0$
(c) $u(0,t) = 0, \quad u_x(L,t) = 0$
(d) $u_x(0,t) = 0, \quad u(L,t) = 0$

2.4 Non-homogeneous Boundary Conditions

We now study the case where the boundary conditions are non-homogeneous. We restrict our attention to the case where the non-homogeneous terms are constants. We also treat the case where the source term is a function of x but not a functions of x and t.

2.4.1 *Steady State Solutions*

We introduce the steady state solutions for the heat equation. They are also called the equilibrium solutions. They are the time independent solutions to the heat equation satisfying the boundary conditions. It is important to know that they do not have to satisfy the initial conditions. For example, the steady state solution $u(x,t) = v(x)$ to

$$u_t - a^2 u_{xx} = 0, \quad 0 < x < L, \quad 0 < t, \tag{2.4.1}$$

$$u(0,t) = T_1, \quad u(L,t) = T_2, \tag{2.4.2}$$

$$u(x,0) = f(x), \tag{2.4.3}$$

satisfies

$$v_{xx} = 0, \tag{2.4.4}$$

$$v(0) = T_1, \quad v(L) = T_2. \tag{2.4.5}$$

It is expected that the solution to (2.4.1) to (2.4.3) approaches the steady state solution v as t approaches infinity. It is not difficult to find the steady state solution. Integrating (2.4.4) twice in x leads to

$$v = c_1 + c_2 x.$$

We choose c_1 and c_2 so that the boundary conditions are satisfied.

$$v(0) = c_1 = T_1, \quad v(L) = c_1 + c_2 L = T_2.$$

Therefore,

$$v(x) = T_1 + \frac{T_2 - T_1}{L} x. \tag{2.4.6}$$

Example 2.4.1. Find the steady state solution to

$$u_t - a^2 u_{xx} = 0, \quad 0 < x < L, \quad 0 < t, \tag{2.4.7}$$

$$u_x(0,t) = T_1, \quad u(L,t) = T_2. \tag{2.4.8}$$

$$u(x,0) = f(x). \tag{2.4.9}$$

Solution: The steady state solution satisfies

$$v_{xx} = 0, \tag{2.4.10}$$

$$v_x(0) = T_1, \quad v(L) = T_2. \tag{2.4.11}$$

As in the previous example,

$$v = c_1 + c_2 x.$$

We choose c_1 and c_2 so that the boundary conditions are satisfied.

$$v_x(0) = c_2 = T_1, \quad v(L) = c_1 + c_2 L = T_2.$$

Therefore,

$$v(x) = T_2 - T_1 L + T_1 x.$$

If the both boundaries are insulated, a special care is necessary. As an example, consider the steady state solution for

$$u_t - a^2 u_{xx} = h(x),$$

$$u_x(0,t) = T_1, \quad u_x(L,t) = T_2,$$

$$u(x,0) = f(x).$$

The steady state solution $u = v(x)$ satisfies

$$-a^2 v_{xx}(x) = h(x), \tag{2.4.12}$$

$$v'(0) = T_1, \quad v'(L) = T_2. \tag{2.4.13}$$

Integrating (2.4.12) from $x = 0$ to L, we find that T_1, T_2, and h must satisfy

$$-a^2(v'(L) - v'(0)) = -a^2(T_2 - T_1) = \int_0^L h(s)ds. \tag{2.4.14}$$

This condition is called the solvability condition. Therefore, to have a steady state solution, T_1 and T_2 cannot be specified arbitrarily. In particular if $h \equiv 0$, $T_1 = T_2$ must hold to have a steady state solution. In the next example, the steady state solution is computed for the case where $h \equiv 0$ and $T_1 = T_2$ is satisfied.

Example 2.4.2. Find the steady state solution to

$$u_t - a^2 u_{xx} = 0, \quad 0 < x < L, \quad 0 < t,$$

$$u_x(0,t) = T_1, \quad u_x(L,t) = T_1,$$

$$u(x,0) = f(x).$$

Solution: The steady state solution $u = v(x)$ satisfies

$$v_{xx} = 0,$$

$$v'(0) = T_1, \quad v'(L) = T_1.$$

Integrating $v_{xx} = 0$, we have $v = c_1 x + c_2$. Using the boundary conditions, we see that

$$v' = c_1 = T_1.$$

Therefore, $v = T_1x + c_2$, where so far c_2 is an arbitrary constant, is a steady state solution. If the initial condition is given, we can determine c_2. Integrating the equation both in x and t and expecting that $\lim_{t\to\infty} u(x,t) = v(x) = T_1x + c_2$, we have

$$\int_0^\infty \int_0^L (u_t - a^2u_{xx})dxdt$$
$$= \int_0^L [\lim_{t\to\infty} u(x,t) - u(x,0)]dx - \int_0^\infty a^2[u_x(L,t) - u_x(0,t)]dt = 0.$$

Substituting $u_x(0,t) = u_x(L,t) = T_1$, $\lim_{t\to\infty} u(x,t) = T_1x + c_2$, and $u(x,0) = f(x)$, we obtain

$$\int_0^L \lim_{t\to\infty} u(x,t)dx = \int_0^L [T_1x + c_2]dx$$
$$= \frac{L^2}{2}T_1 + Lc_2 = \int_0^L f(x)dx.$$

$$c_2 = \frac{1}{L}[\int_0^L f(x)dx - \frac{L^2}{2}T_1].$$

2.4.2 *Non-homogeneous Boundary Conditions*

Now consider the case where the boundary conditions are non-homogeneous. The procedure can be thought of as an application of the superposition principle discussed in Section 1.2. As an example consider the initial boundary value problem (2.4.1) to (2.4.3). To solve the problem, we assume that the solution is given by

$$u(x,t) = w(x,t) + v(x), \tag{2.4.15}$$

where v is the steady state solution (2.4.6) to the problem given in (2.4.4) and (2.4.5). We choose w so that the initial boundary value problem for w is something we can handle. Substituting (2.4.15) to (2.4.1), (2.4.3), and (2.4.2) and using (2.4.6), we see that w satisfies

$$w_t - a^2w_{xx} = 0, \tag{2.4.16}$$

$$w(0,t) = u(0,t) - v(0) = 0, \quad w(L,t) = u(L,t) - v(L) = 0, \tag{2.4.17}$$

$$w(x,0) = f(x) - v(x). \tag{2.4.18}$$

This is the same problem as we did in Section 2.3.1. We can handle the other types of problems in a similar way. The above procedure is explained in the following example. Note that the boundary conditions are different.

Example 2.4.3. Consider

$$u_t - u_{xx} = 0,\ 0 < x < L,\ \ 0 < t, \tag{2.4.19}$$

$$u(0,t) = T_1,\ \ u_x(L,t) = T_2,\ \ 0 < t, \tag{2.4.20}$$

$$u(x,0) = T_2 x + T_3. \tag{2.4.21}$$

(a) Find the steady state solution $v(x)$.

(b) Assume $u(x,t) = w(x,t) + v(x)$ to find the initial boundary value problem for $w(x,t)$.

(c) Solve for $w(x,t)$ and find the solution to the above initial boundary value problem.

Solution: (a) The steady state solution $v(x)$ satisfies

$$v_{xx} = 0,$$

$$v(0) = T_1, \quad v_x(L) = T_2.$$

Integrating the equation, we have

$$v(x) = c_1 x + c_2,$$

where c_1 and c_2 are constants determined from the boundary conditions. Applying the boundary conditions, we see that

$$v(0) = c_2 = T_1, \quad v_x(L) = c_1 = T_2.$$

Therefore,the steady state solution is $v = T_2 x + T_1$.

(b) We assume that the solution is given by

$$u(x,t) = w(x,t) + v(x)$$

and find the equation and the initial and boundary conditions for w. Substituting the above u to (2.4.19)-(2.4.21), we obtain

$$w_t - w_{xx} - v_{xx} = 0 \Rightarrow w_t - w_{xx} = 0, \tag{2.4.22}$$

$$w(0,t) + v(0) = w(0,t) + T_1 = T_1 \Rightarrow w(0,t) = 0,$$

$$w_x(L,t) + v_x(L) = w(L,t) + T_2 = T_2 \Rightarrow w_x(L,t) = 0,$$

$$w(x,0) + v(x) = w(x,0) + T_2 x + T_1 = f(x) \Rightarrow w(x,0) = T_2 x + T_3 - T_2 x - T_1.$$

Therefore, w satisfies

$$w_t - w_{xx} = 0, \quad 0 < x < L,\ 0 < t,$$

$$w(0,t) = 0, \quad w_x(L,t) = 0,$$

$$w(x,0) = T_3 - T_1.$$

(c) The initial boundary value problem is the same as the one discussed in Subsection 2.3.4. Therefore,

$$w(x,t) = \sum_{n=1}^{\infty} a_n e^{-(\frac{a(n-\frac{1}{2})\pi}{L})^2 t} \sin \frac{(n-\frac{1}{2})\pi x}{L},$$

where a_n is given by

$$\begin{aligned} a_n &= \frac{2}{L}\int_0^L w(x,0) \sin \frac{(n-\frac{1}{2})\pi x}{L} dx \\ &= \frac{2}{L}\int_0^L (T_3 - T_1) \sin \frac{(n-\frac{1}{2})\pi x}{L} dx \\ &= \frac{2}{L}[-(T_3 - T_1)\frac{L}{(n-\frac{1}{2})\pi} \cos \frac{(n-\frac{1}{2})\pi x}{L}]_0^L \\ &= \frac{2(T_3 - T_1)}{(n-\frac{1}{2})\pi}, \quad n = 1, 2, \ldots. \end{aligned}$$

The solution is

$$\begin{aligned} u(x,t) &= w(x,t) + v(x) \\ &= \sum_{n=1}^{\infty} \frac{2(T_3 - T_1)}{(n-\frac{1}{2})\pi} e^{-(\frac{a(n-\frac{1}{2})\pi}{L})^2 t} \sin \frac{(n-\frac{1}{2})\pi x}{L} + T_2 x + T_1. \end{aligned}$$

Exercises

1. Find the steady state solutions.

   (a) $\frac{d^2u}{dx^2} + 1 = 0, \quad u(0) = T_1, \quad u(L) = T_2.$
   (b) $\frac{d^2u}{dx^2} + x = 0, \quad u(0) = T_1, \quad u_x(L) = T_2.$
   (c) $\frac{d^2u}{dx^2} = 0, \quad u(0) = T, \quad \frac{du}{dx}(L) + u(L) = 0.$
   (d) $\frac{d^2u}{dx^2} + 1 = 0, \quad u(0) + \frac{du}{dx}(0) = T_1, \quad u(L) = T_2.$

2. For each problem find the value of b for which there is an equilibrium solution and determine the equilibrium solutions.

   (a) $u_t = u_{xx} + x, \quad u(x,0) = f(x), \quad u_x(0,t) = b, \quad u_x(L,t) = 2.$
   (b) $u_t = u_{xx} + b, \quad u(x,0) = f(x), \quad u_x(0,t) = 1, \quad u_x(L,t) = 2.$

3. Consider

$$u_t - a^2 u_{xx} = 0,\ 0 < x < L,\quad t > 0,$$
$$u_x(0,t) = 1,\quad u(L,t) = 0,$$
$$u(x,0) = x.$$

(a) Find the steady state solution $v(x)$.
(b) Assume $u(x,t) = w(x,t) + v(x)$ and find the initial boundary value problem for $w(x,t)$.
(c) Find $w(x,t)$ and find the solution to the above initial boundary value problem.

4. Consider

$$u_t = u_{xx} + x,\quad 0 < x < L,\ t > 0,$$
$$u_x(0,t) = b,\ u_x(L,t) = 7,$$
$$u(x,0) = -\frac{x^3}{6} + (\frac{L^2}{2} + 8)x.$$

(a) Determine the value(s) of b for which there is an equilibrium temperature distribution.
(b) Find the equilibrium solution(s) for the above value(s) of b.
(c) For the value(s) of b found in (a), find the solution $u(x,t)$.

2.5 Robin Boundary Conditions

The Robin boundary conditions are given by (2.2.1). If we restrict to the case where the ambient temperature is zero, the boundary conditions for the Robin problems are generally given by

$$u_x(0,t) - a_0 u(0,t) = 0, \tag{2.5.1}$$

$$u_x(L,t) + a_L u(L,t) = 0, \tag{2.5.2}$$

where a_0 and a_L are constants. We choose the their signs so that the boundary conditions are consistent with the Fourier law (2.1.6). If the temperature is positive at $x = 0$, the heat leaks out from the boundary since ambient temperature is zero. This means that the heat moves to the left and by the Fourier law $u_x(0,t) > 0$. This implies that $a_0 > 0$ is the physically reasonable sign. The case when $u(0,t) < 0$ leads to the same conclusion. Using the similar argument for (2.5.2), we can show that $a_L > 0$.

As an example of the Robin boundary conditions, we consider

Example 2.5.1. Solve the heat equation

$$u_t - a^2 u_{xx} = 0, \quad 0 < x < L, \ 0 < t, \tag{2.5.3}$$

with the boundary conditions

$$u(0,t) = 0, \quad u_x(L,t) + a_L u(L,t) = 0 \tag{2.5.4}$$

and the initial condition

$$u(x,0) = f(x). \tag{2.5.5}$$

Solution: As you see the Robin boundary condition at $x = 0$ is changed to the Dirichlet boundary condition. We discuss the case where $a_L > 0$. The case where $a_L < 0$ will be treated in Exercises in this section.

Step 1: This step is the same as the previous problems. Assuming $u(x,t) = X(x)T(t)$, we obtain two ODE's

$$X'' + \lambda X = 0, \tag{2.5.6}$$

$$T' + \lambda a^2 T = 0, \tag{2.5.7}$$

and we need to determine the values of λ for which we have nontrivial solutions.

Step 2: We solve the set of ODE's obtained in Step 1 starting from (2.5.6). The boundary conditions for X are

$$u(0,t) = 0 \Rightarrow X(0) = 0, \tag{2.5.8}$$

$$u_x(L,t) + a_L u(L,t) = 0 \Rightarrow X'(L) + a_L X(L) = 0. \tag{2.5.9}$$

The general solutions of (2.5.6) are given by

$$\lambda > 0 : \quad X = c_1 \cos\sqrt{\lambda}x + c_2 \sin\sqrt{\lambda}x, \tag{2.5.10}$$

$$\lambda = 0 : \quad X = c_1 + c_2 x, \tag{2.5.11}$$

$$\lambda < 0 : \quad X = c_1 e^{\sqrt{-\lambda}x} + c_2 e^{-\sqrt{-\lambda}x}. \tag{2.5.12}$$

We use the boundary conditions to figure out what values of λ are eigenvalues. For $\lambda > 0$, (2.5.10) with the boundary conditions (2.5.8) and (2.5.9) implies

$$X(0) = c_1 = 0, \tag{2.5.13}$$

$$\begin{aligned} &X'(L) + a_L X(L) \\ &= -\sqrt{\lambda}c_1 \sin\sqrt{\lambda}L + \sqrt{\lambda}c_2 \cos\sqrt{\lambda}L \\ &\quad + a_L(c_1 \cos\sqrt{\lambda}L + c_2 \sin\sqrt{\lambda}L) = 0. \end{aligned} \tag{2.5.14}$$

Since $c_1 = 0$, (2.5.14) leads to

$$[\sqrt{\lambda}\cos\sqrt{\lambda}L + a_L \sin\sqrt{\lambda}L]c_2 = 0.$$

To have a nontrivial solution, $c_2 \neq 0$. This forces us to choose λ such that

$$\tan\sqrt{\lambda}L = -\frac{\sqrt{\lambda}}{a_L}.$$

Unlike the previous examples, it is difficult to find the explicit values of eigenvalues. However, if we set $x = \sqrt{\lambda}L$, it is possible to find eigenvalues approximately by graphing the two functions $y = \tan x$ and $y = -\frac{x}{La_L}$, and finding the intersection points. The case where $La_L = 1$ is illustrated in Figure 2.7.

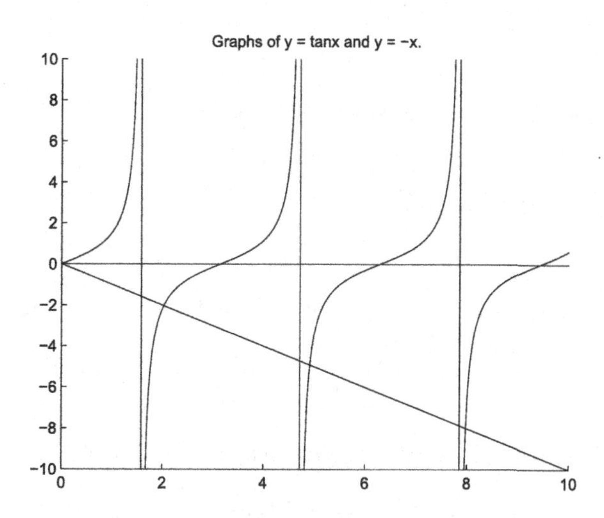

Fig. 2.7 Graphs of $y = \tan x$ and $y = -x$.

For $\lambda = 0$, from (2.5.11) with the boundary conditions (2.5.8) and (2.5.9), we obtain

$$X(0) = c_1 = 0,$$

$$X'(L) + a_L X(L) = c_2 + a_L(c_1 + Lc_2) = 0.$$

Therefore, $c_1 = c_2 = 0$. So, $\lambda = 0$ is not an eigenvalue.

If $\lambda < 0$, from (2.5.12) with the boundary conditions (2.5.8) and (2.5.9), we obtain

$$X(0) = c_1 + c_2 = 0, \tag{2.5.15}$$

$$
\begin{aligned}
&X'(L) + a_L X(L) \\
&= \sqrt{-\lambda}(c_1 e^{\sqrt{-\lambda}L} - c_2 e^{-\sqrt{-\lambda}L}) + a_L(c_1 e^{\sqrt{-\lambda}L} + c_2 e^{-\sqrt{-\lambda}L}) \\
&= 0. \qquad (2.5.16)
\end{aligned}
$$

Substitution of $c_2 = -c_1$in (2.5.15) to (2.5.16) leads to

$$
\begin{aligned}
&\sqrt{-\lambda}c_1(e^{\sqrt{-\lambda}L} + e^{-\sqrt{-\lambda}L}) + a_L c_1(e^{\sqrt{-\lambda}L} - e^{-\sqrt{-\lambda}L}) \\
&= \frac{1}{2}c_1(\sqrt{-\lambda}\cosh\sqrt{-\lambda}L + a_L \sinh\sqrt{-\lambda}L) = 0.
\end{aligned}
$$

Since both $\cosh\sqrt{-\lambda}L$ and $\sinh\sqrt{-\lambda}L$ are positive for $\lambda < 0$, there is no non-trivial solution for $\lambda < 0$.

In summary, the eigenvalues are positive. Unfortunately, we do not know the exact values. So, let's denote them as

$$0 < \lambda_1 < \cdots < \lambda_n < \cdots$$

and denote the corresponding eigenfunctions as

$$X_n = \sin\sqrt{\lambda_n}x, \quad n = 1, 2, \ldots.$$

However, from Figure 2.7 we see that

$$(\frac{\pi}{2L})^2 < \lambda_1 < (\frac{3\pi}{2L})^2 < \cdots < (\frac{(2n-1)\pi}{2L})^2 < \lambda_n < (\frac{(2n+1)\pi}{2L})^2 < \cdots.$$

This relation does not change even if $La_L \neq 1$. Also, we observe that as $n \to \infty$ the eigenvalues behave like

$$\lambda_n \approx (\frac{(2n-1)\pi}{2L})^2.$$

Now, solving (2.5.7) for the eigenvalues of λ, we have

$$\frac{dT_n}{T_n} = -a^2\lambda_n dt \Rightarrow \ln|T_n| = -a^2\lambda_n t + c_n \Rightarrow T_n = b_n e^{-a^2\lambda_n t},$$

where $b_n = e^{c_n}$.

Step 3: Since we assume that $u = XT$, the solutions are

$$u_n = X_n T_n = b_n e^{-a^2\lambda_n t}\sin\sqrt{\lambda_n}x, \quad n = 1, 2, \ldots.$$

Using the superposition principle, we take

$$u(x,t) = \textstyle\sum_{n=1}^{\infty} u_n = \displaystyle\sum_{n=1}^{\infty} b_n e^{-a^2\lambda_n t}\sin\sqrt{\lambda_n}x \qquad (2.5.17)$$

as the solution to the problem. Our task now is to find b_n from the initial condition. At $t = 0$,

$$u(x,0) = \sum_{n=1}^{\infty} b_n \sin\sqrt{\lambda_n}x = f(x). \qquad (2.5.18)$$

To find b_m we multiply (2.5.18) by $\sin\sqrt{\lambda_m}x$ and integrate over $0 < x < L$. Using the orthogonality relation similar to (2.3.18), we obtain

$$b_m = \frac{2}{L}\int_0^L f(x)\sin\sqrt{\lambda_m}x dx. \qquad (2.5.19)$$

In (2.5.17) n starts from one. This shows that u_n exponentially decays to zero for all $n \geq 1$ as t approaches infinity. Therefore,

$$\lim_{t\to\infty} u(x,t) = 0.$$

Exercises

1. Follow the argument for a_0 to show that in (2.5.2) $a_L > 0$ is the physically reasonable sign.
2. Solve the initial boundary value problem (2.5.3), (2.5.4), and (2.5.5) when $a_L < 0$.
3. Suppose λ_m and λ_n are two different eigenvalues and $X_m(x)$ and $X_n(x)$ are the corresponding eigenfunctions for the Robin boundary value problem
$$X'' - \lambda X = 0, \quad 0 < x < L,$$
$$X(0) = 0, \quad X'(L) + a_L X(L) = 0.$$
Show that X_m and X_n are orthogonal, *i.e.*,
$$\int_0^L X_m X_n dx = 0.$$
4. Find the solution to the initial boundary value problem
$$u_t - a^2 u_{xx} = 0, \quad 0 < x < L,\ 0 < t \qquad (2.5.20)$$
$$u_x(0,t) = 0, \quad u_x(L,t) + u(L,t) = 0, \quad 0 < t \qquad (2.5.21)$$
$$u(x,0) = f(x). \qquad (2.5.22)$$
5. Find the solution to the initial boundary value problem
$$u_t - a^2 u_{xx} = 0, \quad 0 < x < L,\ 0 < t \qquad (2.5.23)$$
$$u_x(0,t) - u(0,t) = 0, \quad u_x(L,t) = 0, \quad 0 < t \qquad (2.5.24)$$
$$u(x,0) = f(x). \qquad (2.5.25)$$

2.6 Infinite Domain Problems

There are two types of infinite domain problems. One type is the initial value problems where the problems are specified in the whole real line $-\infty < x < \infty$. In Subsections 2.6.1 and 2.6.2 we derive the solutions based on the change of variables and the Fourier transform, respectively. Another type is the problems where the domain is semi-infinite. A typical example is $0 < x < \infty$. In Subsection 2.6.3 we treat the cases where we can reduce the semi-infinite problems to the initial value problems.

2.6.1 *Initial Value Problems*

In the initial value problems we solve the heat equation in the whole real line $-\infty < x < \infty$. This is a drastic simplification compared to the initial boundary value problems in the sense that the boundary condition will not play an explicit role, though we often look for a solution which vanishes as x approaches infinity.

We solve the heat equation

$$u_t - a^2 u_{xx} = 0, \quad -\infty < x < \infty,\ 0 < t, \tag{2.6.1}$$

with the initial condition

$$u(x, 0) = f(x). \tag{2.6.2}$$

Physically, this problem corresponds to finding the temperature distribution in an infinitely long bar. The initial value problems are justifiable if the boundaries are far away.

We derive the solution to (2.6.1) and (2.6.2) in a heuristic way. We assume that the solution is given in the form $u(x,t) = v(x^2/t)$. One rationale for this assumption is that in the heat equation u is differentiated twice in x and once in t. Setting $z = x^2/t$ and differentiating, we have

$$v_t = v'(z)(-\frac{x^2}{t^2}), \quad v_{xx} = \frac{d}{dx}(v'(z)\frac{2x}{t}) = v''(z)(\frac{2x}{t})^2 + v'(z)\frac{2}{t}.$$

Substituting them in the equation, we obtain

$$4a^2 z v''(z) + (2a^2 + z)v'(z) = 0.$$

This is a first order ODE in v' and separable. Therefore,

$$\int \frac{v''(z)}{v'(z)} dz = -\int (\frac{1}{4a^2} + \frac{1}{2z}) dz \Rightarrow$$

$$\ln|v'| = -(\frac{1}{4a^2}z + \frac{1}{2}\ln|z|) + c_1 \Rightarrow v'(z) = cz^{-\frac{1}{2}}e^{-\frac{1}{4a^2}z},$$

where $c = \pm e^{c_1}$. The integration yields

$$v(z) = c\int^z s^{-\frac{1}{2}}e^{-\frac{1}{4a^2}s}ds + d. \tag{2.6.3}$$

We do not use this v since it is not easy to manipulate. Instead knowing that v_x is also a solution to (2.6.1), we use it to construct the solution satisfying the initial data. Performing the differentiation in x leads to

$$\frac{\partial}{\partial x}v(\frac{x^2}{t}) = \frac{2c}{\sqrt{t}}e^{-\frac{x^2}{4a^2t}}.$$

We choose c so that

$$\int_{-\infty}^{\infty}\frac{2c}{\sqrt{t}}e^{-\frac{x^2}{4a^2t}}dx = 1.$$

Then, $2c = 1/2a\sqrt{\pi}$ and $G(x-y,t) = \exp[-(x-y)^2/(4a^2t)]/(2a\sqrt{\pi t})$ is a solution to (2.6.1) for every y. Here, $\exp(a) = e^a$. How to find this c is given in the Exercise. Therefore,

$$u(x,t) = \frac{1}{2a\sqrt{\pi t}}\int_{-\infty}^{\infty}e^{-\frac{(x-y)^2}{4a^2t}}f(y)dy \tag{2.6.4}$$

is also a solution (2.6.1) as we can verify it. The question is if (2.6.4) satisfies the initial condition. To see this set $z = (y-x)/(2a\sqrt{t})$ and change the variable from y to z. Then,

$$u(x,t) = \frac{1}{\sqrt{\pi}}\int_{-\infty}^{\infty}e^{-z^2}f(x+2a\sqrt{t}z)dz. \tag{2.6.5}$$

As $t \to 0$ we could expect that

$$u(x,0) = \frac{1}{\sqrt{\pi}}\int_{-\infty}^{\infty}e^{-z^2}f(x)dz = f(x). \tag{2.6.6}$$

Remark 2.6.1. The derivation of the solution (2.6.4) to the initial value problem (2.6.1) and (2.6.2) is carried out in a heuristic way. It is rather natural to feel puzzled. In the next subsection and Subsection 9.4.1 we derive the solution to the initial value problems of the heat equation in a more systematic way using the Fourier transform.

The function

$$G(x,t) = \frac{1}{2a\sqrt{\pi t}}e^{-\frac{x^2}{4a^2t}} \tag{2.6.7}$$

is sometimes called a heat kernel, the Green's function, or a fundamental solution. G is symmetric with respect to $x = 0$, *i.e.*, $G(-x,t) = G(x,t)$, and

$$\int_{-\infty}^{\infty} G(x,t)dx = 1.$$

In statistics, G is called the normal (Gaussian) distribution and is one of the most well-known probability distributions.

It is often convenient to define the error function by

$$\mathrm{erf}(x) = \frac{2}{\sqrt{\pi}} \int_0^x e^{-z^2} dz. \tag{2.6.8}$$

The coefficient is chosen so that $\lim_{x\to\infty} \mathrm{erf}(\infty) = 1$. The error function is an odd function, *i.e.*, $\mathrm{erf}(-x) = -\mathrm{erf}(x)$. This function often appears in probability theory as the cumulative distribution function of the normal distribution and is very well tabulated.

Example 2.6.2. Solve

$$u_t - a^2 u_{xx} = 0, \quad -\infty < x < \infty,\ 0 < t,$$

$$u(x,0) = e^x.$$

Solution: This example is meant to illustrate how the formula works. We simply substitute $u(x,0)$ in (2.6.4). Then,

$$u(x,t) = \frac{1}{2a\sqrt{\pi t}} \int_{-\infty}^{\infty} e^{-\frac{(x-y)^2}{4a^2t}} e^y dy = \frac{1}{2a\sqrt{\pi t}} \int_{-\infty}^{\infty} e^{-\frac{(x-y)^2-4a^2ty}{4a^2t}} dy. \tag{2.6.9}$$

Consider the power of exponent. To integrate we complete the square in y. Then,

$$\frac{(x-y)^2 - 4a^2ty}{4a^2t} = \frac{(y-(x+2a^2t))^2}{4a^2t} - a^2t - x. \tag{2.6.10}$$

Therefore, using (2.6.10) and the substitution $z = [y - (x+2a^2t)]/(2a\sqrt{t})$, from (2.6.9) we obtain

$$\begin{aligned} u(x,t) &= \frac{1}{2a\sqrt{\pi t}} \int_{-\infty}^{\infty} e^{-\frac{(y-(x+2a^2t))^2}{4a^2t}+a^2t+x} dy \\ &= e^{a^2t+x} \frac{1}{2a\sqrt{\pi t}} \int_{-\infty}^{\infty} e^{-z^2} 2a\sqrt{t} dz = e^{a^2t+x}. \end{aligned}$$

Example 2.6.3. Express the solution to the initial value problem using the error functions.

$$u_t - ku_{xx} = 0, \quad -\infty < x < \infty, \; 0 < t, \tag{2.6.11}$$

$$u(x,0) = f(x) = \begin{cases} 0, & x < 3 \\ T, & x > 3. \end{cases} \tag{2.6.12}$$

Solution: We use the solution formula (2.6.4). Then,

$$\begin{aligned} u(x,t) &= \frac{1}{2\sqrt{k\pi t}} \int_{-\infty}^{\infty} e^{-\frac{(x-y)^2}{4kt}} f(y)dy \\ &= \frac{1}{2\sqrt{k\pi t}} \int_{-\infty}^{3} e^{-\frac{(x-y)^2}{4kt}} 0dy + \frac{1}{2\sqrt{k\pi t}} \int_{3}^{\infty} e^{-\frac{(x-y)^2}{4kt}} Tdy. \end{aligned}$$

We use the error function to the second term. Once we write the solution in the above form, we choose z in the error function (2.6.8) to be the square root of the exponent. In this problem we use the substitution $z = (y-x)/(2\sqrt{kt})$. Then,

$$u(x,t) = \frac{1}{\sqrt{\pi}} \int_{\frac{3-x}{2\sqrt{kt}}}^{\infty} e^{-z^2} dz = \frac{1}{2}[1 - \mathrm{erf}(\frac{3-x}{2\sqrt{kt}})].$$

2.6.2 *Initial Value Problems via Fourier Transform*

This subsection is an alternative to the previous subsection 2.6.1 where the initial value problems of the heat equation are solved somewhat heuristically. We study the initial value problems for the heat equation by a different method, *i.e.*, the Fourier transform. This is an extension of the Fourier series for the initial value problems and a useful tool. We could regard this subsection as a preview of Chapter 9 where we study more details of the Fourier transform. It is instructive to see what it is and how it is applied to the initial value problems. We make a minimal preparation and then solve the heat equation with the Fourier transform.

Definition 2.6.4. The Fourier transform $\mathcal{F}(f)$ of $f(x)$ is defined by

$$\mathcal{F}(f) = F(\omega) = \int_{-\infty}^{\infty} f(y)e^{-i\omega y}dy$$

and the inverse Fourier transform by

$$\mathcal{F}^{-1}(F) = f(x) = \frac{1}{2\pi} \int_{-\infty}^{\infty} F(\omega)e^{i\omega x}d\omega.$$

The Fourier transform of the derivatives of functions are computed as follows.

$$\mathcal{F}(\frac{\partial f}{\partial x}) = \int_{-\infty}^{\infty} \frac{\partial f}{\partial x} e^{-i\omega x} dx$$
$$= [fe^{-i\omega x}]_{-\infty}^{\infty} + i\omega \int_{R^n} fe^{-i\omega x} dx = i\omega \mathcal{F}(f).$$

Therefore, if $f \to 0$ as $x \to \pm\infty$, there is no contribution from $x = \pm\infty$. Hence,

$$\mathcal{F}(\frac{\partial f}{\partial x}) = i\omega \mathcal{F}(f).$$

To find the Fourier transform of the second derivatives, we repeat the procedure.

$$\mathcal{F}(\frac{\partial^2 f}{\partial x^2}) = \int_{-\infty}^{\infty} \frac{\partial^2 f}{\partial x^2} e^{-i\omega x} dx$$
$$= [\frac{\partial f}{\partial x} e^{-i\omega x}]_{-\infty}^{\infty} + i\omega \int_{R^n} \frac{\partial f}{\partial x} e^{-i\omega x} dx = i\omega \mathcal{F}(\frac{\partial f}{\partial x}).$$

Using the previous result, we have

$$\mathcal{F}(\frac{\partial^2 f}{\partial x^2}) = (i\omega)^2 \mathcal{F}(f). \tag{2.6.13}$$

We are now ready to solve the initial value problems for the heat equation using the Fourier transform. The basic idea is as follows. First apply the Fourier transform to a partial differential equations and convert to an ODE in time where ω appears passively as a parameter. Integrate the ODE in t to find the Fourier transform of solution at t. Then, use the inverse Fourier transform to go back to the original variables and obtain the solution in the original variables. As an example we consider the initial value problem.

$$u_t - a^2 u_{xx} = 0, \quad -\infty < x < \infty,\ 0 < t, \tag{2.6.14}$$

$$u(x, 0) = f(x). \tag{2.6.15}$$

Step 1: We denote the Fourier transform of u by

$$U(\omega, t) = \int_{-\infty}^{\infty} u(y, t) e^{-i\omega y} dy.$$

As explained before the problem, first we apply the Fourier transform, *i.e.*, we multiply the equation (2.6.14) and the initial condition (2.6.15) by $e^{-i\omega x}$

and integrate over R. Then, after performing the integration by parts or using the formula (2.6.13), we obtain

$$U_t(\omega,t)+a^2\omega^2U(\omega,t)=0, \tag{2.6.16}$$

$$U(\omega,0)=F(\omega)=\int_{-\infty}^{\infty} f(y)e^{-i\omega y}dy.$$

This is an ODE in t. Integrating (2.6.13) in t, we obtain

$$U(\omega,t)=F(\omega)e^{-a^2\omega^2t}.$$

Step 2: Now we apply the inverse Fourier transform. Then,

$$\begin{aligned} u(x,t) &= \frac{1}{2\pi}\int_{-\infty}^{\infty} F(\omega)e^{-a^2\omega^2t}e^{i\omega x}d\omega \\ &= \frac{1}{2\pi}\int_{-\infty}^{\infty}(\int_{-\infty}^{\infty} f(y)e^{-i\omega y}dy)e^{-a^2\omega^2t}e^{i\omega x}d\omega \\ &= \frac{1}{2\pi}\int_{-\infty}^{\infty} f(y)\int_{-\infty}^{\infty} e^{-a^2\omega^2t}e^{i\omega(x-y)}d\omega dy. \end{aligned} \tag{2.6.17}$$

In (2.6.17) ω variable in the inside the exponents is combined as

$$-a^2t\omega^2+i\omega(x-y)=-a^2t[\omega-\frac{i(x-y)}{2a^2t}]^2-\frac{1}{4a^2t}(x-y)^2.$$

Therefore, (2.6.17) is rewritten as

$$u(x,t)=\frac{1}{2\pi}[\int_{-\infty}^{\infty} f(y)e^{-\frac{1}{4a^2t}(x-y)^2}\int_{-\infty}^{\infty} e^{-a^2t[\omega-\frac{i(x-y)}{2a^2t}]^2}d\omega dy] \tag{2.6.18}$$

In (2.6.18) $z=\omega-i(x-y)/(2a^2t)$ is a horizontal line in the complex plane. We use the Cauchy theorem in the complex analysis to effectively "shift" the integration path to the real line (See Remark 2.6.5). Using also the result of Exercise 2 in Section 2.6, we obtain

$$\int_{-\infty}^{\infty} e^{-a^2t[\omega-\frac{i(x-y)}{2a^2t}]^2}d\omega=\int_{-\infty}^{\infty} e^{-a^2tz^2}dz=\sqrt{\frac{\pi}{a^2t}}. \tag{2.6.19}$$

Therefore, (2.6.18) is computed as

$$u(x,t)=\frac{1}{2a\sqrt{\pi t}}\int_{-\infty}^{\infty} f(y)e^{-\frac{1}{4a^2t}(x-y)^2}dy=\int_{-\infty}^{\infty} G(x-y,t)f(y)dy, \tag{2.6.20}$$

where G is the heat kernel defined in (2.6.7). The formula (2.6.20) is precisely the same as (2.6.4) in the previous subsection.

Remark 2.6.5. It may be useful to explain the rationale for (2.6.19). We use the Cauchy theorem for complex integrals which is stated roughly as follows. The line integral of f around a closed path C in the complex plane is zero, *i.e.*,

$$\int_C f(z)dz = 0 \tag{2.6.21}$$

provided that f is not singular in the region surrounded by C. An example of a singular function is $f(z) = 1/z$ if the origin of the complex plane is inside C. For the integral (2.6.19) we choose $f(z) = e^{-a^2tz^2}$, apply (2.6.21) to a path C depicted in Figure 2.8, and take the limit as $A \to \infty$. Then, since f does not have any singularity inside C, by the Cauchy theorem

$$\int_C e^{-a^2tz^2}dz = \{\int_{C_1} + \int_{C_2} + \int_{C_3} + \int_{C_4}\}e^{-a^2tz^2}dz = 0.$$

Here,

$$\int_{C_1} e^{-a^2tz^2}dz = \lim_{A\to\infty}\int_{-A}^{A} e^{-a^2t[\omega-\frac{i(x_i-y_i)}{2a^2t}]^2}d\omega, \quad z = \omega - \frac{i(x_i-y_i)}{2a^2t},$$

$$\int_{C_2} e^{-a^2tz^2}dz = \lim_{A\to\infty}\int_{-\frac{(x-y)}{2a^2t}}^{0} e^{-a^2t(A+i\omega)^2}d\omega = 0, \quad z = A + i\omega,$$

$$\int_{C_3} e^{-a^2tz^2}dz = \lim_{A\to\infty}\int_{A}^{-A} e^{-a^2t\omega^2}d\omega = -\sqrt{\frac{\pi}{a^2t}}, \quad z = \omega,$$

$$\int_{C_4} e^{-a^2tz^2}dz = \lim_{A\to\infty}\int_{0}^{-\frac{(x-y)}{2a^2t}} e^{-a^2t(-A+i\omega)^2}d\omega = 0, \quad z = -A + i\omega.$$

Combining the above integrals, we obtain (2.6.19).

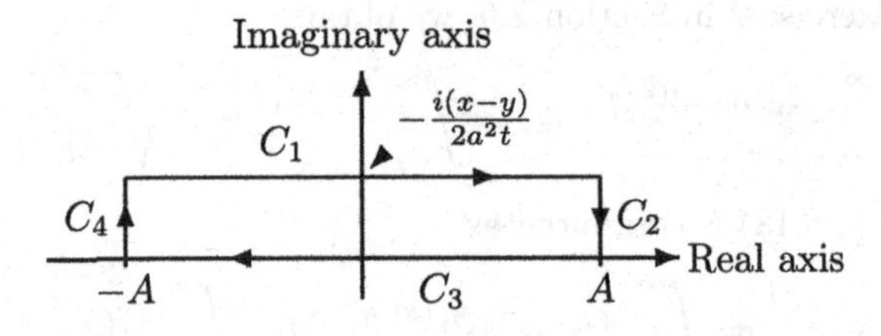

Fig. 2.8 Path integral in the complex plane.

2.6.3 *Semi-infinite Domains*

In the heat equation this case corresponds to the heat conduction in a semi-infinite rod on the interval $[0, \infty)$. We specify the boundary condition at $x = 0$. Typical conditions are the ones discussed in the initial boundary value problems. As an example we treat the case where the temperature at $x = 0$ is fixed at zero. Then, the initial boundary value problem is given by

$$u_t - a^2 u_{xx} = 0, \quad 0 < x < \infty,\ 0 < t, \tag{2.6.22}$$

$$u(0,t) = 0, \tag{2.6.23}$$

$$u(x,0) = f(x). \tag{2.6.24}$$

To solve the above problem, we consider the initial value problem where the initial condition has a special structure.

$$u_t - a^2 u_{xx} = 0, \quad -\infty < x < \infty,\ 0 < t, \tag{2.6.25}$$

$$u(x,0) = \begin{cases} -f(-x) & -\infty < x < 0 \\ f(x) & 0 < x < \infty \end{cases}. \tag{2.6.26}$$

As the form of the initial condition suggests, we extended the initial condition as an odd function to the interval $-\infty < x < 0$. The solution to (2.6.25) and (2.6.26) is given by

$$u(x,t) = \frac{1}{2a\sqrt{\pi t}} \int_{-\infty}^{0} e^{-\frac{(x-y)^2}{4a^2 t}} (-f(-y))dy + \frac{1}{2a\sqrt{\pi t}} \int_{0}^{\infty} e^{-\frac{(x-y)^2}{4a^2 t}} f(y)dy.$$

The first term is rewritten as follows. Set $s = -y$. Then,

$$\begin{aligned} &\frac{1}{2a\sqrt{\pi t}} \int_{-\infty}^{0} e^{-\frac{(x-y)^2}{4a^2 t}} (-f(-y))dy \\ &= \frac{1}{2a\sqrt{\pi t}} \int_{\infty}^{0} e^{-\frac{(x+y)^2}{4a^2 t}} (-f(s))(-ds) \\ &= -\frac{1}{2a\sqrt{\pi t}} \int_{0}^{\infty} e^{-\frac{(x+s)^2}{4a^2 t}} f(s)ds. \end{aligned}$$

This implies that

$$u(x,t) = \frac{1}{2a\sqrt{\pi t}} \int_{0}^{\infty} \{e^{-\frac{(x-y)^2}{4a^2 t}} - e^{-\frac{(x+y)^2}{4a^2 t}}\} f(y)dy. \tag{2.6.27}$$

Next, we show that u in (2.6.27) satisfies the boundary condition (2.6.23). This is easy. Evaluating u at $x = 0$, we find out that

$$u(0,t) = \frac{1}{2a\sqrt{\pi t}} \int_{0}^{\infty} \{e^{-\frac{y^2}{4a^2 t}} - e^{-\frac{y^2}{4a^2 t}}\} f(y)dy = 0.$$

This shows that u in (2.6.27) satisfies both the initial and boundary conditions and hence it is a solution to (2.6.22) to (2.6.24).

Example 2.6.6. Use the error function to express the solution to (2.6.22) to (2.6.24) if $f(x) = 1$.
Solution: From (2.6.27) we have

$$u(x,t) = \frac{1}{2a\sqrt{\pi t}} \int_0^\infty \{e^{-\frac{(x-y)^2}{4a^2t}} - e^{-\frac{(x+y)^2}{4a^2t}}\}dy.$$

Use the change of variables $z = \frac{y-x}{2a\sqrt{t}}$ and $w = \frac{y+x}{2a\sqrt{t}}$ to the first and second integrals, respectively. Then, since the error function is odd, we have

$$\begin{aligned}
u(x,t) &= \frac{1}{2a\sqrt{\pi t}} \int_{-\frac{x}{2a\sqrt{t}}}^\infty e^{-z^2} 2a\sqrt{t}dz - \frac{1}{2a\sqrt{\pi t}} \int_{\frac{x}{2a\sqrt{t}}}^\infty e^{-w^2} 2a\sqrt{t}dw \\
&= \frac{1}{\sqrt{\pi}}[\int_0^\infty e^{-z^2}dz + \int_{-\frac{x}{2a\sqrt{t}}}^0 e^{-z^2}dz] \\
&\quad - \frac{1}{\sqrt{\pi}}[\int_0^\infty e^{-z^2}dz + \int_{\frac{x}{2a\sqrt{t}}}^0 e^{-z^2}dz] \\
&= \frac{1}{2}[-\text{erf}(-\frac{x}{2a\sqrt{t}}) + \text{erf}(\frac{x}{2a\sqrt{t}})] = \text{erf}(\frac{x}{2a\sqrt{t}}).
\end{aligned}$$

Exercises

1. Verify that G defined in (2.6.7) and u defined in (2.6.4) satisfy the heat equation (2.6.1).
2. Compute $I = \int_0^\infty e^{-ax^2}dx$ $(a > 0)$ in the following way.
   (a) Change the double integral $I^2 = \int_0^\infty e^{-ax^2}dx \int_0^\infty e^{-ay^2}dy$ to the one in the Polar coordinates.
   (b) Integrate I^2 in the Polar coordinates to show that $I^2 = \pi/(4a)$.
   (c) Evaluate $J = \int_{-\infty}^\infty e^{-ax^2}dx$ $(a > 0)$.
3. Find the solution. You may have to use the error function.

$$u_t - a^2u_{xx} = 0, \quad -\infty < x < \infty, \ 0 < t,$$

$$u(x,0) = \begin{cases} 0 & -\infty < x < 0, \\ x & 0 < x < 1, \\ 0 & 0 < x < \infty. \end{cases}$$

4. Solve

$$u_t - a^2u_{xx} = 0, \quad -\infty < x < \infty, \ 0 < t,$$

$$u(x,0) = e^{-2x}.$$

5. Express the solution to the initial value problem using the error functions.

$$u_t - 4u_{xx} = 0, \quad -\infty < x < \infty,\ 0 < t,$$

$$u(x,0) = \begin{cases} T_1, & x < 0 \\ T_2, & x > 0. \end{cases}$$

6. Express the solution using the error function.

$$u_t - a^2 u_{xx} = 0,$$

$$u(x,0) = \begin{cases} 0, & x < -1, \\ 1, & -1 \le x \le 1, \\ 0, & 1 < x. \end{cases}$$

7. Consider a semi-infinite problem where the bar at $x = 0$ is insulated. Then, the initial boundary value problem is given by

$$u_t - a^2 u_{xx} = 0, \quad 0 < x < \infty,\ 0 < t, \tag{2.6.28}$$

$$u_x(0,t) = 0, \tag{2.6.29}$$

$$u(x,0) = f(x). \tag{2.6.30}$$

(a) Find the solution to

$$u_t - a^2 u_{xx} = 0, \quad -\infty < x < \infty,\ 0 < t,$$

$$u(x,0) = \begin{cases} f(-x) & -\infty < x < 0 \\ f(x) & 0 < x < \infty. \end{cases}$$

(b) Show that the solution in (a) satisfies the boundary condition (2.6.29). This and (a) show that the solution to (2.6.28) to (2.6.30) is given by (a).

(c) Express the solution to (2.6.28) to (2.6.30) using the error function(s) if

$$f(x) = \begin{cases} 0, & 0 \le x < 1, \\ 1, & 1 \le x < \infty. \end{cases}$$

8. We extend the derivation of the heat kernel to the multi-dimensional case. We proceed in the same way as we did in the one-dimensional case. We assume that the solution to the heat equation in R^n

$$u_t = a^2 \Delta u = a^2 (u_{x_1x_1} + u_{x_2x_2} + \cdots + u_{x_nx_n}) \tag{2.6.31}$$

is given by $u = v(|\mathbf{x}|^2/t)/t^\alpha$, where $|\mathbf{x}|^2 = x_1^2 + \cdots + x_n^2$.

(a) Set $y = |\mathbf{x}|^2/t$ and show that

$$u_t = -\alpha t^{-\alpha-1}v - t^{-\alpha}v'|\mathbf{x}|^2t^{-2} = -\alpha t^{-\alpha-1}v - t^{-\alpha-1}yv',$$

$$\Delta u = t^{-\alpha}[v''\sum(\frac{2x_i}{t})^2 + v'\frac{2n}{t}] = t^{-\alpha-1}[4yv'' + 2nv'].$$

(b) Substitute $u = v(y)/t^\alpha$ in (2.6.31) to show that v satisfies.

$$4a^2(yv'' + \frac{n}{2}v') + (yv' + \alpha v) = 0. \qquad (2.6.32)$$

(c) Choose $\alpha = n/2$ and multiply (2.6.32) by the integrating factor $y^{n/2-1}$ to show that v satisfies

$$4a^2(y^{n/2}v')' + (y^{n/2}v)' = 0 \Rightarrow y^{n/2}(4a^2v' + v) = c_1.$$

(d) We look for a solution v decaying to zero faster than $y^{n/2}$ as $|y| \to \infty$, in which case $c_1 = 0$. Then, show that v is given by

$$(v' + \frac{1}{4a^2}v) = 0 \Rightarrow (e^{y/(4a^2)}v)' = 0 \Rightarrow v = ce^{-y/(4a^2)}.$$

(e) We choose the constant c so that $\int_{R^n} u d\mathbf{x} = 1$. Then, show that u is given by

$$u = \frac{1}{(4\pi a^2t)^{n/2}}e^{-|\mathbf{x}|^2/(4a^2t)}.$$

(f) Verify that the solution to (2.6.31) with the initial condition $u(\mathbf{x}, 0) = f(\mathbf{x})$ is

$$u(\mathbf{x}, t) = \int_{R^n} \frac{1}{(4\pi a^2t)^{n/2}}e^{-|\mathbf{x}-\mathbf{y}|^2/(4a^2t)}f(\mathbf{y})d\mathbf{y}. \qquad (2.6.33)$$

As in one dimensional case,

$$G(\mathbf{x}, t) = \frac{1}{(4\pi a^2t)^{n/2}}e^{-|\mathbf{x}|^2/(4a^2t)} \qquad (2.6.34)$$

is called the heat kernel.

2.7 Maximum Principle, Energy Method, and Uniqueness of Solutions

2.7.1 *Maximum Principle*

The maximum principle is a useful tool for the heat equation. It basically says that the maximum (or minimum) of the solution is attained on the initial line or on the boundary. We study this principle in one-dimension and discuss its applications. Among others we prove the uniqueness of the solutions using the maximum principle.

Let $\Omega = \{(x,t) \mid 0 < x < L,\ 0 < t < T\}$ be a rectangular region in the xt-plane, $\partial\Omega_U = \{(x,t) \mid 0 < x < L,\ t = T\}$ be the upper boundary of Ω, and $\partial\Omega_L$ be the lower boundary given by

$$\partial\Omega_L = \{(x,t) \mid 0 \le x \le L,\ t = 0\} \cup \{(x,t) \mid x = 0, x = L,\ 0 \le t \le T\}.$$

They are illustrated in Figure 2.9.

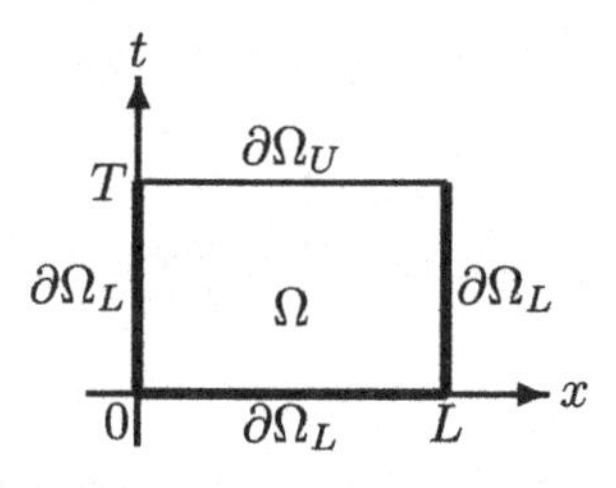

Fig. 2.9 Maximum Principle.

Theorem 2.7.1. ***(Maximum Principle)*** *If $u(x,t)$ satisfies the heat equation on a rectangular region Ω in xt-space, the maximum value of $u(x,t)$ is attained either on the initial line ($t = 0$) or on the boundaries (on the vertical sides, $x = 0$ or $x = L$), i.e.,*

$$\max_{\bar{\Omega}} u = \max_{\partial\Omega_L} u. \tag{2.7.1}$$

Proof. The idea of proof is calculus. At the maximum the first derivative u_t is zero and the second derivative u_{xx} is negative or zero. Therefore, $u_t = a^2 u_{xx}$ does not hold except when $u_{xx} = 0$. Since u_{xx} could be zero, this makes proof more delicate.

First consider the case where $u_t - a^2 u_{xx} < 0$ in Ω. We use the contradiction to show that the maximum won't take place in $\Omega \cup \partial\Omega_U$. Let Ω_ε $(0 < \varepsilon < T)$ be the set

$$\Omega_\varepsilon = \{(x,t) \mid 0 < x < L,\ 0 < t < T - \varepsilon\}.$$

There exists a point $(x,t) \in \bar{\Omega}_\varepsilon$ with

$$u(x,t) = \max_{\bar{\Omega}_\varepsilon} u.$$

If $(x,t) \in \Omega_\varepsilon$, the necessary relations

$$u_t = 0, \quad u_{xx} \le 0$$

would contradict $u_t - a^2 u_{xx} < 0$. If $(x,t) \in \partial\Omega_{\varepsilon U}$, we would have

$$u_t \geq 0, \ u_{xx} \leq 0$$

leading to the same contradiction. Thus $(x,t) \in \partial\Omega_{\varepsilon L}$, and

$$\max_{\bar{\Omega}_\varepsilon} u = \max_{\partial\Omega_{\varepsilon L}} u \leq \max_{\partial\Omega_L} u.$$

Since this holds for every Ω_ε and u is continuous in $\bar{\Omega}$, (2.7.1) follows.

Next, let $u_t - a^2 u_{xx} \leq 0$ in Ω. Introduce

$$v(x,t) = u(x,t) - kt,$$

where k is a positive constant. Then, $v_t - v_{xx} = u_t - a^2 u_{xx} - k < 0$ and

$$\max_{\bar{\Omega}} u = \max_{\bar{\Omega}}(v + kt) \leq \max_{\bar{\Omega}} v + kT = \max_{\partial\Omega_L} v + kT \leq \max_{\partial\Omega_L} u + kT.$$

As $k \to 0$ we obtain (2.7.1). □

Remark 2.7.2. If u satisfies the heat equation, we can apply the maximum principle to $-u$ to obtain

$$\min_{\bar{\Omega}} u = \min_{\partial\Omega_L} u.$$

The maximum principle can be used to show the uniqueness of solutions to the Dirichlet problem for the heat equation.

Theorem 2.7.3. *There is at most one solution to*

$$u_t - a^2 u_{xx} = \phi(x,t), \quad -\infty < x < \infty, \ 0 < t,$$

$$u(0,t) = g(t), \quad u(L,t) = h(t),$$

$$u(x,0) = f(x).$$

Proof. Suppose u_1 and u_2 are two different solutions. Let $w = u_1 - u_2$. Then w satisfies $w_t - a^2 w_{xx} = 0$, $w(x,0) = 0$, $u(0,t) = u(L,t) = 0$. Applying the maximum principle for w and $-w$, we obtain

$$\max_{\bar{\Omega}} w = \min_{\bar{\Omega}} w = 0.$$

Therefore, $w = 0$ in $\bar{\Omega}$. This implies that $u_1 = u_2$. □

Remark 2.7.4. The maximum principle can be used to verify whether your numerical computation of heat equation is consistent with the theoretical result. For example, if the maximum or minimum of the numerical result happens in Ω but not on $\partial\Omega_L$, there could be a mistake in the code.

Example 2.7.5. (a) Verify that $u = 1 - x^2 - 2a^2 t$ satisfies the heat equation $u_t - a^2 u_{xx} = 0$.

(b) Find the maximum and minimum of u and their locations in the closed rectangle $\Omega = \{(x,t) \mid 0 \leq x \leq 1,\ 0 \leq t \leq T\}$.

Solution: (b) We examine the maximum and minimum of u on $t = 0$, $x = 0$, and $x = 1$. Then,

$$\text{On } t = 0, \quad \begin{cases} \max u = 1 & \text{at } x = 0, \\ \min u = 0 & \text{at } x = 1. \end{cases}$$

$$\text{On } x = 0, \quad \begin{cases} \max u = 1 & \text{at } t = 0, \\ \min u = 1 - 2a^2 T & \text{at } t = T. \end{cases}$$

$$\text{On } x = 1, \quad \begin{cases} \max u = 0 & \text{at } t = 0, \\ \min u = -2a^2 T & \text{at } t = T. \end{cases}$$

Therefore, the maximum of u is one at $(x,t) = (0,0)$ and the minimum of u is $-2a^2T$ at $(x,t) = (1, -2a^2T)$.

2.7.2 *Energy Method*

Before we start the energy method, we need a few definitions.

Definition 2.7.6. We say that a function $f(x)$ belongs to $L^2(a,b)$ if $\int_a^b f^2(x)dx < \infty$. We define L^2-norm as $\|f\| = [\int_a^b f^2(x)dx]^{1/2}$.

Remark 2.7.7. The word "norm" needs some explanation. This is a generalization of length, magnitude, or distance. For example, the L^2-distance (or difference) between two functions f and g in $L^2(a,b)$ is given by

$$\|f - g\| = [\int_a^b (f(x) - g(x))^2 dx]^{1/2}.$$

The energy method is a useful tool in PDE's. The name comes from the observation that the term $\int_0^L u^2(x,T)dx$ appearing in (2.7.6) below resembles the energy. To describe the basic idea, consider

$$u_t - a^2 u_{xx} = 0, \quad 0 < x < L,\ 0 < t, \tag{2.7.2}$$

$$u(0,t) = 0, \quad u(L,t) = 0, \quad 0 < t, \tag{2.7.3}$$

$$u(x,0) = f(x). \tag{2.7.4}$$

Multiply (2.7.2) by u and integrate over $0 < x < L$. Then,

$$\frac{1}{2}\int_0^L (u^2)_t dtdx - a^2\int_0^L u_{xx}udxdt = 0.$$

Using the integration by parts and the boundary conditions to the second term, we see

$$a^2\int_0^L u_{xx}udxdt = -a^2\int_0^L u_x^2 dxdt.$$

Therefore, after interchanging the order of integration and differentiation in the first term, we obtain

$$\frac{1}{2}\frac{d}{dt}\int_0^L u^2(x,t)dx = -a^2\int_0^L u_x^2 dxdt. \tag{2.7.5}$$

Integrating (2.7.5) over $0 < t < T$ where T is a positive constant, we obtain

$$\frac{1}{2}\int_0^L u^2(x,T)dx + a^2\int_0^T\int_0^L u_x^2 dxdt = \frac{1}{2}\int_0^L f^2(x)dx. \tag{2.7.6}$$

From (2.7.5) and (2.7.6), we see that the L^2 energy of the solution is decreasing and bounded by that of the initial data.

As an application of the energy method, we consider the uniqueness of the solutions to

$$w_t - a^2 w_{xx} = h(x,t), \quad 0 < x < L,\ 0 < t, \tag{2.7.7}$$

$$w(0,t) = g_1(t), \quad w(L,t) = g_2(t), \quad 0 < t, \tag{2.7.8}$$

$$w(x,0) = f(x). \tag{2.7.9}$$

Suppose that there are more than one solution and denote two of them by w_1 and w_2. Then, $u = w_1 - w_2$ satisfies (2.7.2), (2.7.3), and the initial condition

$$u(x,0) = 0, \quad 0 < x < L.$$

So, instead (2.7.6), we obtain

$$\frac{1}{2}\int_0^L u^2(x,T)dx + a^2\int_0^T\int_0^L u_x^2 dxdt = 0.$$

Since both terms are nonnegative, $\int_0^L u^2(x,T)dx = 0$ for all $T > 0$. This implies that $w_1 = w_2$ for all $T \geq 0$.

We can use the energy method to compare two solutions. Suppose w_1 and w_2 satisfy (2.7.7), (2.7.8), but the different initial data $w_1(x,0) = f_1(x)$ and $w_2(x,0) = f_2(x)$. Then, from (2.7.6) $u = w_1 - w_2$ satisfies

$$\frac{1}{2}\int_0^L u^2(x,T)dx + a^2\int_0^T\int_0^L u_x^2 dxdt = \frac{1}{2}\int_0^L [f_1(x) - f_2(x)]^2 dx.$$

This shows that the L^2-distance of $u = w_1 - w_2$ is bounded by the L^2-distance of the initial data.

Exercises

1. (a) Consider $f(x) = 1/x^p$, $p > 0$. For what values of p, does $f(x)$ belong to $L^2(0,1)$?
   (b) Does $f(x) = \ln x$ belong to $L^2(0,1)$?
2. Prove the comparison principle for the heat equation: If u and v are two solutions, and if $u \leq v$ for $t = 0$, for $x = 0$, and for $x = L$, then $u \leq v$ for $0 \leq t < \infty$, $0 < x < L$.
3. Consider the heat equation

$$u_t = a^2 u_{xx}, \quad 0 < x < 1, \ 0 < t,$$

$$u(0,t) = 0, \quad u(1,t) = 0,$$

$$u(x,0) = 4x(1-x).$$

   (a) Show that $0 < u(x,t) < 1$ for all $t > 0$ and $0 < x < 1$.
   (b) Show that $u(x,t) = u(1-x,t)$ for all $t \geq 0$ and $0 \leq x \leq 1$.
   (c) Use the energy method to show that $\int_0^t u^2 dx$ is a strictly decreasing function of t.

4. Obtain the L^2 energy for the Neumann boundary value problem.

$$u_t - a^2 u_{xx} = 0, \quad 0 < x < L, \ 0 < t,$$

$$u_x(0,t) = 0, \quad u_x(L,t) = 0, \quad 0 < t,$$

$$u(x,0) = f(x).$$

5. Obtain the L^2 energy for the Robin boundary value problem. Also, show the uniqueness. In this problem a_0 and a_L are both positive constants.

$$u_t - a^2 u_{xx} = 0, \quad 0 < x < L, \ 0 < t,$$

$$u_x(0,t) - a_0 u(0,t) = 0, \quad u_x(L,t) + a_L u(0,t) = 0, \quad 0 < t,$$

$$u(x,0) = f(x).$$

Chapter 3

Wave Equation

In this chapter we study the wave equation in details. This equation or its extension explains the phenomena of propagation and oscillations (or vibrations). We all know that the sounds propagate in air or in other media. We also know that the lights propagate through the universe. The Internet we enjoy uses the fiber optics where the information (actually pulses carrying information) propagates at the speed of light. Equations modeling these phenomena are more complicated than the wave equation that we study in this chapter. Nevertheless, by studying the wave equation, we are able to learn the basic idea of how the information such as sound propagates.

In Section 3.1 we derive the wave equation. We also discuss the meaning of boundary conditions. In Section 3.2 we study the initial value problems. We introduce the characteristics and use them to see how the wave propagates in space and time. We study the reflection problems in Section 3.3. We hear echo. We also use mirrors to see ourselves or see if there are cars behind us while changing the lane. We are benefiting from the wave reflection in our daily life. In Section 3.4 we discuss the initial boundary value problems where we use the separation of variables.

3.1 Derivation of Wave Equation

3.1.1 *One-dimensional Case*

We derive one-dimensional wave equation. A typical example is the vibration of string. To derive the wave equation we use the Newton's Law, which says that the mass times acceleration (the time rate of change of linear momentum) is equal to the force applied, *i.e.*, $\mathbf{F} = m\mathbf{a}$. We also make various simplifying assumptions. We need to figure out how $\mathbf{F} = m\mathbf{a}$ is applied and what kind of simplifying assumptions are made. Now, let $u(x,t)$ be the vertical displacement of string from its equilibrium position and let $\mathbf{T}(x,t)$ and $T(x,t)$ be the tension vector and its magnitude. The following are the main assumptions.

1. The string has a constant density ρ per unit length.
2. The direction of tension is tangent to the string.
3. The horizontal movement of the string is negligible compared to the vertical movement.
4. The vertical motion is small so that the slope of the string $u_x(x,t)$ is small.
5. The magnitude $T(x,t)$ of the tension will be assumed to be a constant.

The tension at (x,t) is given by

$$\mathbf{T}(x,t) = \left[T\cos\theta, T\sin\theta\right] = \left[\frac{T}{\sqrt{1+u_x^2}}, \frac{Tu_x}{\sqrt{1+u_x^2}}\right].$$

The change of linear momentum in a segment $[x_1, x_2]$ in the string is given by

$$m\mathbf{a} = \left[0, \frac{d}{dt}\int_{x_1}^{x_2} \rho u_t dx\right] = \left[0, \int_{x_1}^{x_2} \rho u_{tt} dx\right],$$

where u_t and u_{tt} are the velocity and acceleration of string in the vertical direction. Note that we exchanged the order of integration and differentiation. See Appendix A.1 for the details of when it is possible. The x component of $m\mathbf{a}$ is zero by Assumption 3. The force applied to this segment is given by

$$\mathbf{F} = \mathbf{T}(x_2,t) - \mathbf{T}(x_1,t) = \int_{x_1}^{x_2} \frac{\partial}{\partial x}\left[\frac{T}{\sqrt{1+u_x^2}}, \frac{Tu_x}{\sqrt{1+u_x^2}}\right] dx. \tag{3.1.1}$$

Using $\mathbf{F} = m\mathbf{a}$ and the fact that x_1 and x_2 are arbitrary, we arrive at

$$\left[0, \rho u_{tt}\right] = \left[\frac{\partial}{\partial x}\frac{T}{\sqrt{1+u_x^2}}, \frac{\partial}{\partial x}\frac{Tu_x}{\sqrt{1+u_x^2}}\right]. \tag{3.1.2}$$

Since u_x is small, we use the approximation $\sqrt{1+u_x^2} \approx 1$. This together with the x component in (3.1.2) yields

$$0 = \frac{\partial}{\partial x}\frac{T}{\sqrt{1+u_x^2}} \approx \frac{\partial T}{\partial x}.$$

This justifies Assumption 5 and we may assume that T is constant. Then, the vertical force is Tu_x and the y components in (3.1.2) lead to

$$u_{tt} = c^2 u_{xx}, \tag{3.1.3}$$

where $c = \sqrt{T/\rho}$.

We often specify how the two ends of string are attached. They are called the boundary conditions. For example, in a violin or guitar both ends are firmly fixed and there is no vertical movement, *i.e.*, the vertical displacement is zero at both ends. Suppose the string is attached to a violin at $x = 0$ and at $x = L$, then the appropriate boundary conditions are

$$u(0,t) = 0, \quad u(L,t) = 0.$$

Another example is the case where the end $x = 0$ is fixed and the other end is free to move, we have

$$u(0,t) = 0, \quad u_x(L,t) = 0.$$

The phrase "free to move" means that no force is applied. Since the vertical force is Tu_x, this implies that u_x is zero at $x = L$.

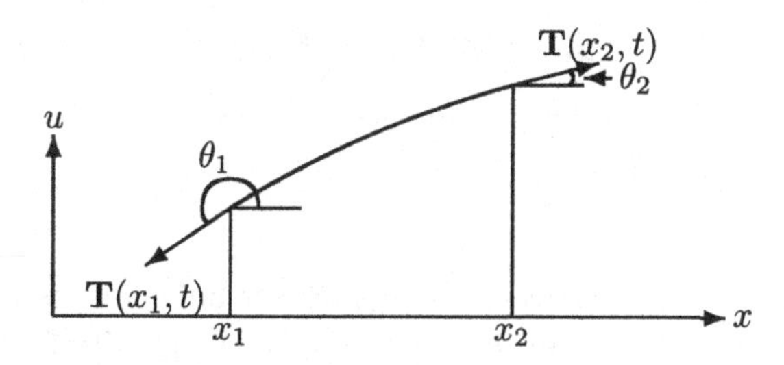

Fig. 3.1 String motion.

3.1.2 *Multi-dimensional Case*

We derive the wave equation in two dimensional case. We apply the Newton's law to a displacement u in some direction. This law says that the rate of change of linear momentum in a closed region Ω is equal to the force applied on the boundary. Therefore, we have

$$\frac{d}{dt}\iint_{\Omega} \rho u_t dxdy = \iint_{\Omega} (\rho u_t)_t dxdy = F = -\int_{\partial\Omega} \mathbf{T}\cdot\mathbf{n}ds, \tag{3.1.4}$$

where $\mathbf{T}$ is the force applied on the boundary $\partial\Omega$ of Ω and $\mathbf{n}$ is the outward normal vector to $\partial\Omega$. An example of such a force is the tension or the pressure applied on the boundary. Applying the divergence theorem, we find that

$$\int_{\partial\Omega} \mathbf{T} \cdot \mathbf{n} ds = \iint_{\Omega} \nabla \cdot \mathbf{T} dx dy. \tag{3.1.5}$$

Combining (3.1.4) and (3.1.5), we have

$$(\rho u_t)_t = -\nabla \cdot \mathbf{T}. \tag{3.1.6}$$

For simplicity, if we assume that the force is proportional to the gradient of the displacement, the force is given by

$$\mathbf{T} = -T\nabla u.$$

Using this in (3.1.6), we obtain

$$(\rho u_t)_t = \nabla \cdot (T\nabla u).$$

For simplicity we assume that both ρ and T are constant. Then, the above equation is simplified to

$$u_{tt} = c^2 \nabla \cdot \nabla u = c^2 \Delta u,$$

where $c = \sqrt{T/\rho}$. This is the two dimensional wave equation.

Exercises

1. Suppose that a vertical force $h(x,t)$ per unit mass is acting on a string. An example of the vertical force is a gravity force.
   (a) What modification do you need in (3.1.1)?
   (b) Derive an equation corresponding to (3.1.3).

3.2 Initial Value Problems

3.2.1 *Homogeneous Wave Equation*

In this section we discuss the initial value problems for the wave equation in one space dimension. The multi-dimensional case will be treated in Section 11.1. To understand the wave propagation, it is easier to start from the initial value problems where we do not have to consider the effects of boundary conditions to the wave propagation. The problem is given by

$$u_{tt}(x,t) - c^2 u_{xx}(x,t) = 0, \quad -\infty < x < \infty,\ 0 < t, \tag{3.2.1}$$

$$u(x,0) = f(x),\ u_t(x,0) = g(x). \tag{3.2.2}$$

The conditions (3.2.2) are called the initial conditions. The line $t = 0$ is called the initial line. We find the solution to the wave equation (3.2.1) on the interval $-\infty < x < \infty$ for positive t satisfying the initial conditions (3.2.2). In terms of the string problem, there is an infinitely long string and we find the vertical displacement $u(x,t)$ at x and t satisfying the initial displacement $f(x)$ and the initial velocity $g(x)$.

Equation (3.2.1) is hyperbolic. There are several ways to solve the above initial value problem. Well-known ones are the methods based on the first order linear equations (Exercise 4), the Divergence Theorem, etc. Here, we show the method based on Section 1.4. We perform the coordinate transform or the change of variables of the form

$$\xi = x + ct, \quad \eta = x - ct \tag{3.2.3}$$

and find the equation satisfied by u when we use (ξ, η) as the independent variables. As in Section 1.4.3 using the chain rule, we obtain

$$u_x = u_\xi \xi_x + u_\eta \eta_x = u_\xi + u_\eta,$$

$$u_t = u_\xi \xi_t + u_\eta \eta_t = c(u_\xi - u_\eta),$$

$$u_{xx} = u_{\xi\xi}\xi_x + u_{\xi\eta}\eta_x + u_{\eta\xi}\xi_x + u_{\eta\eta}\eta_x = u_{\xi\xi} + 2u_{\xi\eta} + u_{\eta\eta},$$

$$u_{xt} = u_{\xi\xi}\xi_t + u_{\xi\eta}\eta_t + u_{\eta\xi}\xi_t + u_{\eta\eta}\eta_t = c(u_{\xi\xi} - u_{\eta\eta}),$$

$$u_{tt} = c(u_{\xi\xi}\xi_t + u_{\xi\eta}\eta_t - u_{\eta\xi}\xi_t - u_{\eta\eta}\eta_t) = c^2(u_{\xi\xi} - 2u_{\eta\xi} + u_{\eta\eta}).$$

Substitution of u_{xx} and u_{tt} to (3.2.1) yields

$$-4c^2 u_{\xi\eta} = 0.$$

Integration in η implies

$$u_\xi = h(\xi)$$

and the integration in ξ leads to

$$u = F(\xi) + G(\eta),$$

where $F(\xi) = \int^\xi h(s)ds$ and G will be determined by the initial conditions. Going back to the original variables (x,t), we see that the solution is given by

$$u(x,t) = F(x+ct) + G(x-ct). \tag{3.2.4}$$

Along $x - ct = s$, $G(x-ct) = G(s)$ and along $x + ct = s$, $F(x+ct) = F(s)$. Here s is an arbitrary position on the initial line. So, the solution consists

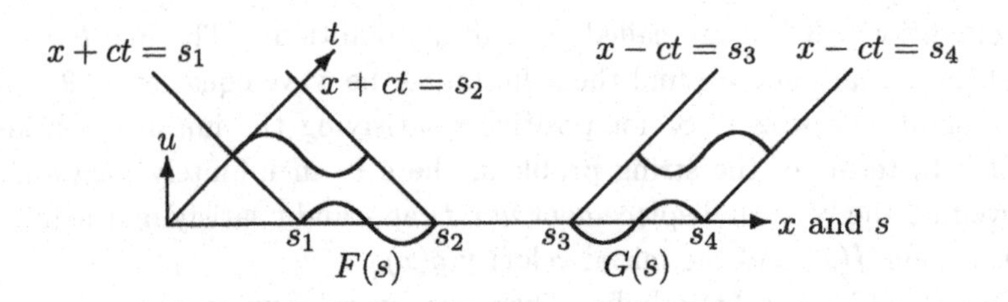

Fig. 3.2 Characteristics and wave propagation.

of a profile G moving to the right with speed c and a profile F moving to the left with speed c. The lines $x - ct =$ constant and $x + ct =$ constant are called characteristics or characteristic curves. Characteristics and the way waves propagate are shown in Figure 3.2. The important observation is that the profiles F and G move with the negative and positive speeds c, respectively, but they do not change their shape.

Now, we find F and G satisfying the initial data (3.2.2). At $t = 0$,

$$u(x, 0) = F(x) + G(x) = f(x). \tag{3.2.5}$$

Since $u_t = cF'(x + ct) - cG'(x - ct)$,

$$u_t(x, 0) = cF'(x) - cG'(x) = g(x). \tag{3.2.6}$$

Integrating (3.2.6), we have

$$F(x) - G(x) = \frac{1}{c}\int_0^x g(s)ds + K, \tag{3.2.7}$$

where K is a constant of integration. From (3.2.5) and (3.2.7), we see that

$$F(x) = \frac{1}{2}[f(x) + \frac{1}{c}\int_0^x g(s)ds + K],$$

$$G(x) = \frac{1}{2}[f(x) - \frac{1}{c}\int_0^x g(s)ds - K]. \tag{3.2.8}$$

Therefore,

$$\begin{aligned} u(x,t) &= F(x + ct) + G(x - ct) \\ &= \frac{1}{2}[f(x - ct) + f(x + ct)] + \frac{1}{2c}\int_{x-ct}^{x+ct} g(s)ds. \end{aligned} \tag{3.2.9}$$

To figure out what the formula (3.2.9) says, let's fix a point (x_1, t_1) in the xt-plane and find $u(x_1, t_1)$. We notice that along the line $x - ct = x_1 - ct_1$, $f(x - ct) = f(x_1 - ct_1)$ and along the line $x + ct = x_1 + ct_1$, $f(x + ct) = f(x_1 +$

$ct_1)$. As depicted in Figure 3.3 (Left), these are characteristic curves passing through (x_1, t_1) with positive and negative slopes, respectively. Also, we notice that

$$\frac{1}{2c}\int_{x-ct}^{x+ct} g(s)ds = \frac{1}{2c}\int_{x_1-ct_1}^{x_1+ct_1} g(s)ds.$$

This shows that the value $u(x_1, t_1)$ is affected by the initial data on the interval $[x_1 - ct_1, x_1 + ct_1]$. This interval is called the domain of dependence of the solution at (x_1, t_1). If we draw the lines $x + ct = s$ and $x - ct = s$ as in Figure 3.3 (Right), where s is the x-coordinate of the initial line, we see that the initial data at $(s, 0)$ will influence the solution in the triangular region given by

$$|x - s| \leq ct,\ t \geq 0.$$

This region is called the range of influence of initial data at $(s, 0)$.

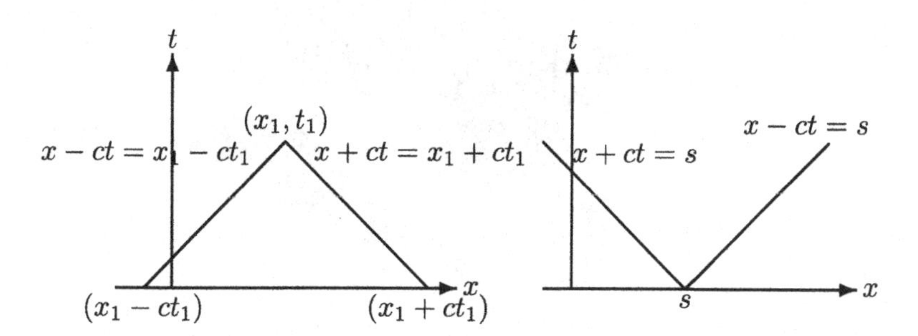

Fig. 3.3 Domain of dependence (Left) and Range of influence (Right).

To illustrate the meaning of (3.2.9), we consider the examples of the initial value problems.

Example 3.2.1. Find the solution $u(x, t)$ to

$$u_{tt}(x,t) - c^2 u_{xx}(x,t) = 0, \quad -\infty < x < \infty,\ 0 < t, \tag{3.2.10}$$

$$u(x, 0) = \cos(2x),\ u_t(x, 0) = \sin(3x). \tag{3.2.11}$$

Solution: We replace x in (3.2.11) with $x - ct$ and $x + ct$ and apply the formula (3.2.9). Then,

$$u(x,t) = \frac{1}{2}\{\cos[2(x - ct)] + \cos[2(x + ct)]\} + \frac{1}{2c}\int_{x-ct}^{x+ct} \sin(3s)ds.$$

Example 3.2.2. Consider the wave equation

$$u_{tt}(x,t) - c^2 u_{xx}(x,t) = 0, \quad -\infty < x < \infty,\ 0 < t$$

with the initial data

$$u(x,0) = f(x) = \begin{cases} 0, & x < -1, \\ 1, & -1 < x < 1, \\ 0, & 1 < x. \end{cases} \quad u_t(x,0) = 0,$$

(a) Find the solution.
(b) Sketch the profiles of solution at $t = 1/(2c)$, $1/c$, $2/c$.

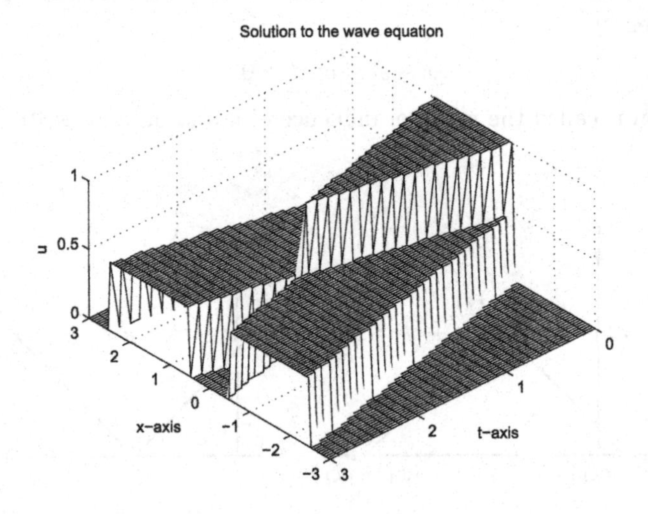

Fig. 3.4 Solution to Example 3.2.2 with $c = 1/2$.

Solution: (a) The solution is given by

$$u(x,t) = \frac{1}{2}[f(x-ct) + f(x+ct)].$$

The solution is the sum of the half of the initial profile f moving to the right with speed c and the other half of the initial profile f moving to the left with speed c as shown in Figure 3.4. Since the initial profile changes at $x = -1$ and $x = 1$, it is useful to keep track them by drawing the characteristic curves $x \pm ct = -1$ and $x \pm ct = 1$. These four characteristics divide the xt-plane into six regions. The four characteristics with $c = 1/2$ are depicted in Figure 3.5 for illustrative purpose.

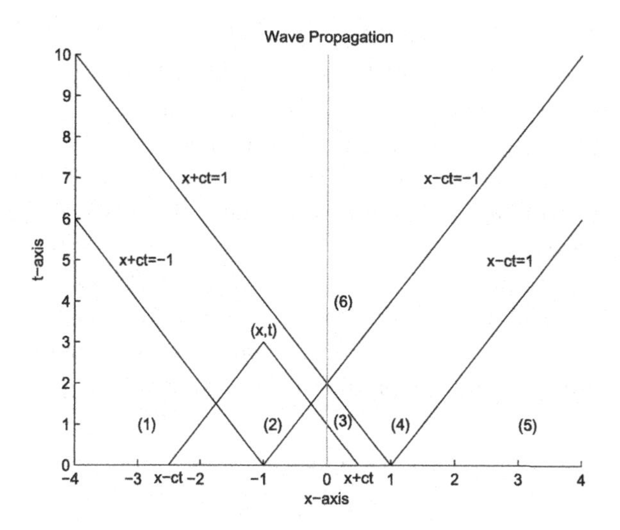

Fig. 3.5 Wave propagation ($c = 1/2$).

There are two ways to find the solution. From Figure 3.5 we see that in the band between $x + ct = -1$ and $x + ct = 1$ $f(x + ct) = 1/2$, and in the band between $x - ct = -1$ and $x - ct = 1$ $f(x - ct) = \frac{1}{2}$. We add (or superimpose) them to find the solution u. For example the region (2) is in the band between $x + ct = -1$ and $x + ct = 1$ but not in the band between $x - ct = -1$ and $x - ct = 1$. Therefore, $u(x, t) = [0 + 1]/2 = 1/2$ in (2). In the region (3) where the two bands overlap u is 1.

Another way is to draw the characteristic curves from the point (x, t) to the initial line and see where they intersect with the initial line. In Figure 3.5 the point (x, t) is placed in the region (2). Then, the two characteristics through this point hit the initial line at $x - ct$ and $x + ct$ as depicted in Figure 3.5. Since $x - ct$ is less than -1, $f(x - ct) = 0$ and since $x + ct$ is between -1 and 1, $f(x + ct) = 1$. Therefore, $u(x, t) = [0 + 1]/2 = 1/2$ in (2). The solution $u(x, t)$ in the other regions can be found similarly and summarized as follows.

Region (1): $x + ct < -1$ and $x - ct < -1$.
$u(x, t) = [f(x - ct) + f(x + ct)]/2 = [0 + 0]/2 = 0$.

Region (2): $x - ct < -1$ and $-1 < x + ct < 1$. $u(x, t) = [0 + 1]/1 = 1/2$.

Region (3): $-1 < x - ct < 1$ and $-1 < x + ct < 1$. $u(x, t) = [1 + 1]/2 = 1$.

Region (4): $-1 < x - ct < 1$ and $1 < x + ct$. $u(x, t) = [1 + 0]/2 = 1/2$.

Region (5): $1 < x - ct$ and $1 < x + ct$. $u(x,t) = [0 + 0]/2 = 0$.

Region (6): $x - ct < -1$ and $1 < x + ct$. $u(x,t) = [0 + 0]/2 = 0$.

(b) One way is to use the fact that $f(x + ct)/2$ and $f(x - ct)/2$ are the profile $f(x)/2$ moving to the left and right at speed c, respectively. At $t = 1/(2c), 1/c, 2/c$ $f(x+ct)$ is $f(x+\frac{1}{2})$, $f(x+1)$, $f(x+2)$. So, we shift the profile $f(x)$ by 1/2, 1, 2 to the left. For $f(x-ct)$ we shift the profile $f(x)$ by 1/2, 1, 2 to the right. For example at $t = 1/c$, since $u(x,t) = [f(x-ct)+f(x+ct)]/2$, we superimpose the profiles $f(x + 1)/2$ and $f(x - 1)/2$. Also, it is a good idea to draw horizontal lines $t = 1/(2c)$, $1/c$, $2/c$ in Figure 3.5 and see how they intersect with regions (1) to (6). The profiles of solutions are given in Figure 3.6. Also, compare the profiles in Figure 3.6 with Figure 3.4.

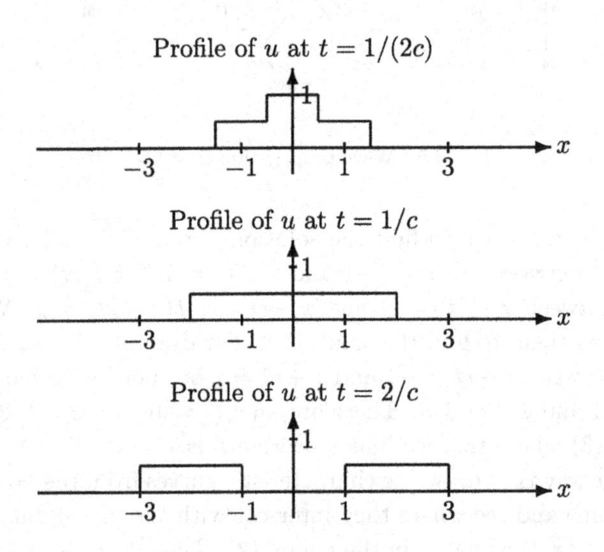

Fig. 3.6 Profiles of u at $t = 1/(2c)$, $1/c$, $2/c$ in Example 3.2.2.

In the next example, we study the effect of g.

Example 3.2.3. Consider the wave equation

$$u_{tt}(x,t) - c^2 u_{xx}(x,t) = 0, \quad -\infty < x < \infty,\ t > 0$$

with the initial data

$$u(x,0) = 0,\ u_t(x,0) = g(x) = \begin{cases} 0, & x < -h, \\ 1, & -h < x < h, \\ 0, & h < x, \end{cases}$$

where h is a positive constant.

(a) Find the solution.

(b) Sketch the profiles of solution at $t = h/(2c)$, h/c, $2h/c$ if $c = 1/2$ and $h = 1$.

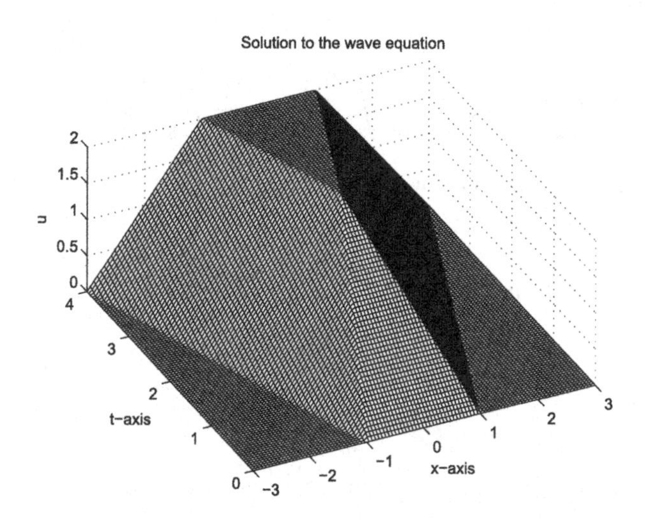

Fig. 3.7 Solution to Example 3.2.3 with $c = 1/2$ and $h = 1$.

Solution: (a) It is a good idea to draw characteristics through $(-h, 0)$ and $(h, 0)$ and the domain of dependence of $u(x,t)$ for a given (x,t). The solution is given by

$$u(x,t) = \frac{1}{2c}\int_{x-ct}^{x+ct} g(s)ds. \tag{3.2.12}$$

First, find the time at which the characteristic curves $x + ct = h$ and $x - ct = -h$ intersect. Setting x equal, we have

$$h - ct = -h + ct \Rightarrow t = \frac{h}{c}.$$

To determine the integral (3.2.12) it is important to know how the two intervals $[x - ct, x + ct]$ and $[-h, h]$ overlap since we integrate g over $[x - ct, x+ct]$, and $g = 1$ on $[-h, h]$ and zero otherwise. As we did in the previous example, we need to locate the interval $[x - ct, x + ct]$ on the initial line for a given (x,t). For this purpose, we draw the characteristics through (x,t) and see where they intersect with the initial line. For example if (x,t) is in

(2) as depicted in Figure 3.5 (need to change 1 to h and -1 to $-h$), we see how the two intervals overlap. We evaluate (3.2.12) as follows.

In Region (1) $x+ct < -h$. Hence, the interval $[x-ct, x+ct]$ is to the left of the interval $[-h, h]$. Therefore, g is zero on $[x-ct, x+ct]$. So, $u(x,t) = 0$.
In Region (2) $-h < x + ct < h$ and $x - ct < -h$. The two intervals overlap on $[-h, x + ct]$.

$$u(x,t) = \frac{1}{2c}\int_{x-ct}^{x+ct} g(s)ds = \frac{1}{2c}\int_{x-ct}^{-h} 0ds + \frac{1}{2c}\int_{-h}^{x+ct} ds = \frac{x+ct+h}{2c}.$$

In Region (3) $-h < x - ct$ and $x + ct < h$. The interval $[x - ct, x + ct]$ is contained in $[-h, h]$.

$$u(x,t) = \frac{1}{2c}\int_{x-ct}^{x+ct} ds = \frac{2ct}{2c} = t.$$

In Region (4) $h < x + ct$ and $-h < x - ct < h$. The two intervals overlap on $[x - ct, h]$.

$$u(x,t) = \frac{1}{2c}\int_{x-ct}^{x+ct} g(s)ds = \frac{1}{2c}\int_{x-ct}^{h} ds + \frac{1}{2c}\int_{-h}^{x+ct} 0ds = \frac{h-x+ct}{2c}.$$

In Region (5) $h < x - ct$. The interval $[x - ct, x + ct]$ is to the right of $[-h, h]$. $u(x,t) = 0$.
In Region (6) $h < x+ct$ and $x - ct < -h$. The interval $[-h, h]$ is contained in $[x - ct, x + ct]$.

$$\begin{aligned} u(x,t) &= \frac{1}{2c}\int_{x-ct}^{x+ct} g(s)ds \\ &= \frac{1}{2c}\int_{x-ct}^{-h} 0ds + \frac{1}{2c}\int_{-h}^{h} ds + \frac{1}{2c}\int_{-h}^{x+ct} 0ds = \frac{h}{c}. \end{aligned}$$

The solution of the problem for $0 \le t \le 4$ are depicted in Figure 3.7.
(b) The profiles for $c = 1/2$ and $h = 1$ are given in Figure 3.8. It is interesting to compare these two figures.

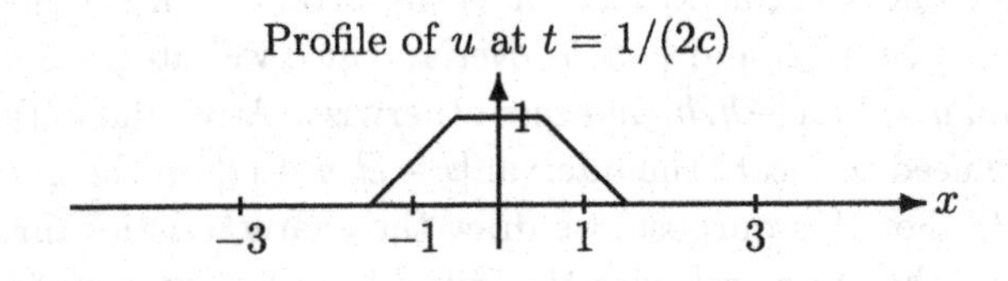

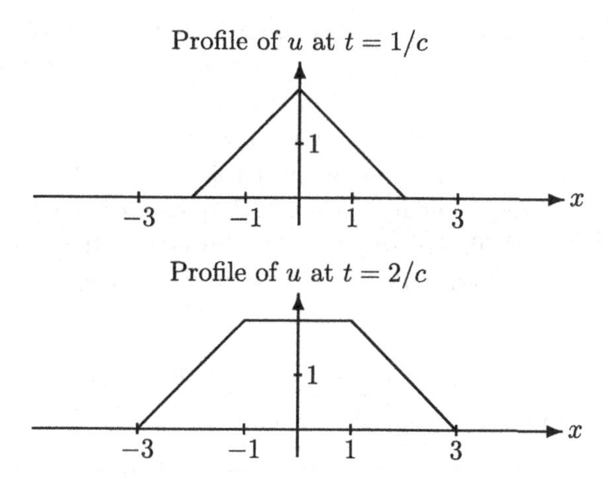

Fig. 3.8 Profiles of u at $t = 1/(2c)$, $1/c$, $2/c$ in Example 3.2.3.

3.2.2 *Non-homogeneous Wave Equation*

We now consider a non-homogeneous equation

$$u_{tt} - c^2 u_{xx} = h(x,t), \quad -\infty < x < \infty, \ 0 < t \tag{3.2.13}$$

with the initial condition

$$u(x,0) = f(x), \quad u_t(x,0) = g(x).$$

Unlike the heat equation, the non-homogeneous term h is often called the forcing term.

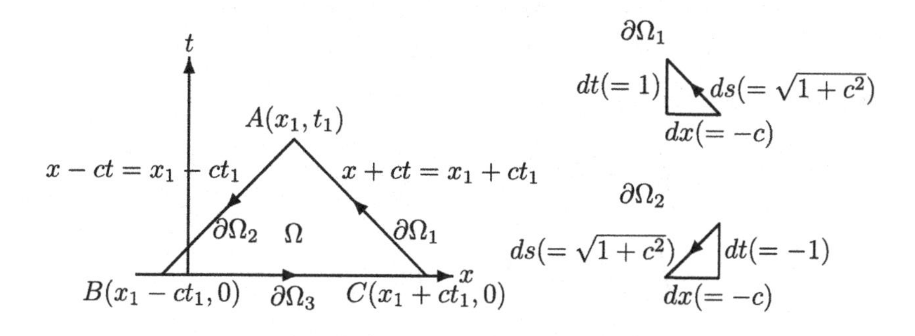

Fig. 3.9 Divergence theorem for non-homogeneous wave equation.

One interesting approach to this problem is to apply the divergence theorem to find the solution $u(x_1, t_1)$. We define $\nabla = [\partial/\partial x, \partial/\partial t]$ and set $\mathbf{u} = [-c^2\partial/\partial x, \partial/\partial t]$. As illustrated in Figure 3.9, we apply the divergence theorem to the region Ω $(= \triangle ABC)$ bounded by the characteristics $x+ct = x_1 + ct_1$, $x - ct = x_1 - ct_1$, and the initial line $t = 0$. The outward unit normal $\mathbf{n}$ to the three boundary lines $\partial\Omega_1$, $\partial\Omega_2$, and $\partial\Omega_3$ are $[1, c]/\sqrt{1+c^2}$, $[-1, c]/\sqrt{1+c^2}$, and $[0, -1]$, respectively. Since the left hand side of the wave equation can be written in the divergence form as $u_{tt} - c^2u_{xx} = \nabla\cdot\mathbf{u}$, we have

$$\iint_{\Omega} \nabla\cdot\mathbf{u}dxdt = \int_{\partial\Omega} \mathbf{u}\cdot\mathbf{n}ds = \iint_{\Omega} h(x,t)dxdt, \tag{3.2.14}$$

where the middle integral is the line integral going counter clockwise along the boundary $\partial\Omega$ consisting of the three sides of $\triangle ABC$.

$$\begin{aligned}\int_{\partial\Omega} \mathbf{u}\cdot\mathbf{n}ds = &\int_{\partial\Omega_1} \frac{1}{\sqrt{1+c^2}}[1, c]\cdot[-c^2\frac{\partial u}{\partial x}, \frac{\partial u}{\partial t}]ds \\ &+ \int_{\partial\Omega_2} \frac{1}{\sqrt{1+c^2}}[-1, c]\cdot[-c^2\frac{\partial u}{\partial x}, \frac{\partial u}{\partial t}]ds \\ &+ \int_{\partial\Omega_3} [0, -1]\cdot[-c^2\frac{\partial u}{\partial x}, \frac{\partial u}{\partial t}]ds. \end{aligned} \tag{3.2.15}$$

We evaluate the each integral by the method of characteristics (or the chain rule) as follows. In the first integral, from the triangle for $\partial\Omega_1$ in Figure 3.9 (upper right) we find the ratios of the differentials $\frac{dt}{ds} = 1/\sqrt{1+c^2}$ and $\frac{dx}{ds} = c/\sqrt{1+c^2}$. Therefore,

$$\frac{du}{ds} = \frac{\partial u}{\partial t}\frac{dt}{ds} + \frac{\partial u}{\partial x}\frac{dx}{ds} = \frac{1}{\sqrt{1+c^2}}(\frac{\partial u}{\partial t} - c\frac{\partial u}{\partial x}).$$

Using this, along the boundary $\partial\Omega_1$ we obtain

$$\begin{aligned}&\int_{\partial\Omega_1} \frac{1}{\sqrt{1+c^2}}[1, c]\cdot[-c^2\frac{\partial u}{\partial x}, \frac{\partial u}{\partial t}]ds \\ &= \int_{\partial\Omega_1} \frac{c}{\sqrt{1+c^2}}(-c\frac{\partial u}{\partial x} + \frac{\partial u}{\partial t})ds \\ &= \int_C^A c\frac{du}{ds}ds = c[u(x_1, t_1) - u(x_1 + ct_1, 0)]. \end{aligned} \tag{3.2.16}$$

Similarly, along the boundary $\partial\Omega_2$ we obtain

$$\int\limits_{\partial\Omega_2} \frac{1}{\sqrt{1+c^2}}[-1,c]\cdot[-c^2\frac{\partial u}{\partial x},\frac{\partial u}{\partial t}]ds$$

$$= \int\limits_{\partial\Omega_2} \frac{c}{\sqrt{1+c^2}}(c\frac{\partial u}{\partial x}+\frac{\partial u}{\partial t})ds$$

$$= -\int_A^B c\frac{du}{ds}ds = c[u(x_1,t_1)-u(x_1-ct_1,0)]. \tag{3.2.17}$$

The boundary $\partial\Omega_3$ is on the initial line and therefore,

$$\int\limits_{\partial\Omega_3} [0,-1]\cdot[-c^2\frac{\partial u}{\partial x},\frac{\partial u}{\partial t}]ds$$

$$= -\int_B^C \frac{\partial u}{\partial t}(x,0)dx = -\int_{x_1-ct_1}^{x_1+ct_1} g(x)dx. \tag{3.2.18}$$

We also need to evaluate the integral on the right hand side of (3.2.14). To find the limits of integration we note that for each t, x changes from $x_1-c(t_1-t)$ to $x_1+c(t_1-t)$ and t changes from 0 to t_1. So,

$$\frac{1}{2c}\iint\limits_{\Omega} h(x,t)dxdt = \frac{1}{2c}\int_0^{t_1}\int_{x_1-c(t_1-t)}^{x_1+c(t_1-t)} h(x,t)dxdt. \tag{3.2.19}$$

Combining (3.2.16), (3.2.17), (3.2.18), and (3.2.19), we obtain

$$u(x_1,t_1) = \frac{1}{2}[f(x_1+ct_1)+f(x_1-ct_1)]$$
$$+\frac{1}{2c}\int_{x_1-ct_1}^{x_1+ct_1} g(x)dx + \frac{1}{2c}\iint\limits_{\Omega} h(x,t)dxdt. \tag{3.2.20}$$

Note that the above formula reduces to (3.2.9) if $h=0$. In other words, we could have used the divergence theorem to find the solution (3.2.9) in the homogeneous case.

Example 3.2.4. Solve

$$u_{tt}-c^2u_{xx}=e^{-t}, \quad -\infty<x<\infty,\ 0<t,$$
$$u(x,0)=\cos x, \quad u_t(x,0)=1.$$

Solution: Using (3.2.20), we have

$$u(x,t) = \frac{1}{2}[\cos(x+ct)+\cos(x-ct)] + \frac{1}{2c}\int_{x-ct}^{x+ct} dy$$
$$+\frac{1}{2c}\int_0^t\int_{x-c(t-s)}^{x+c(t-s)} e^{-s}dyds$$
$$= \frac{1}{2}[\cos(x+ct)+\cos(x-ct)] + 2t + e^{-t} - 1.$$

Exercises

1. Consider the wave equation

$$u_{tt}(x,t) - (\frac{1}{2})^2 u_{xx}(x,t) = 0, \quad -\infty < x < \infty,\ 0 < t$$

with the initial data

$$u(x,0) = f(x) = \begin{cases} 0, & x \le -1, \\ 1 - x^2, & -1 < x \le 1, \\ 0, & 1 < x. \end{cases} \quad u_t(x,0) = 0,$$

(a) Find the solution.
(b) Sketch the profiles of solution at $t = 1,\ 2,\ 3$.

2. Consider the initial value problem for the wave equation.

$$u_{tt} - (\frac{1}{2})^2 u_{xx} = 0, \quad -\infty < x < \infty,\ t > 0,$$

$$u(x,0) = f(x) = 0 \quad u_t(x,0) = g(x) = \begin{cases} 0, & x < -1, \\ 1 - x^2, & -1 \le x < 1, \\ 0, & 1 \le x. \end{cases}$$

(a) Find the solution.
(b) Sketch the profiles of solution at $t = 1,\ 2,\ 3$. Indicate the height of solution and the locations where the height changes.

3. For (3.2.17) find dt/ds, dx/ds, du/ds, and derive (3.2.17).
4. We solve the wave equation using the method of characteristics. Define $\mathcal{L}_1 = \partial/\partial t - c\partial/\partial x$ and $\mathcal{L}_2 = \partial/\partial t + c\partial/\partial x$.

(a) Show that the wave equation is written as

$$u_{tt} - c^2 u_{xx} = (\frac{\partial}{\partial t} - c\frac{\partial}{\partial x})(\frac{\partial}{\partial t} + c\frac{\partial}{\partial x})u = \mathcal{L}_1 \mathcal{L}_2 u = 0.$$

(b) Set $v = (\partial/\partial t + c\partial/\partial x)u$. Show that u and v satisfy

$$u_t + cu_x = v, \qquad (3.2.21)$$

$$v_t - cv_x = 0. \qquad (3.2.22)$$

(c) Find the initial conditions for u and v if $u(x,0) = f(x)$ and $u_t(x,0) = g(x)$.
(d) Solve (3.2.22) and then (3.2.21).

5. Find the solution to the wave equation

$$u_{tt}(x,t) - (\frac{1}{2})^2 u_{xx}(x,t) = \sin x, \quad -\infty < x < \infty, \ 0 < t$$

with the initial data

$$u(x,0) = f(x) = \cos(2x), \ u_t(x,0) = 0.$$

6. Solve

$$u_{tt} - c^2 u_{xx} = e^{-t}, \quad -\infty < x < \infty, \ 0 < t,$$

$$u(x,0) = 0, \quad u_t(x,0) = \cos x.$$

3.3 Wave Reflection Problems

In this section we discuss the wave reflection on the boundaries. These problems are formulated as the initial boundary problems with semi-infinite domains. In the string problem this is a semi-infinite string problem where the one end of the string is at $x = 0$. We discuss both homogeneous and non-homogeneous boundary conditions and in each case the Dirichlet and the Neumann boundary conditions. As described in the introduction, the wave reflection is an important phenomenon in physics. We study the mechanism behind the wave reflection.

3.3.1 *Homogeneous Boundary Conditions*

Fixed End

We assume that the boundary is located at $x = 0$. The case where the end is fixed is formulated as an initial boundary value problem

$$u_{tt}(x,t) - c^2 u_{xx}(x,t) = 0, \quad 0 < x < \infty, \ 0 < t, \tag{3.3.1}$$

where the boundary condition is

$$u(0,t) = 0 \tag{3.3.2}$$

and the initial conditions are

$$u(x,0) = f(x), \ u_t(x,0) = g(x). \tag{3.3.3}$$

In the string problem, this is the case where the end of a semi-infinite string is attached to the boundary at $x = 0$.

There are two typical ways to solve the problem, the methods based on the divergence theorem and on the extension. The first method will be

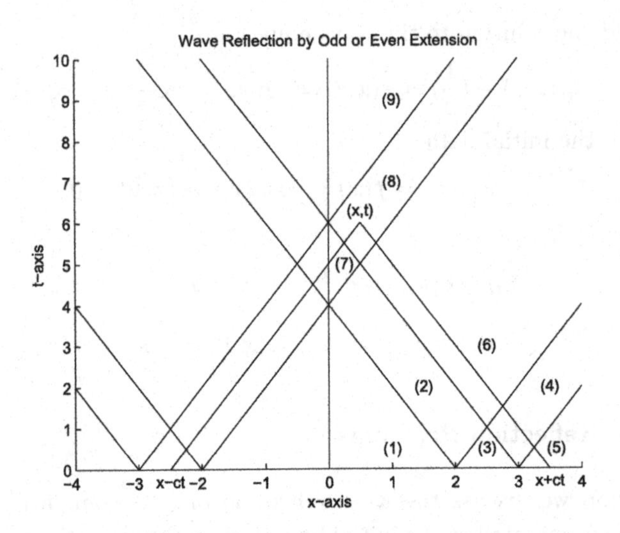

Fig. 3.10 Wave reflection by odd or even extension. $c = 1/2$.

treated in Exercises. Here, we continue the extension method developed in the heat equation. Since we have the homogeneous Dirichlet boundary condition, we extend the initial data into $x < 0$ as odd functions so that the solution on $x \geq 0$ of the resulting initial value problem will be the solution to the above initial boundary value problem. Therefore, we consider the following initial value problem

$$u_{tt}(x,t) - c^2 u_{xx}(x,t) = 0, \quad -\infty < x < \infty,\ 0 < t,$$

$$u(x,0) = p(x) = \begin{cases} -f(-x), & x < 0, \\ f(x), & 0 < x, \end{cases}$$

$$u_t(x,0) = q(x) = \begin{cases} -g(-x), & x < 0, \\ g(x), & 0 < x. \end{cases} \tag{3.3.4}$$

The solution is given by

$$u(x,t) = \frac{1}{2}[p(x-ct) + p(x+ct)] + \frac{1}{2c}\int_{x-ct}^{x+ct} q(s)ds. \tag{3.3.5}$$

We need to examine if (3.3.5) satisfies the initial conditions and the boundary condition. At $t = 0$, for $x > 0$,

$$u(x,0) = \frac{1}{2}[p(x) + p(x)] = f(x),$$

and

$$u_t(x,0) = \frac{1}{2}[-cp'(x) + cp'(x)] + \frac{1}{2c}[q(x)c + q(x)c] = g(x).$$

At $x = 0$, since

$$\frac{1}{2c}\int_{-ct}^{0} q(s)ds = -\frac{1}{2c}\int_{-ct}^{0} g(-s)ds = -\frac{1}{2c}\int_{0}^{ct} g(\xi)d\xi,$$

(3.3.5) is

$$\begin{aligned} u(0,t) &= \frac{1}{2}[p(-ct) + p(ct)] + \frac{1}{2c}\int_{-ct}^{ct} q(s)ds \\ &= \frac{1}{2}[-f(ct) + f(ct)] + \frac{1}{2c}\int_{0}^{ct} g(s)ds - \frac{1}{2c}\int_{0}^{ct} g(\xi)d\xi = 0. \end{aligned}$$

Therefore, (3.3.5) satisfies both the initial and boundary conditions and for $x > 0$ it is a solution to (3.3.1)-(3.3.3). The uniqueness of solutions, which validates the solution (3.3.5), is discussed in Exercises 3 in Section 3.5.

We can rewrite (3.3.5) using (3.3.4). If $ct < x$, the solution is the same as the initial value problem and therefore,

$$u(x,t) = \frac{1}{2}[f(x-ct) + f(x+ct)] + \frac{1}{2c}\int_{x-ct}^{x+ct} g(s)ds.$$

On the other hand, if $x < ct$, since q is an odd function, we have

$$\frac{1}{2c}\int_{x-ct}^{0} q(s)ds = -\frac{1}{2c}\int_{x-ct}^{0} g(-s)ds = \frac{1}{2c}\int_{ct-x}^{0} g(\xi)d\xi,$$

where $\xi = -s$. Therefore, after replacing ξ with s, we obtain

$$\begin{aligned} u(x,t) &= \frac{1}{2}[p(x-ct) + f(x+ct)] + \frac{1}{2c}\int_{0}^{x+ct} g(s)ds + \frac{1}{2c}\int_{x-ct}^{0} q(s)ds \\ &= \frac{1}{2}[-f(ct-x) + f(x+ct)] + \frac{1}{2c}\int_{ct-x}^{x+ct} g(s)ds. \end{aligned} \tag{3.3.6}$$

Combining two cases, we have

$$u(x,t) = \begin{cases} \frac{1}{2}[f(x+ct) + f(x-ct)] + \frac{1}{2c}\int_{x-ct}^{x+ct} g(s)ds, & x - ct > 0 \\ \frac{1}{2}[f(x+ct) - f(ct-x)] + \frac{1}{2c}\int_{ct-x}^{x+ct} g(s)ds, & x - ct < 0. \end{cases} \tag{3.3.7}$$

The first case in (3.3.7) is the same as the solution to the initial value problem. To understand the second case, we consider the solution u at $(x,t) = (x_1,t_1)$ satisfying $x_1 < ct_1$. It is given by

$$u(x_1,t_1) = \frac{1}{2}[f(x_1+ct_1) - f(ct_1-x_1)] + \frac{1}{2c}\int_{ct_1-x_1}^{x_1+ct_1} g(s)ds.$$

We examine the characteristics through the point. Since $x_1 < ct_1$, the characteristic $x - ct = x_1 - ct_1$ intersects the t-axis at $(0, (ct_1 - x_1)/c)$. Consider the characteristic $x + ct$ = constant through this point. It is given by $x + ct = ct_1 - x_1$. This characteristic intersects with the x-axis at $(ct_1 - x_1, 0)$. The domain of influence is $ct_1 - x_1 \leq x \leq x_1 + ct_1$. An interpretation of the term $-f(ct_1 - x_1)$ is the following. The value $f(ct_1 - x_1)$ propagates along $x + ct = ct_1 - x_1$ till it hits the boundary $x = 0$. Then, it changes the sign and propagates along $x - ct = ct_1 - x_1$ to reach (x_1, t_1).

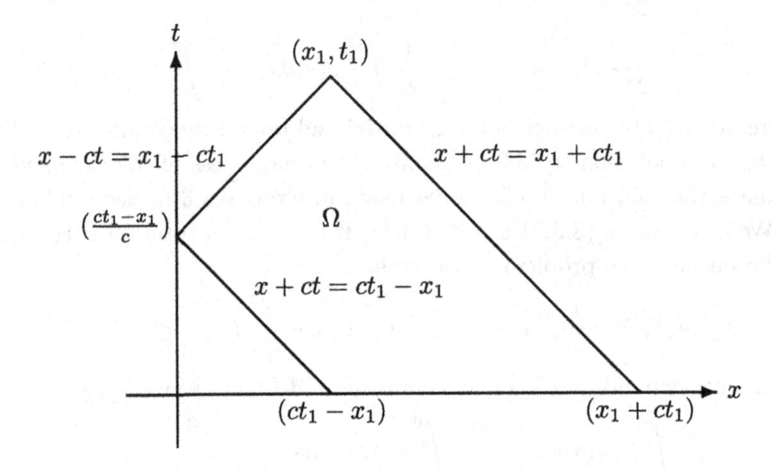

Fig. 3.11 Wave reflection.

The following example illustrates how we solve the reflection problem with the Dirichlet boundary condition.

Example 3.3.1. Consider an initial boundary value problem

$$u_{tt}(x,t) - (\frac{1}{2})^2 u_{xx}(x,t) = 0, \quad 0 < x < \infty,\ 0 < t,$$

where the boundary condition is

$$u(0,t) = 0$$

and the initial conditions are given by

$$u(x,0) = f(x) = \begin{cases} 0, & 0 < x < 2, \\ 1, & 2 < x < 3, \\ 0, & 3 < x. \end{cases} \quad u_t(x,0) = 0,$$

(a) Find the solution.

(b) Sketch the profiles of solution at $t = 4,\ 5,\ 6$. Indicate the height of solution and the locations where the height changes.

Solution: There are two ways to find the solution. One way is to extend the initial data as odd functions and use the formula (3.3.5). It is a good idea to draw a figure like Figure 3.10 where the characteristics from $x = 2,\ 3$ are drawn. The characteristics with negative and positive slopes are $x + t/2 =$ constant and $x - t/2 =$ constant, respectively. Since $c = 1/2$ and $g = 0$, the solution is given by

$$u(x,t) = \frac{1}{2}[p(x - \frac{t}{2}) + p(x + \frac{t}{2})],$$

where p is the odd extension of f. In this problem

$$p(x) = \begin{cases} 0, & x < -3, \\ -1, & -3 < x < -2, \\ 0, & -2 < x < 2, \\ 1, & 2 < x < 3, \\ 0, & 3 < x. \end{cases}$$

In Figure 3.10 an example of the characteristics through (x,t) located in Region (8) is given. In this case the characteristic through (x,t) with negative slope hits $x + t/2$ on the initial line. This value of $x + t/2$ is larger than 3. Therefore, $p(x + t/2) = 0$. Also, the characteristic through (x,t) with positive slope hits $x - t/2$ on the initial line. This value of $x - t/2$ is between -3 and -2. Therefore, $p(x - t/2) = -1$. So, if (x,t) is located in Region (8),

$$u(x,t) = \frac{1}{2}[p(x - \frac{t}{2}) + p(x + \frac{t}{2})] = -\frac{1}{2}.$$

The other cases are treated similarly.

Also, we can use the reflection. From (3.3.7) the solution is given by

$$u(x,t) = \begin{cases} \frac{1}{2}[f(x + \frac{t}{2}) + f(x - \frac{t}{2})], & x - \frac{t}{2} > 0, \\ \frac{1}{2}[f(x + \frac{t}{2}) - f(\frac{t}{2} - x)], & x - \frac{t}{2} < 0. \end{cases}$$

It is a good idea to draw a figure like Figure 3.12 where the characteristics from $x = 2,\ 3$ and their reflections are drawn. The characteristics with negative and positive slopes are $x + t/2 =$ constant and $x - t/2 =$ constant, respectively. In Figure 3.12 an example of the characteristics through (x,t) located in Region (8) is given. In this case the characteristic through (x,t)

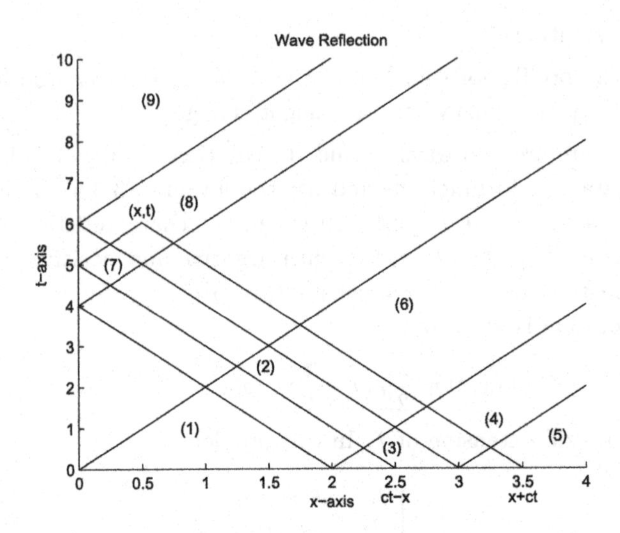

Fig. 3.12 Characteristics for wave reflections.

with negative slope hits $x + t/2$ on the initial line. This value of $x + t/2$ is larger than 3. Therefore, $f(x + t/2) = 0$. On the other hand, the characteristic through (x, t) with positive slope reflects on the boundary $x = 0$ and hits $t/2 - x$ on the initial line. This value of $t/2 - x$ is between 2 and 3. Therefore, $f(t/2 - x) = 1$. So,

$$u(x,t) = \frac{1}{2}[f(x + \frac{t}{2}) - f(\frac{t}{2} - x)] = -\frac{1}{2}.$$

The solution in the regions (1) through (9) are given below. In each case we do the similar thing as we did for the case (8), *i.e.*, draw backward characteristics from a point (x, y) in each region and observe where they reach on the initial line. Either way there are nine regions to consider.

Region (1): $0 < x < -t/2 + 2$. $u(x,t) = 0$.
Region (2): $-t/2 + 2 < x < -t/2 + 3$ and $t/2 - 2 < x < t/2 + 2$.
  $u(x, t) = 1/2$.
Region (3): $t/2 + 2 < x < -t/2 + 3$. $u(x,t) = 1$.
Region (4): $t/2 + 2 < x < t/2 + 3$ and $-t/2 + 3 < x$. $u(x,t) = 1/2$.
Region (5): $t/2 + 3 < x$. $u(x,t) = 0$.
Region (6): $t/2 - 2 < x < t/2 + 2$ and $-t/2 + 3 < x$. $u(x,t) = 0$.
Region (7): $0 < x < t/2 - 2$ and $0 < x < -t/2 + 3$. $u(x,t) = 0$.
Region (8): $t/2 - 3 < x < t/2 - 2$ and $-t/2 + 3 < x$. $u(x,t) = -1/2$.

Region (9): $0 < x < t/2 - 3$. $u(x,t) = 0$.

In Region (7) the incoming wave $f(x + t/2)/2$ and the reflecting wave $-f(t/2 - x)/2$ cancel out.

(b) The profiles are given in Figure 3.13.

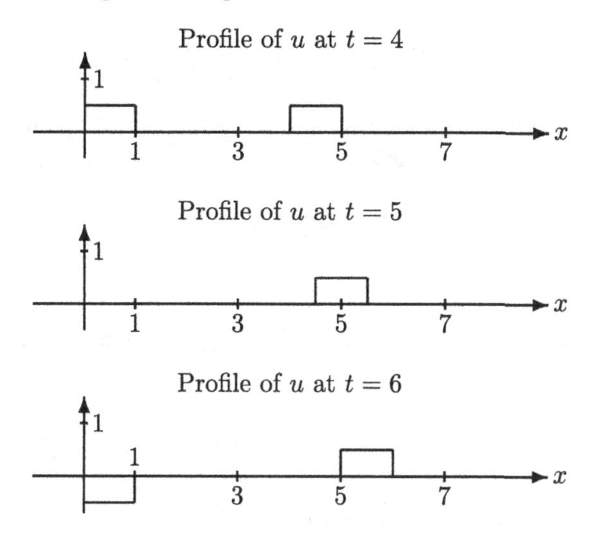

Fig. 3.13 Wave reflection: Dirichlet boundary conditions.

Free End

Another case is when one end of string is free to move vertically at $x = 0$, *i.e.*, there is no vertical force on the boundary. Since the vertical force is Tu_x, this is formulated as an initial boundary value problem

$$u_{tt}(x,t) - c^2 u_{xx}(x,t) = 0, \quad 0 < x < \infty,\ 0 < t, \tag{3.3.8}$$

where the boundary condition is

$$u_x(0,t) = 0 \tag{3.3.9}$$

and the initial conditions are given by

$$u(x,0) = f(x),\ u_t(x,0) = g(x). \tag{3.3.10}$$

To solve the above problem, consider the initial value problem on the interval $-\infty < x < \infty$, where the initial data (3.3.10) are extended as even functions to $x < 0$

$$u_{tt}(x,t) - c^2 u_{xx}(x,t) = 0, \quad -\infty < x < \infty,\ 0 < t,$$

$$u(x,0) = p(x) = \begin{cases} f(-x), & x < 0, \\ f(x), & 0 \leq x, \end{cases} \quad u_t(x,0) = q(x) = \begin{cases} g(-x), & x < 0, \\ g(x), & 0 \leq x. \end{cases} \tag{3.3.11}$$

The solution is given by

$$u(x,t) = \frac{1}{2}[p(x-ct) + p(x+ct)] + \frac{1}{2c}\int_{x-ct}^{x+ct} q(s)ds. \tag{3.3.12}$$

We need to examine if (3.3.12) satisfies the initial condition and the boundary condition. At $t = 0$, for $x \geq 0$,

$$u(x,0) = \frac{1}{2}[p(x) + p(x)] = f(x),$$

and

$$u_t(x,0) = \frac{1}{2}[-cp'(x) + cp'(x)] + \frac{1}{2c}[q(x)c + q(x)c] = g(x).$$

Since

$$u_x = \frac{1}{2}[p'(x-ct) + p'(x+ct)] + \frac{1}{2c}[q(x+ct) - q(x-ct)]$$

and $p'(x) = -f'(-x)$ if $x < 0$, we see at $x = 0$

$$u_x(0,t) = \frac{1}{2}[-f'(ct) + f'(ct)] + \frac{1}{2c}[g(ct) - g(-ct)] = 0.$$

Therefore, (3.3.12) satisfies both the initial and boundary conditions and for $x > 0$ it is a solution to (3.3.8)-(3.3.10).

We can rewrite (3.3.12) using (3.3.11). If $ct \leq x$, the solution is the same as the initial value problem and therefore,

$$u(x,t) = \frac{1}{2}[f(x-ct) + f(x+ct)] + \frac{1}{2c}\int_{x-ct}^{x+ct} g(s)ds.$$

On the other hand, if $x < ct$, since q is an even function, we have

$$\frac{1}{2c}\int_{x-ct}^{0} q(s)ds = \frac{1}{2c}\int_{x-ct}^{0} g(-s)ds = \frac{1}{2c}\int_{0}^{ct-x} g(\xi)d\xi,$$

where $\xi = -s$. Therefore,

$$\begin{aligned} u(x,t) &= \frac{1}{2}[p(x-ct) + f(x+ct)] + \frac{1}{2c}\int_{0}^{x+ct} g(s)ds + \frac{1}{2c}\int_{x-ct}^{0} q(s)ds \\ &= \frac{1}{2}[f(ct-x) + f(x+ct)] + \frac{1}{2c}\int_{0}^{x+ct} g(s)ds + \frac{1}{2c}\int_{0}^{ct-x} g(\xi)d\xi. \end{aligned}$$

Combining two cases and replacing ξ with s, we obtain

$$u(x,t) = \begin{cases} \frac{1}{2}[f(x+ct)+f(x-ct)] + \frac{1}{2c}\int_{x-ct}^{x+ct} g(s)ds, & x-ct>0, \\ \frac{1}{2}[f(x+ct)+f(ct-x)] & \\ \quad +\frac{1}{2c}[\int_0^{x+ct} g(s)ds + \int_0^{ct-x} g(s)ds], & x-ct<0. \end{cases} \tag{3.3.13}$$

It is interesting to observe how the difference in the boundary conditions affects the form of the solution in the region where $x - ct < 0$. With the boundary condition $u(0,t) = 0$, $f(ct - x)$ changes sign after the reflection and there is a cancellation in the integral $\int g(s)ds$. On the other hand, with the boundary condition $u_x(0,t) = 0$, $f(ct - x)$ does not change sign after the reflection and there is no cancellation in the integral $\int g(s)ds$. The following examples illustrates some of the differences.

Example 3.3.2. Consider the initial boundary value problem for the wave equation.

$$u_{tt} - (\frac{1}{2})^2 u_{xx} = 0, \quad 0 < x < \infty,\ 0 < t,$$

$$u_x(0,t) = 0,$$

$$u(x,0) = f(x) = \begin{cases} 0,\ 0 < x < 2, \\ 1,\ 2 < x < 3, \\ 0, \quad 3 < x, \end{cases} \quad u_t(x,0) = 0.$$

(a) Find the solution.

(b) Sketch the profiles of solution at $t = 4,\ 5,\ 6$. Indicate the height of solution and the locations where the height changes.

Solution: (a) One way is to extend the initial data as even functions and use the formula (3.3.12). It is a good idea to draw a figure like Figure 3.10 where the characteristics from $x = 2,\ 3$ are drawn. The characteristics with negative and positive slopes are $x + t/2 =$ constant and $x - t/2 =$ constant, respectively. Since $c = 1/2$ and $g = 0$, the solution is given by

$$u(x,t) = \frac{1}{2}[p(x - \frac{t}{2}) + p(x + \frac{t}{2})],$$

where p is the even extension of f. In this problem

$$p(x) = \begin{cases} 0, & x < -3, \\ 1 & -3 < x < -2, \\ 0, & -2 < x < 2, \\ 1, & 2 < x < 3, \\ 0, & 3 < x. \end{cases}$$

In Figure 3.10 an example of the characteristics through (x,t) located in Region (8) is given. In this case the characteristic through (x,t) with negative slope hits $x+t/2$ on the initial line. This value of $x+t/2$ is larger than 3. Therefore, $p(x+t/2)=0$. Also, the characteristic through (x,t) with positive slope hits $x-t/2$ on the initial line. This value of $x-t/2$ is between -3 and -2. Therefore, $p(x-t/2)=1$. So, if (x,t) is located in Region (8),

$$u(x,t)=\frac{1}{2}[p(x-\frac{t}{2})+p(x+\frac{t}{2})]=\frac{1}{2}.$$

The other cases are treated similarly.

Also, we can use the reflection. From (3.3.13) the solution is given by

$$u(x,t)=\begin{cases}\frac{1}{2}[f(x+\frac{t}{2})+f(x-\frac{t}{2})], & x-\frac{t}{2}>0,\\ \frac{1}{2}[f(x+\frac{t}{2})+f(\frac{t}{2}-x)], & x-\frac{t}{2}<0.\end{cases}$$

As in Example 3.3.1 it is a good idea to draw a figure like Figure 3.12 where the characteristics from $x=2,\ 3$ and their reflections are drawn. The characteristics with negative and positive slopes are $x+t/2=$ constant and $x-t/2=$ constant, respectively. In Figure 3.12 an example of the characteristics through (x,t) located in Region (8) is given. In this case the characteristic through (x,t) with negative slope hits $x+t/2$ on the initial line. This value of $x+t/2$ is larger than 3. Therefore, $f(x+t/2)=0$. On the other hand, the characteristic through (x,t) with positive slope reflects on the boundary $x=0$ and hits $t/2-x$ on the initial line. This value of $t/2-x$ is between 2 and 3. Therefore, $f(t/2-x)=1$. So,

$$u(x,t)=\frac{1}{2}[f(x+\frac{t}{2})+f(\frac{t}{2}-x)]=\frac{1}{2}.$$

The other cases are similarly computed.

Region (1): $0<x<-t/2+2$. $u(x,t)=0$.

Region (2): $-t/2+2<x<-t/2+3$ and $t/2-2<x<t/2+2$.
$u(x,t)=1/2$.

Region (3): $t/2+2<x<-t/2+3$. $u(x,t)=1$.

Region (4): $t/2+2<x<t/2+3$ and $-t/2+3<x$. $u(x,t)=1/2$.

Region (5): $t/2+3<x$. $u(x,t)=0$.

Region (6): $t/2-2<x<t/2+2$ and $-t/2+3<x$. $u(x,t)=0$.

Region (7): $0<x<t/2-2$ and $0<x<-t/2+3$. $u(x,t)=1$.

Region (8): $t/2-3<x<t/2-2$ and $-t/2+3<x$. $u(x,t)=1/2$.

Region (9): $0<x<t/2-3$. $u(x,t)=0$.

In Region (7) the incoming wave $f(x+t/2)/2$ and the reflecting wave $f(t/2-x)/2$ add up.

(b) The profiles of solution are given in Figure 3.14. It is interesting to compare Figures 3.14 and 3.13 to see the effects of different boundary conditions.

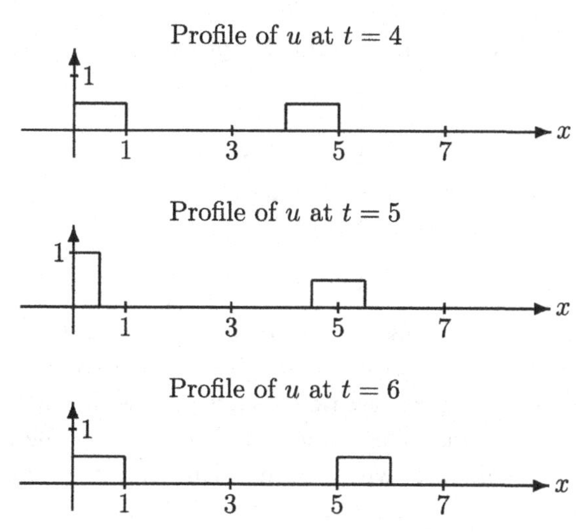

Fig. 3.14 Wave reflection: Neumann boundary conditions.

Example 3.3.3. Consider the initial boundary value problem for the wave equation.

$$u_{tt}-(\frac{1}{2})^2u_{xx}=0, \quad 0<x<\infty,\ 0<t,$$

$$u_x(0,t)=0,$$

$$u(x,0)=f(x)=0 \quad u_t(x,0)=g(x)=\begin{cases}0,\ 0<x<2,\\ 1,\ 2<x<3,\\ 0, \quad 3<x.\end{cases}$$

(a) Find the solution.

(b) Sketch the profiles of solution at $t=4,\ 5,\ 6$. Indicate the height of solution and the locations where the height changes.

Solution: (a) One way is to extend the initial data as even functions and use the formula (3.3.12). It is a good idea to draw a figure like Figure 3.10

where the characteristics from $x = 2,\ 3$ are drawn. The characteristics with negative and positive slopes are $x + t/2 =$ constant and $x - t/2 =$ constant, respectively. Since $c = 1/2$ and $f = 0$, the solution is given by

$$u(x,t) = \frac{1}{2c}\int_{x-ct}^{x+ct} q(s)ds,$$

where q is the even extension of g. In this problem

$$q(x) = \begin{cases} 0, & x < -3, \\ 1 & -3 < x < -2, \\ 0, & -2 < x < 2, \\ 1, & 2 < x < 3, \\ 0, & 3 < x. \end{cases}$$

There are nine regions to distinguish in the region $x > 0$. In Figure 3.10 an example of the characteristics through (x, t) located in Region (8) is given. In this case the characteristic through (x, t) with negative slope hits $x + t/2$ on the initial line. This value of $x + t/2$ is larger than 3. Also, the characteristic through (x, t) with positive slope hits $x - t/2$ on the initial line. This value of $x - t/2$ is between -3 and -2. So, if (x, t) is located in Region (8),

$$\begin{aligned} u(x,t) &= \int_{x-t/2}^{x+t/2} q(s)ds \\ &= \int_{x-t/2}^{-2} ds + \int_{-2}^{2} 0ds + \int_{2}^{3} ds + \int_{3}^{x+t/2} 0ds = \frac{t}{2} - x - 1. \end{aligned}$$

The other cases are treated similarly.

Also, we can use the reflection. From (3.3.13) the solution is given by

$$u(x,t) = \begin{cases} \frac{1}{2c}\int_{x-ct}^{x+ct} g(s)ds, & x - ct > 0 \\ \frac{1}{2c}[\int_0^{x+ct} g(s)ds + \int_0^{ct-x} g(s)ds], & x - ct < 0. \end{cases}$$

As in Example 3.3.1 it is a good idea to draw a figure like Figure 3.12 where the characteristics from $x = 2,\ 3$ and their reflections are drawn. The characteristics with negative and positive slopes are $x + t/2 =$ constant and $x - t/2 =$ constant, respectively. In Figure 3.12 an example of the characteristics through (x, t) located in Region (8) is given. In this case the characteristic through (x, t) with negative slope hits $x + t/2$ on the initial line. This value of $x + t/2$ is larger than 3. The characteristic through (x, t)

with positive slope reflects on the boundary $x = 0$ and hits $t/2 - x$ on the initial line. This value of $t/2 - x$ is between 2 and 3. So,

$$u(x,t) = \int_0^{x+t/2} g(s)ds + \int_0^{t/2-x} g(s)ds$$
$$= \int_0^2 0ds + \int_2^3 ds + \int_3^{x+t/2} 0ds + \int_0^2 0ds + \int_2^{t/2-x} ds = \frac{t}{2} - x - 1.$$

The other cases can be calculated similarly.

Region (1): $0 < x < -t/2 + 2$. $u(x,t) = 0$.

Region (2): $-t/2 + 2 < x < -t/2 + 3$ and $t/2 - 2 < x < t/2 + 2$.
$u(x,t) = \int_2^{x+t/2} ds = x + t/2 - 2$.

Region (3): $t/2 + 2 < x < -t/2 + 3$. $u(x,t) = \int_{x-t/2}^{x+t/2} ds = t$.

Region (4): $t/2 + 2 < x < t/2 + 3$ and $-t/2 + 3 < x$.
$u(x,t) = \int_{x-t/2}^{3} ds = 3 - x + t/2$.

Region (5): $t/2 + 3 < x$. $u(x,t) = 0$.

Region (6): $t/2 - 2 < x < t/2 + 2$ and $-t/2 + 3 < x$. $u(x,t) = \int_2^3 ds = 1$.

Region (7): $0 < x < t/2 - 2$ and $0 < x < -t/2 + 3$.
$u(x,t) = \int_2^{x+t/2} ds + \int_2^{t/2-x} ds = t - 4$.

Region (8): $t/2 - 3 < x < t/2 - 2$ and $-t/2 + 3 < x$.
$u(x,t) = \int_2^3 ds + \int_2^{t/2-x} ds = t/2 - x - 1$.

Region (9): $0 < x < t/2 - 3$. $u(x,t) = \int_2^3 ds + \int_2^3 ds = 2$.

(b) The profiles of solution are given in Figure 3.15.

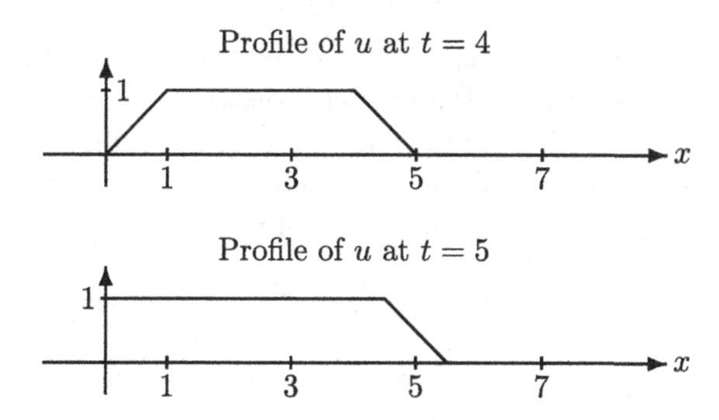

Fig. 3.15 Wave reflection: Neumann boundary conditions.

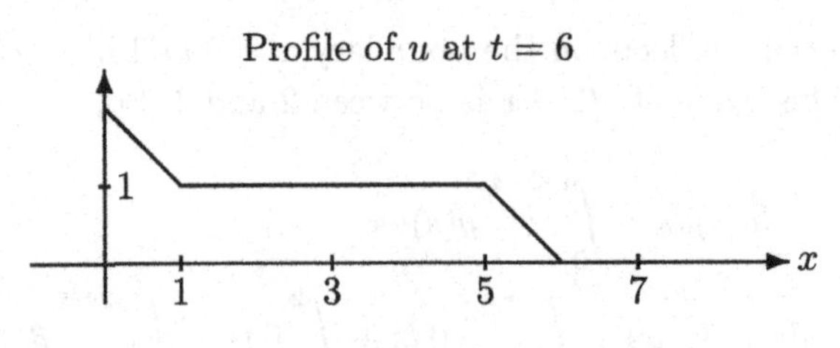

Fig. 3.15 (Continued)

3.3.2 *Non-homogeneous Boundary Conditions*

Now we discuss the wave reflection problems where the boundary conditions are not homogeneous. In the case of sound propagation this corresponds to the case where the sound is not only reflected but also is generated on the boundary. For example a speaker can be placed on a wall at $x = 0$. There are two typical problems depending on the Dirichlet or Neumann type boundary conditions. In the following example, we consider the Dirichlet the boundary condition. The Neumann problem will be discussed in Exercises.

Example 3.3.4. Find the solution to

$$u_{tt}(x,t) - c^2 u_{xx}(x,t) = 0, \quad 0 < x < \infty,\ 0 < t, \tag{3.3.14}$$

where the boundary condition is given by

$$u(0,t) = b(t) \tag{3.3.15}$$

and the initial conditions are given by

$$u(x,0) = f(x), \quad u_t(x,0) = g(x), \tag{3.3.16}$$

Solution: Extending the initial data to $x < 0$ as odd functions may not be so easy. Instead we treat the problem based on the method of characteristics using the solution obtained in (3.2.4) and (3.2.8), *i.e.*, we find the appropriate expression for $F(x+ct)$ and $G(x-ct)$. If $ct < x$, the solution is the same as the initial value problem and therefore,

$$u(x,t) = F(x+ct) + G(x-ct) = \frac{1}{2}[f(x-ct) + f(x+ct)] + \frac{1}{2c}\int_{x-ct}^{x+ct} g(s)ds.$$

On the other hand, if $x - ct < 0$, we can still use $F(x+ct) = [f(x+ct) + \frac{1}{c}\int_0^{x+ct} g(s)ds + K]/2$, but we cannot use $G(x-ct) = [f(x-ct) - \frac{1}{c}\int_0^{x-ct} g(s)ds - K]/2$ since f and g are defined only for $x - ct > 0$. We need to find the expression for $G(x-ct)$. The solution must satisfy (3.3.15). Therefore, on the boundary $x=0$, we have

$$u(0,t) = F(ct) + G(-ct) = b(t).$$

If we set $z = ct$, then

$$F(z) + G(-z) = b(\frac{z}{c}).$$

Therefore, setting $z = ct - x$, we have

$$\begin{aligned} G(x-ct) &= b(\frac{ct-x}{c}) - F(ct-x) \\ &= b(\frac{ct-x}{c}) - \frac{1}{2}[f(ct-x) + \frac{1}{c}\int_0^{ct-x} g(s)ds + K]. \end{aligned}$$

Therefore, for $x - ct < 0$, the solution is given by

$$\begin{aligned} &u(x,t) \\ &= F(x+ct) + G(x-ct) \\ &= \frac{1}{2}[f(x+ct) + \frac{1}{c}\int_0^{x+ct} g(s)ds + K] \\ &\quad -\frac{1}{2}[f(ct-x) + \frac{1}{c}\int_0^{ct-x} g(s)ds + K] + b(\frac{ct-x}{c}) \\ &= \frac{1}{2}[f(x+ct) - f(ct-x)] + \frac{1}{2c}\int_{ct-x}^{x+ct} g(s)ds + b(\frac{ct-x}{c}). \quad (3.3.17) \end{aligned}$$

For $x - ct < 0$, it is interesting to compare the solution (3.3.6) in the homogeneous case and (3.3.17). The difference is the non-homogeneous term $b((ct-x)/c)$. This shows that the method in this section is also applicable to the homogeneous problems.

Exercises

1. Draw the profiles of solutions for Examples 3.3.1, 3.3.2, and 3.3.3 at $t = 8$.
2. Consider the initial boundary value problem

$$u_{tt}(x,t) - (\frac{1}{2})^2 u_{xx}(x,t) = 0, \quad 0 < x < \infty,\ 0 < t,$$

$$u(0,t) = 0$$

$$u(x,0) = f(x) = \begin{cases} 0, & x \le 1, \\ 1 - (x-2)^2, & 1 < x \le 3, \\ 0, & 3 < x, \end{cases} \quad u_t(x,0) = 0.$$

(a) Find the solution.
(b) Sketch the profiles of solution at $t = 2,\ 4,\ 8$.

3. Consider the initial boundary value problem

$$u_{tt}(x,t) - (\frac{1}{2})^2 u_{xx}(x,t) = 0, \quad 0 < x < \infty,\ 0 < t,$$

$$u(0,t) = 0,$$

$$u(x,0) = f(x) = 0, \quad u_t(x,0) = \begin{cases} 0, & x \leq 1, \\ 1 - (x-2)^2, & 1 < x \leq 3, \\ 0, & 3 < x. \end{cases}$$

(a) Find the solution.
(b) Sketch the profiles of solution at $t = 2,\ 4,\ 8$.

4. In Exercises 2 and 3, change the boundary conditions to $u_x(0,t) = 0$ and solve the problems.
5. Find the solution formula for

$$u_{tt}(x,t) - c^2 u_{xx}(x,t) = 0, \quad 0 < x < \infty,\ 0 < t,$$

$$u_x(0,t) = b(t),$$

$$u(x,0) = f(x), \quad u_t(x,0) = g(x).$$

6. Apply the divergence theorem to the region Ω in Figure 3.11 to find the solution formula.

$$u_{tt}(x,t) - c^2 u_{xx}(x,t) = h(x,t), \quad 0 < x < \infty,\ 0 < t,$$

$$u(0,t) = b(t),$$

$$u(x,0) = f(x), \quad u_t(x,0) = g(x).$$

3.4 Initial Boundary Value Problems

It is possible to use the method of characteristics to handle the initial boundary value problems. However, we need to consider multiple wave reflections, which makes the problem complicated. Instead, we apply the separation of variables introduced in the heat equation.

We consider a vertical motion of string with length L with one end at $x = 0$ and the other end at $x = L$. Depending on the fixed or free ends, as in the heat equation, there are four typical boundary conditions given below.

$$u(0,t) = 0, \quad u(L,t) = 0, \tag{3.4.1}$$

$$u_x(0,t) = 0, \quad u_x(L,t) = 0, \tag{3.4.2}$$

$$u(0,t) = 0, \quad u_x(L,t) = 0, \tag{3.4.3}$$

$$u_x(0,t) = 0, \quad u(L,t) = 0. \tag{3.4.4}$$

For example, (3.4.3) says that the end $x = 0$ is fixed and $x = L$ is free. Also, the Robin boundary conditions could be discussed.

As an example we consider the following initial boundary value problem with the boundary conditions (3.4.2) to illustrate how the separation of variables works for the wave equation.

$$u_{tt} - c^2 u_{xx} = 0, \quad 0 < x < L, \quad 0 < t, \tag{3.4.5}$$

$$u_x(0,t) = 0, \quad u_x(L,t) = 0, \tag{3.4.6}$$

$$u(x,0) = f(x), \quad u_t(x,0) = g(x). \tag{3.4.7}$$

Step 1 (Separating Variables): As we did in the heat equation, we assume

$$u(x,t) = X(x)T(t)$$

and derive ODE's for X and T. Substitute this in (3.4.5). Then,

$$XT'' - c^2 X''T = 0.$$

Dividing this by $c^2 XT$, we obtain

$$\frac{T''}{c^2 T} = \frac{X''}{X} = -\lambda,$$

where λ is a constant to be determined. We need to solve ODE's

$$X'' + \lambda X = 0, \tag{3.4.8}$$

$$T'' + \lambda c^2 T = 0. \tag{3.4.9}$$

Step 2 (Solving ODE's): As we did in the heat equation first we solve the eigenvalue problem for (3.4.8) and then solve the initial value problem for (3.4.9). The boundary conditions for (3.4.8) are

$$u_x(0,t) = 0 \Rightarrow X'(0)T(t) = 0 \Rightarrow X'(0) = 0, \tag{3.4.10}$$

$$u_x(L,t) = 0 \Rightarrow X'(L)T(t) = 0 \Rightarrow X'(L) = 0. \tag{3.4.11}$$

We notice that the eigenvalue problem for X is the same as the one discussed in Subsection 2.3.3. The eigenvalues are

$$\lambda_0 = 0, \quad \lambda_n = (\frac{n\pi}{L})^2, \ n = 1, 2, \ldots$$

and the corresponding eigenfunctions are

$$X_0 = 1, \quad X_n = \cos\frac{n\pi x}{L}, \quad n = 1, 2, \ldots.$$

Now, we solve (3.4.9) for the eigenvalues of λ. If $\lambda = \lambda_0 = 0$,

$$T_0 = \frac{a_0}{2} + \frac{b_0}{2}t$$

and if $\lambda = \lambda_n = (n\pi/L)^2$,

$$T_n = a_n \cos\frac{n\pi ct}{L} + b_n \sin\frac{n\pi ct}{L}.$$

Step 3 (Finding the Solution): Since we assume that $u = XT$, the solutions are

$$u_0 = X_0 T_0 = \frac{a_0}{2} + \frac{b_0}{2}t,$$

$$u_n = X_n T_n = (a_n \cos\frac{n\pi ct}{L} + b_n \sin\frac{n\pi ct}{L})\cos\frac{n\pi x}{L}, \quad n = 1, 2, \ldots.$$

Using the superposition principle, we take

$$\begin{aligned} u(x,t) &= \sum_{n=0}^{\infty} u_n \\ &= \frac{a_0}{2} + \frac{b_0}{2}t + \sum_{n=1}^{\infty}(a_n \cos\frac{n\pi ct}{L} + b_n \sin\frac{n\pi ct}{L})\cos\frac{n\pi x}{L} \end{aligned} \tag{3.4.12}$$

as the solution to the problem.

Our task now is to find a_n and b_n from the initial conditions. At $t = 0$,

$$u(x,0) = \frac{a_0}{2} + \sum_{n=1}^{\infty} a_n \cos\frac{n\pi x}{L} = f(x), \tag{3.4.13}$$

$$u_t(x,0) = \frac{b_0}{2} + \sum_{n=1}^{\infty} \frac{n\pi c}{L} b_n \cos\frac{n\pi x}{L} = g(x). \tag{3.4.14}$$

To find a_0 and b_0 we integrate (3.4.13) and (3.4.14) over $0 < x < L$. Then,

$$a_0 = \frac{2}{L}\int_0^L f(x)dx, \quad b_0 = \frac{2}{L}\int_0^L g(x)dx. \tag{3.4.15}$$

To find a_m and b_m we multiply (3.4.13) and (3.4.14) by $\cos(m\pi/L)$ and integrate over $0 < x < L$. Using the orthogonality relation

$$\int_0^L \cos\frac{m\pi x}{L}\cos\frac{n\pi x}{L}dx = \begin{cases} \frac{L}{2} & m = n \\ 0 & m \neq n, \end{cases}$$

we obtain

$$a_m = \frac{2}{L}\int_0^L f(x)\cos\frac{m\pi x}{L}dx,$$

$$b_m = \frac{L}{m\pi c}\frac{2}{L}\int_0^L g(x)\cos\frac{m\pi x}{L}dx, \ m = 1, 2, \ldots. \tag{3.4.16}$$

Comparing (3.4.15) and (3.4.16), we see that (3.4.16) is also valid for $m = 0$.

Example 3.4.1. Find the solution to the initial boundary value problem

$$\begin{aligned} &u_{tt} - c^2 u_{xx} = 0, \quad 0 < x < L, \quad 0 < t, \\ &u_x(0,t) = 0, \quad u_x(L,t) = 0, \\ &u(x,0) = x, \quad u_t(x,0) = 0. \end{aligned}$$

Solution: From (3.4.12) and (3.4.16), we see that $b_n = 0$ $(n = 0, 1, \ldots)$ and

$$\begin{aligned} a_0 &= \frac{2}{L}\int_0^L f(x)dx = \frac{2}{L}\int_0^L x dx \\ &= \frac{2}{L}[\frac{x^2}{2}]_0^L = L, \end{aligned}$$

$$\begin{aligned} a_n &= \frac{2}{L}\int_0^L f(x)\cos\frac{n\pi x}{L}dx = \frac{2}{L}\int_0^L x\cos\frac{n\pi x}{L}dx \\ &= \frac{2}{L}[x\frac{L}{n\pi}\sin\frac{n\pi x}{L}]_0^L - \frac{2}{L}\int_0^L \frac{L}{n\pi}\sin\frac{n\pi x}{L}dx \\ &= \frac{2}{L}(\frac{L}{n\pi})^2[\cos\frac{n\pi x}{L}]_0^L = \frac{2}{L}(\frac{L}{n\pi})^2((-1)^n - 1). \end{aligned}$$

Therefore, the solution is

$$u(x,t) = \frac{L}{2} + \sum_{n=1}^{\infty} \frac{2}{L}(\frac{L}{n\pi})^2((-1)^n - 1)\cos\frac{n\pi x}{L}\cos\frac{n\pi ct}{L}.$$

Exercises

1. Consider the initial boundary value problem.

$$u_{tt} = c^2 u_{xx}, \; 0 < x < 1, \; 0 < t,$$

$$u(0,t) = 0, \quad u(1,t) = 0,$$

$$u(x,0) = f(x), \quad u_t(x,0) = g(x).$$

   (a) Assume $u(x,t) = X(x)T(t)$ to derive ODE's for $X(x)$ and $T(t)$. Put c^2 to the ODE for T.
   (b) Which are the eigenvalues and eigenfunctions of the ODE for $X(x)$?
   (c) Solve the ODE for $T(t)$ to find the solution.
   (d) From the initial data, find the formula for the coefficients of the solution.

2. Find the solutions to the initial boundary value problems

$$u_{tt} = c^2 u_{xx}, \; 0 < x < L, \; 0 < t,$$

$$u(x,0) = f(x), \quad u_t(x,0) = g(x)$$

   with the following boundary conditions.

   (a) $u_x(0,t) = 0, \quad u(1,t) = 0$
   (b) $u(0,t) = 0, \quad u_x(L,t) = 0$
   (c) $u(0,t) = 0, \quad u_x(L,t) + u(L,t) = 0$

3. Find the solutions to the initial boundary value problems

$$u_{tt} = c^2 u_{xx}, \; 0 < x < L, \; 0 < t,$$

$$u(0,t) = 0, \quad u_x(L,t) = 0$$

   with the following initial conditions.

   (a) $u(x,0) = 1, \quad u_t(x,0) = 0$
   (b) $u(x,0) = 0, \quad u_t(x,0) = 1$
   (c) $u(x,0) = 1, \quad u_t(x,0) = 1$
   (d) Explain the relation among (a), (b), and (c).

4. Consider

$$u_{tt} = c^2 u_{xx}, \quad 0 \le x \le L, \quad 0 < t,$$

$$u_x(0,t) = 1, \quad u(L,t) = 0,$$

$$u(x,0) = x, \quad u_t(x,0) = 0.$$

(a) Find the steady state solution $v(x)$.
(b) Assume $u(x,t) = w(x,t) + v(x)$ and find the initial boundary value problem for $w(x,t)$.
(c) Find $w(x,t)$ and find the solution to the above initial boundary value problem.
(d) Does $u(x,t)$ approach $v(x)$ as $t \to \infty$?

5. Consider a damped wave equation

$$u_{tt} = c^2 u_{xx} - r u_t, \quad 0 \le x \le L, \quad 0 < t,$$

$$u(0,t) = 0, \quad u(L,t) = 0,$$

$$u(x,0) = f(x), \quad u_t(x,0) = g(x).$$

(a) Solve the problem if $r^2 < 4\pi^2 c^2/L^2$.
(b) Solve the problem if $4\pi^2 c^2/L^2 < r^2 < 16\pi^2 c^2/L^2$.

6. Consider

$$u_{tt} = c^2 u_{xx} - r u_t, \quad 0 \le x \le L, \quad 0 < t,$$

$$u_x(0,t) = 1, \quad u(L,t) = 0,$$

$$u(x,0) = x, \quad u_t(x,0) = 0.$$

(a) Find the steady state solution $v(x)$.
(b) Assume $u(x,t) = w(x,t) + v(x)$ and find the initial boundary value problem for $w(x,t)$.
(c) Find $w(x,t)$ and find the solution to the above initial boundary value problem. Assume $r^2 < 4\pi^2 c^2/L^2$.
(d) If $r^2 < 4\pi^2 c^2/L^2$, does $u(x,t)$ approach $v(x)$ as $t \to \infty$?

3.5 Energy Method

We have the similar L^2 energy method as in the heat equation. However, unlike the heat equation the energy method for the wave equation has physical meaning as we see it below. As a specific example consider

$$u_{tt} - c^2 u_{xx} = 0, \quad 0 < x < L, \quad 0 < t, \tag{3.5.1}$$

$$u(0,t) = 0, \quad u_x(L,t) = 0, \tag{3.5.2}$$

$$u(x,0) = f(x), \quad u_t(x,0) = g(x). \tag{3.5.3}$$

For the wave equation we multiply (3.5.1) by u_t and integrate the resulting equation by parts over $0 < x < L$. Then, using $u_t u_{tt} = \frac{1}{2}\frac{\partial}{\partial t}(u_t^2)$, we have

$$\int_0^L [\frac{1}{2}\frac{\partial}{\partial t}(u_t^2) - c^2 u_{xx} u_t] dx = 0.$$

Since $u(0,t) = 0$ along the t-axis $u_t(0,t) = 0$. This and the integration by parts in x for the second term lead to

$$-\int_0^L c^2 u_{xx} u_t dx = -[c^2 u_x u_t]|_{x=0}^{x=L} + \int_0^L c^2 u_x u_{xt} dx = \frac{1}{2}\int_0^L c^2 \frac{\partial}{\partial t}(u_x^2) dx.$$

Therefore,

$$\int_0^L [\frac{1}{2}\frac{\partial}{\partial t}(u_t^2) - c^2 u_{xx} u_t] dx = \frac{1}{2}\int_0^L \frac{\partial}{\partial t}[u_t^2 + c^2 u_x^2] dx$$

$$= \frac{1}{2}\frac{d}{dt}\int_0^L [(u_t^2) + c^2(u_x^2)] dx = 0.$$

Integrating over $0 < t < T$, where T is a positive constant, we see that

$$\int_0^L [u_t^2(x,T) + c^2 u_x^2(x,T)] dx = \int_0^L [g_t^2(x) + c^2 f_x^2(x)] dx.$$

The first and second terms on the left hand side are regarded as the kinetic and potential energy, respectively. This shows that the total energy of the string is conserved.

Exercises

1. Consider

$$u_{tt} - c^2 u_{xx} = 0, \quad 0 < x < L, \quad 0 < t, \tag{3.5.4}$$

$$u(0,t) = 0, \quad u(L,t) = 0, \tag{3.5.5}$$

$$u(x,0) = f(x), \quad u_t(x,0) = g(x). \tag{3.5.6}$$

   (a) Show that the total energy is conserved.
   (b) Show that the solution is unique.
   (c) Show that L^2 energy is bounded by that of the initial data.

2. Consider

$$u_{tt} - c^2 u_{xx} = 0, \quad 0 < x < L, \quad 0 < t, \tag{3.5.7}$$

$$u_x(0,t) - u(0,t) = 0, \quad u_x(L,t) + u(L,t) = 0, \quad 0 < t, \tag{3.5.8}$$

$$u(x,0) = f(x), \quad u_t(x,0) = g(x). \tag{3.5.9}$$

(a) Show that the following is conserved.

$$\int_0^L [u_t^2(x,t) + c^2 u_x^2(x,t)]dx + c^2[u^2(L,t) + u^2(0,t)].$$

(b) Show that the solution is unique.

3. We discuss the uniqueness of solutions for the wave reflection problem in Section 3.3. Suppose that f and g below have compact support in x. In other words f and g vanish after finite intervals. Show the following

$$u_{tt} - c^2 u_{xx} = 0, \quad 0 < x < \infty, \quad 0 < t,$$

$$u(0,t) = 0,$$

$$u(x,0) = f(x), \quad u_t(x,0) = g(x).$$

(a) Show that the solution u has a compact support in x for finite t.
(b) Show that the total energy is conserved.
(c) Show that the solution is unique.
(d) Show that L^2 energy is bounded by that of the initial data.

4. If there is a damping due to air resistance, the wave equation is modified to

$$u_{tt} - c^2 u_{xx} + r u_t = 0, \quad 0 < x < L, \quad 0 < t, \tag{3.5.10}$$

where r is a positive constant. This is called the damped wave equation. Consider (3.5.10) with (3.5.5) and (3.5.6).

(a) Show that the total energy decays in time.
(b) Show that the solution is unique.
(c) Show that L^2 energy is bounded by that of the initial data.

Chapter 4

Laplace Equation

In this chapter we consider the Laplace and the Poisson equations. Motivations for this chapter are explained in Section 4.1. In Section 4.2 we continue the separation of variables and we apply the method to the boundary value problems of the Laplace equation. We discuss the cases where the domains are rectangular and circular. In Section 4.3 we study the fundamental solutions. This is a solution to the Laplace equation which depends only on the radial variable and is used to construct the Green's function. This is discussed in Section 4.4 and we have solution representation of the Laplace equation by the Green's functions when the domains are simple. Solutions of the Laplace equation are called the harmonic functions and their properties are discussed in Section 4.5 including the maximum principle and the mean value property. In Section 4.6 one of the well-posedness issues will be studied.

4.1 Motivations

There are several sources where the Laplace equation appears. First, we are often interested in the time independent solutions of wave or heat equations. These solutions are called the steady state solutions. In the heat equation, we would expect that the solution approaches the steady state solutions as t approaches infinity. For example, for the initial boundary value problem of the heat equation

$$u_t = a^2 u_{xx}, \quad 0 < x < 1,\ 0 < t,$$

$$u(0,t) = 0,\ u(1,t) = 1,$$

$$u(x,0) = f(x),$$

the corresponding steady state solution is

$$u_{xx} = 0,$$

$$u(0, t) = 0, \ u(1, t) = 1.$$

The two-dimensional and three-dimensional heat equations are given by

$$u_t = a^2(u_{xx} + u_{yy}), \quad u_t = a^2(u_{xx} + u_{yy} + u_{zz}).$$

The corresponding steady state solutions are

$$\triangle u = u_{xx} + u_{yy} = 0,$$

$$\triangle u = u_{xx} + u_{yy} + u_{zz} = 0,$$

respectively. The above equations are called the Laplace equation and C^2 functions u satisfying $\triangle u = 0$ are called potential functions or harmonic functions.

We also often encounter the equation

$$\triangle u = h.$$

This equation is called the Poisson equation. The following are the examples. Maxwell formulated a system of PDE's called the Maxwell equations

$$\frac{\partial \mathbf{B}}{\partial t} + c\boldsymbol{\nabla} \times \mathbf{E} = 0, \tag{4.1.1}$$

$$\frac{\partial \mathbf{E}}{\partial t} - c\boldsymbol{\nabla} \times \mathbf{B} = -4\pi \mathbf{J}, \tag{4.1.2}$$

$$\boldsymbol{\nabla} \cdot \mathbf{E} = \sigma, \tag{4.1.3}$$

$$\boldsymbol{\nabla} \cdot \mathbf{B} = 0, \tag{4.1.4}$$

governing the electromagneticity in the late 19th century. Here $\mathbf{E}$ and $\mathbf{B}$ are the electric and magnetic fields, respectively, $\mathbf{J}$ is the current, σ is the charge, and c is the speed of light. We introduce the electric potential u and the vector potential $\mathbf{A}$ for the magnetic field satisfying $\mathbf{E} = -\boldsymbol{\nabla} u$ and $\mathbf{B} = \boldsymbol{\nabla} \times \mathbf{A}$, respectively. Then, from (4.1.3) and (4.1.2) we obtain, respectively

$$-\triangle u = \sigma, \tag{4.1.5}$$

$$\frac{1}{c}\frac{\partial^2 \mathbf{A}}{\partial t^2} - c\triangle \mathbf{A} = 4\pi \mathbf{J}. \tag{4.1.6}$$

In the time independent case we see that both u and $\mathbf{A}$ satisfy the Poisson equation.

Exercises

1. Derive (4.1.5) and (4.1.6).

4.2 Boundary Value Problems - Separation of Variables

4.2.1 *Laplace Equation on a Rectangular Domain*

We study the boundary value problems for the Laplace equation. We start with the case where the domain Ω is a rectangle. There are two well-known boundary conditions, the Dirichlet and the Neumann boundary conditions.
In the Dirichlet problem, u is specified on the boundary.

$$\Delta u = 0, \quad 0 < x < L, \quad 0 < y < H, \tag{4.2.1}$$

$$u(0,y) = f_1(y), \quad u(L,y) = f_2(y), \tag{4.2.2}$$

$$u(x,0) = g_1(x), \quad u(x,H) = g_2(x). \tag{4.2.3}$$

In the Neumann problem, the derivative of u is specified on the boundary.

$$\Delta u = 0, \quad 0 < x < L, \quad 0 < y < H,$$

$$u_x(0,y) = f_1(y), \quad u_x(L,y) = f_2(y),$$

$$u_y(x,0) = g_1(x), \quad u_y(x,H) = g_2(x).$$

Using $\Delta u = \nabla \cdot \nabla u$ and the divergence theorem, in the case of the Neumann problem we see that the following solvability condition must hold.

$$\begin{aligned}
0 &= \int_0^L \int_0^H \nabla \cdot \nabla u dy dx = \int_{\partial\Omega} \frac{\partial u}{\partial n} ds \\
&= \int_0^H [1,0] \cdot [\frac{\partial u}{\partial x}, \frac{\partial u}{\partial y}]|_{x=L} dy + \int_L^0 [0,1] \cdot [\frac{\partial u}{\partial x}, \frac{\partial u}{\partial y}]|_{y=H}(-dx) \\
&\quad + \int_H^0 [-1,0] \cdot [\frac{\partial u}{\partial x}, \frac{\partial u}{\partial y}]|_{x=0}(-dy) + \int_0^L [0,-1] \cdot [\frac{\partial u}{\partial x}, \frac{\partial u}{\partial y}]|_{y=0} dx \\
&= \int_0^H f_2(y) dy + \int_L^0 g_2(x)(-dx) \\
&\quad + \int_H^0 (-f_1(y))(-dy) + \int_0^L (-g_1(x)) dx.
\end{aligned} \tag{4.2.4}$$

Here, $\partial u/\partial n$ is a simplified notation for the directional derivative $\nabla u \cdot \mathbf{n}$, where $\mathbf{n}$ is the unit outward normal vector to Ω. We are integrating counter clockwise. That is why $ds = -dy$ and $ds = -dx$ in the first and the last integrals, respectively. This shows that f_i and g_i $(i = 1, 2)$ are not arbitrary. We had a similar situation in the steady state solution for the one-dimensional heat equation with the Neumann boundary conditions.

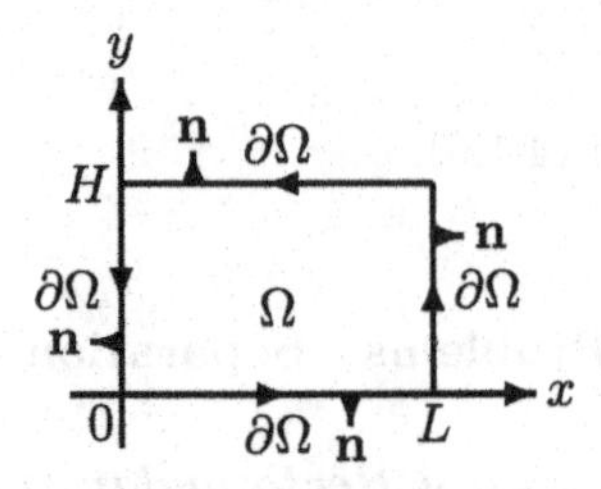

Fig. 4.1 The line integral.

Superposition Principle

To solve the above problems we use the superposition principle to simplify the problems. For example, in the Dirichlet problem the solution to (4.2.1), (4.2.2), and (4.2.3) is the same as the sum of the solutions to four subproblems where the one side of the boundary is nonzero and the other three sides are zero. More precisely, suppose u_1, u_2, u_3, and u_4 are the solutions to (1), (2), (3), and (4), respectively.

$$\begin{aligned}(1):\quad \Delta u &= 0, \quad 0<x<L, \quad 0<y<H,\\ u(0,y) &= f_1(y), \quad u(L,y)=0,\\ u(x,0) &= 0, \quad u(x,H)=0,\end{aligned}$$

$$\begin{aligned}(2):\quad \Delta u &= 0, \quad 0<x<L, \quad 0<y<H,\\ u(0,y) &= 0, \quad u(L,y)=f_2(y),\\ u(x,0) &= 0, \quad u(x,H)=0,\end{aligned}$$

$$\begin{aligned}(3):\quad \Delta u &= 0, \quad 0<x<L, \quad 0<y<H,\\ u(0,y) &= 0, \quad u(L,y)=0,\\ u(x,0) &= g_1(x), \quad u(x,H)=0,\end{aligned}$$

$$\begin{aligned}(4):\quad \Delta u &= 0, \quad 0<x<L, \quad 0<y<H,\\ u(0,y) &= 0, \quad u(L,y)=0,\\ u(x,0) &= 0, \quad u(x,H)=g_2(x).\end{aligned}$$

Then, the solution to (4.2.1), (4.2.2), and (4.2.3) is given by

$$u(x,y) = u_1(x,y) + u_2(x,y) + u_3(x,y) + u_4(x,y).$$

Separation of Variables

The separation of variables that we studied in the heat and wave equations can be applied to each sub-problem. We consider the following problem as an example.

Example 4.2.1. Find the solution to the boundary value problem.

$$\Delta u = 0, \quad 0 < x < L, \quad 0 < y < H, \tag{4.2.5}$$

$$u(0, y) = 0, \quad u(L, y) = 0, \tag{4.2.6}$$

$$u(x, 0) = 0, \quad u(x, H) = g_2(x). \tag{4.2.7}$$

Solution: Step 1 (Separating variables): We assume

$$u(x, y) = X(x)Y(y) \tag{4.2.8}$$

and derive ODE's for X and Y. Substitute this in (4.2.5). Then,

$$XY'' + X''Y = 0.$$

Divide the above equation by XY. Then,

$$\frac{X''}{X} = -\frac{Y''}{Y} = \pm\lambda,$$

where the sign $\pm$ and the constant λ are to be determined. We need to solve ODE's

$$X'' \pm \lambda X = 0. \tag{4.2.9}$$

$$Y'' \mp \lambda Y = 0. \tag{4.2.10}$$

To determine the sign of λ and which equation we solve first, we substitute (4.2.8) to the boundary conditions. Then,

$$u(0, y) = X(0)Y(y) = 0 \Rightarrow X(0) = 0, \tag{4.2.11}$$

$$u(L, y) = X(L)Y(y) = 0 \Rightarrow X(L) = 0, \tag{4.2.12}$$

$$u(x, 0) = X(x)Y(0) = 0 \Rightarrow Y(0) = 0, \tag{4.2.13}$$

$$u(x, H) = X(x)Y(H) = g_2(x) \Rightarrow Y(H) =?.$$

Since the boundary conditions for X is clear but those for Y is not clear, we solve X first. Consequently, it is more convenient to choose the plus sign for λ. Then, the boundary value problem for X has a familiar sign for λ, *i.e.*, we have the same eigenvalue problems as in the heat and wave equations. However unlike $T(t)$ for the heat and wave equations where we solved the initial value problems, we solve the boundary value problems for $Y(y)$.

Step 2 (Solving ODE's): We solve the eigenvalue problem for (4.2.9) with the positive sign for λ and with the boundary conditions (4.2.11) and (4.2.12). We have already discussed this problem in Subsection 2.3.1. The eigenvalues are

$$\lambda_n = (\frac{n\pi}{L})^2, \quad n = 1, 2, \ldots$$

and the corresponding eigenfunctions are

$$X_n(x) = \sin\frac{n\pi x}{L}, \quad n = 1, 2, \ldots.$$

Next, we solve the boundary value problem for (4.2.10) for the eigenvalues of λ. The general solution for $\lambda = \lambda_n = (n\pi/L)^2$ is

$$Y_n(y) = a_n \cosh\frac{n\pi y}{L} + b_n \sinh\frac{n\pi y}{L}.$$

At $y = 0$, from (4.2.13)

$$Y_n(0) = a_n \cosh 0 = a_n = 0.$$

Therefore,

$$Y_n(y) = b_n \sinh\frac{n\pi y}{L}.$$

Step 3 (Finding the solution): Since we assume that $u = XY$, the solutions are

$$u_n = X_n Y_n = b_n \sinh\frac{n\pi y}{L}\sin\frac{n\pi x}{L}, \quad n = 1, 2, \ldots.$$

Using the superposition principle, we take

$$\begin{aligned} u(x,y) &= \sum_{n=0}^{\infty} u_n \\ &= \sum_{n=1}^{\infty} b_n \sinh\frac{n\pi y}{L}\sin\frac{n\pi x}{L} \end{aligned} \tag{4.2.14}$$

as the solution to the problem.

Our task now is to find b_n from the other boundary condition for $Y(y)$. At $y = H$, from $u(x, H) = g_2(x)$ and (4.2.14)

$$u(x,H) = \sum_{n=1}^{\infty} b_n \sinh\frac{n\pi H}{L}\sin\frac{n\pi x}{L} = g_2(x). \tag{4.2.15}$$

To find b_m we multiply by $\sin(m\pi x/L)$ and integrate over $0 < x < L$. Using the orthogonality relation

$$\int_0^L \sin\frac{m\pi x}{L}\sin\frac{n\pi x}{L}dx = \begin{cases} \frac{L}{2} & m = n \\ 0 & m \neq n \end{cases},$$

we obtain

$$b_m = \frac{1}{\sinh\frac{m\pi H}{L}}\frac{2}{L}\int_0^L g_2(x)\sin\frac{m\pi x}{L}dx. \tag{4.2.16}$$

Let's consider a specific case of Example 4.2.1 in the next example.

Example 4.2.2. Find the solution to the boundary value problem (4.2.5) to (4.2.7) if $g_2 = L - x$ in (4.2.7).

Solution: Evaluate (4.2.16) with $g_2 = L - x$. Then, since b_n is given by

$$\begin{aligned}\frac{L}{2}\sinh\frac{n\pi H}{L}b_n &= \int_0^L g_2(x)\sin\frac{n\pi x}{L}dx = \int_0^L (L-x)\sin\frac{n\pi x}{L}dx \\ &= [(L-x)\frac{L}{n\pi}(-1)\cos\frac{n\pi x}{L}]_0^L - \int_0^L \frac{L}{n\pi}\cos\frac{n\pi x}{L}dx \\ &= \frac{L^2}{n\pi},\end{aligned}$$

the solution is

$$u(x,y) = \sum_{n=1}^{\infty}\frac{1}{\sinh\frac{n\pi H}{L}}\frac{2L}{n\pi}\sinh\frac{n\pi y}{L}\sin\frac{n\pi x}{L}.$$

4.2.2 *Laplace Equation on a Circular Disk*

Laplacian in the Polar Coordinates

We consider the Laplace equation on a circular disk. For this purpose we change the variables to the polar coordinates (r, θ). The relation between the rectangular coordinates (x, y) and the polar coordinates (r, θ) is given by

$$x = r\cos\theta, \quad y = r\sin\theta. \tag{4.2.17}$$

Also, we have

$$r^2 = x^2 + y^2, \quad \tan\theta = \frac{y}{x}. \tag{4.2.18}$$

If we use (4.2.18), partial derivatives of (r, θ) with respect to (x, y) are given by

$$2r\frac{\partial r}{\partial x} = 2x, \quad \frac{1}{\cos^2\theta}\frac{\partial\theta}{\partial x} = -\frac{y}{x^2},$$

$$2r\frac{\partial r}{\partial y} = 2y, \quad \frac{1}{\cos^2\theta}\frac{\partial \theta}{\partial y} = \frac{1}{x}.$$

Then, we obtain

$$u_x = u_r\frac{\partial r}{\partial x} + u_\theta\frac{\partial \theta}{\partial x} = u_r\frac{x}{r} + u_\theta(-\frac{y}{x^2}\cos^2\theta),$$
$$u_y = u_r\frac{\partial r}{\partial y} + u_\theta\frac{\partial \theta}{\partial y} = u_r\frac{y}{r} + u_\theta(\frac{1}{x}\cos^2\theta),$$

$$\begin{aligned}
u_{xx} &= (u_{rr}\frac{\partial r}{\partial x} + u_{r\theta}\frac{\partial \theta}{\partial x})\frac{\partial r}{\partial x} + u_r\frac{\partial}{\partial x}(\frac{x}{r}) \\
&+(u_{\theta r}\frac{\partial r}{\partial x} + u_{\theta\theta}\frac{\partial \theta}{\partial x})(-\frac{y}{x^2}\cos^2\theta) + u_\theta\frac{\partial}{\partial x}(-\frac{y}{x^2}\cos^2\theta) \\
&= u_{rr}(\frac{x}{r})^2 + 2u_{r\theta}\frac{x}{r}(-\frac{y}{x^2}\cos^2\theta) + u_r\frac{r^2 - x^2}{r^3} + u_{\theta\theta}(-\frac{y}{x^2}\cos^2\theta)^2 \\
&+u_\theta(2\frac{y}{x^3}\cos^2\theta - (\frac{y}{x^2})^2 2\cos\theta\sin\theta\cos^2\theta),
\end{aligned} \quad (4.2.19)$$

$$\begin{aligned}
u_{yy} &= u_{rr}(\frac{y}{r})^2 + 2u_{r\theta}\frac{y}{r}(\frac{1}{x}\cos^2\theta) + u_r\frac{r^2 - y^2}{r^3} \\
&+u_{\theta\theta}(\frac{1}{x}\cos^2\theta)^2 - u_\theta(\frac{1}{x^2}2\cos\theta\sin\theta\cos^2\theta).
\end{aligned} \quad (4.2.20)$$

Adding (4.2.19) and (4.2.20) and observing that u_θ terms cancel out, we see that

$$u_{xx} + u_{yy} = u_{rr} + \frac{1}{r^2}u_{\theta\theta} + \frac{1}{r}u_r = \frac{1}{r}\frac{\partial}{\partial r}(r\frac{\partial u}{\partial r}) + \frac{1}{r^2}\frac{\partial^2 u}{\partial \theta^2}. \quad (4.2.21)$$

Therefore, we obtain

$$\Delta u = \frac{1}{r}\frac{\partial}{\partial r}(r\frac{\partial u}{\partial r}) + \frac{1}{r^2}\frac{\partial^2 u}{\partial \theta^2}.$$

Separation of Variables

We consider the following problem.

$$\Delta u = \frac{1}{r}\frac{\partial}{\partial r}(r\frac{\partial u}{\partial r}) + \frac{1}{r^2}\frac{\partial^2 u}{\partial \theta^2} = 0, \quad 0 \le r < a, \ -\pi \le \theta \le \pi, \quad (4.2.22)$$

$$u(a,\theta) = f(\theta), \quad (4.2.23)$$

$$u(r,-\pi) = u(r,\pi), \quad u_\theta(r,-\pi) = u_\theta(r,\pi). \quad (4.2.24)$$

The boundary conditions (4.2.24) need some explanation. They are called the periodic boundary conditions. We should note that both $\theta = -\pi$ and $\theta = \pi$ are the negative x-axis. They guarantee that u is smoothly connected across the negative x-axis $\theta = \pm\pi$. We could equally use $0 \le \theta \le 2\pi$ as the domain of θ. Then, the boundary conditions would be

$$u(r,0) = u(r,2\pi), \quad u_\theta(r,0) = u_\theta(r,2\pi)$$

so that the solution is smoothly connected across the positive x-axis. However, we prefer $-\pi \le \theta \le \pi$ because of the symmetry.

Step 1: We use the separation of variables and assume $u = R(r)\Theta(\theta)$. Substituting this in (4.2.22) and dividing the resulting equation by $R(r)\Theta(\theta)/r^2$, we obtain

$$\frac{r(rR')'}{R} + \frac{\Theta''}{\Theta} = 0 \Rightarrow \frac{r(rR')'}{R} = -\frac{\Theta''}{\Theta} = \lambda.$$

$$\Theta'' + \lambda\Theta = 0, \tag{4.2.25}$$

$$r(rR')' - \lambda R = r^2R'' + rR' - \lambda R = 0. \tag{4.2.26}$$

From (4.2.24) the boundary conditions are

$$u(r,-\pi) = u(r,\pi) \Rightarrow R(r)\Theta(-\pi) = R(r)\Theta(\pi) \Rightarrow \Theta(-\pi) = \Theta(\pi), \tag{4.2.27}$$

$$u_\theta(r,-\pi) = u_\theta(r,\pi) \Rightarrow R(r)\Theta'(-\pi) = R(r)\Theta'(\pi) \Rightarrow \Theta'(-\pi) = \Theta'(\pi). \tag{4.2.28}$$

Step 2: We solve the ODE's. The general solution of (4.2.25) is given by

$$\begin{aligned}
\lambda > 0 &: \quad \Theta = c_1\cos\sqrt{\lambda}\theta + c_2\sin\sqrt{\lambda}\theta, \\
\lambda = 0 &: \quad \Theta = c_1 + c_2\theta, \\
\lambda < 0 &: \quad \Theta = c_1e^{\sqrt{-\lambda}\theta} + c_2e^{-\sqrt{-\lambda}\theta}.
\end{aligned}$$

We use the boundary conditions to figure out what values of λ are eigenvalues. We start from $\lambda > 0$. From the boundary condition (4.2.27),

$$\begin{aligned}
\Theta(-\pi) &= c_1\cos\sqrt{\lambda}\pi - c_2\sin\sqrt{\lambda}\pi \\
&= c_1\cos\sqrt{\lambda}\pi + c_2\sin\sqrt{\lambda}\pi = \Theta(\pi).
\end{aligned}$$

This implies that

$$c_2\sin\sqrt{\lambda}\pi = 0. \tag{4.2.29}$$

Also, from (4.2.28)

$$\begin{aligned}
\Theta'(-\pi) &= +c_1\sin\sqrt{\lambda}\pi + c_2\cos\sqrt{\lambda}\pi \\
&= -c_1\sin\sqrt{\lambda}\pi + c_2\cos\sqrt{\lambda}\pi = \Theta'(\pi).
\end{aligned}$$

This implies that

$$c_1 \sin \sqrt{\lambda}\pi = 0. \tag{4.2.30}$$

Equations (4.2.29) and (4.2.30) imply that to have nontrivial solutions

$$\sqrt{\lambda} = n \Rightarrow \lambda = n^2, \quad n = 1, 2, \ldots$$

are the eigenvalues and the corresponding eigenfunctions are

$$\Theta_n = a_n \cos n\theta + b_n \sin n\theta, \quad n = 1, 2, \ldots.$$

If $\lambda = 0$,

$$\Theta(-\pi) = c_1 - c_2\pi = c_1 + c_2\pi = \Theta(\pi) \Rightarrow c_2 = 0,$$

$$\Theta'(-\pi) = c_2 = c_2 = \Theta'(\pi).$$

Therefore, c_1 is arbitrary. So, $\lambda = 0$ is an eigenvalue and as a corresponding eigenfunction we take

$$\Theta_0 = 1.$$

If $\lambda < 0$,

$$\Theta(-\pi) = c_1 e^{-\sqrt{-\lambda}\pi} + c_2 e^{\sqrt{-\lambda}\pi} = c_1 e^{\sqrt{-\lambda}\pi} + c_2 e^{-\sqrt{-\lambda}\pi} = \Theta(\pi),$$

$$\begin{aligned}\Theta'(-\pi) &= \sqrt{-\lambda} c_1 e^{-\sqrt{-\lambda}\pi} - \sqrt{-\lambda} c_2 e^{\sqrt{-\lambda}\pi} \\ &= \sqrt{-\lambda} c_1 e^{\sqrt{-\lambda}\pi} - \sqrt{-\lambda} c_2 e^{-\sqrt{-\lambda}\pi} = \Theta'(\pi).\end{aligned}$$

Rewriting the above relations in a matrix form, we have

$$(e^{\sqrt{-\lambda}\pi} - e^{-\sqrt{-\lambda}\pi}) \begin{bmatrix} 1 & -1 \\ 1 & 1 \end{bmatrix} \begin{bmatrix} c_1 \\ c_2 \end{bmatrix} = \begin{bmatrix} 0 \\ 0 \end{bmatrix}.$$

Since $e^{\sqrt{-\lambda}\pi} - e^{-\sqrt{-\lambda}\pi} \neq 0$ and $\det \begin{bmatrix} 1 & -1 \\ 1 & 1 \end{bmatrix} = 2 \neq 0$, by Appendix A.4 we see that $c_1 = c_2 = 0$ is the only solution. So, there is no nontrivial solution. In summary, the eigenvalues are

$$\lambda_n = n^2, \quad n = 0, 1, 2, \ldots \tag{4.2.31}$$

and the corresponding eigenfunctions are

$$\Theta_0(x) = a_0, \quad \Theta_n(x) = a_n \cos n\theta + b_n \sin n\theta, \quad n = 1, 2, \ldots. \tag{4.2.32}$$

Now we solve the ODE for R for each eigenvalue.

$$r^2 R'' + rR' - \lambda R = 0.$$

If $\lambda = \lambda_0 = 0$,

$$\frac{R''}{R'} = -\frac{1}{r} \Rightarrow \ln|R'| = -\ln r + c \Rightarrow R' = \frac{c_2}{r} \Rightarrow R = c_1 + c_2 \ln r.$$

If $c_2 \neq 0$, R becomes infinite at the origin. So, we choose $c_2 = 0$. If $\lambda = \lambda_n = n^2$, assuming $R = r^k$, we have

$$k^2 - n^2 = 0 \Rightarrow R_n = c_1 r^n + c_2 r^{-n}.$$

If $c_2 \neq 0$, R becomes infinite at the origin. So, we choose $c_2 = 0$.

Step 3: Next, we use the superposition principle to take

$$u = \frac{a_0}{2} + \sum_{n=1}^{\infty}(a_n \cos n\theta + b_n \sin n\theta)r^n$$

as a general form of solution. To find the coefficients, we use the boundary condition (4.2.23). On $r = a$, we have

$$u(a,\theta) = \frac{a_0}{2} + \sum_{n=1}^{\infty}(a_n \cos n\theta + b_n \sin n\theta)a^n = f(\theta). \qquad (4.2.33)$$

As we have done in the previous chapters, we use the mutual orthogonality to find the coefficients.

For a_0 we integrate (4.2.33) over $[-\pi, \pi]$. Then,

$$\int_{-\pi}^{\pi} \frac{a_0}{2} d\theta = \int_{-\pi}^{\pi} f(\theta)d\theta \Rightarrow a_0 = \frac{1}{\pi}\int_{-\pi}^{\pi} f(\theta)d\theta.$$

For a_m $(m = 1, 2, \ldots)$ multiplying (4.2.33) by $\phi_m(\theta) = \cos m\theta$ and integrating over $[-\pi, \pi]$, we have

$$a_m = \frac{1}{a^m}\frac{\int_{-\pi}^{\pi} f(\theta)\phi_m(\theta)d\theta}{\int_{-\pi}^{\pi}\phi_m^2(\theta)d\theta} = \frac{1}{\pi a^m}\int_{-\pi}^{\pi} f(\theta)\cos m\theta d\theta.$$

For b_m $(m = 1, 2, \ldots)$ if we set $\phi_m(\theta) = \sin m\theta$ this time, we obtain similarly

$$b_m = \frac{1}{a^m}\frac{\int_{-\pi}^{\pi} f(\theta)\phi_m(\theta)d\theta}{\int_{-\pi}^{\pi}\phi_m^2(\theta)d\theta} = \frac{1}{\pi a^m}\int_{-\pi}^{\pi} f(\theta)\sin m\theta d\theta.$$

Remark 4.2.3. Notice that in (4.2.33) we represent $f(x)$ given on $-L \leq x \leq L$ $(L = \pi)$ using trigonometric functions as follows.

$$f(x) = \frac{a_0}{2} + \sum_{n=1}^{\infty}(a_n \cos\frac{n\pi x}{L} + b_n \sin\frac{n\pi x}{L}), \quad -L \leq x \leq L, \qquad (4.2.34)$$

where

$$a_n = \frac{1}{L}\int_{-L}^{L} f(x)\cos\frac{n\pi x}{L}dx, \quad b_n = \frac{1}{L}\int_{-L}^{L} f(x)\sin\frac{n\pi x}{L}dx.$$

The right hand side of (4.2.34) is called the Fourier series of f. The coefficients a_n and b_n are called the Fourier coefficients of f. We learn more about the Fourier series in Chapter 6.

Example 4.2.4. Find the solution to the boundary value problem (4.2.22) to (4.2.24) if the boundary condition is given by

$$u(a,\theta) = |\theta|, \quad -\pi \le \theta \le \pi.$$

Solution: We can use the fact that $|\theta|$ is an even function to simplify the integration.

$$a_0 = \frac{1}{\pi}\int_{-\pi}^{\pi} |\theta| d\theta = \frac{2}{\pi}\int_0^{\pi} |\theta| d\theta = \frac{2}{\pi}\int_0^{\pi} \theta d\theta = \pi.$$

$$\begin{aligned}
a_m &= \frac{1}{\pi a^m}\int_{-\pi}^{\pi} |\theta| \cos m\theta d\theta \\
&= \frac{2}{\pi a^m}\int_0^{\pi} \theta \cos m\theta d\theta \\
&= \frac{2}{\pi a^m}[\frac{\theta}{m}\sin m\theta]_0^{\pi} - \frac{2}{\pi a^m}\int_0^{\pi} \frac{1}{m}\sin m\theta d\theta \\
&= \frac{2}{\pi a^m}[\frac{1}{m^2}\cos m\theta]_0^{\pi} = \frac{2}{m^2 \pi a^m}[(-1)^m - 1].
\end{aligned}$$

$$b_m = \frac{1}{\pi a^m}\int_{-\pi}^{\pi} |\theta| \sin m\theta d\theta = 0.$$

Therefore,

$$\begin{aligned}
u(r,\theta) &= \frac{a_0}{2} + \sum_{n=1}^{\infty}(a_n \cos n\theta + b_n \sin n\theta) r^n \\
&= \frac{\pi}{2} + \sum_{n=1}^{\infty} \frac{2}{\pi}\frac{[(-1)^n - 1]}{n^2}(\frac{r}{a})^n \cos n\theta.
\end{aligned}$$

Exercises

1. (a) Use the divergence theorem to derive (4.2.4).
   (b) Show that the same result can be obtained by directly integrating once each integral on the right hand side of (4.2.35) and using the boundary conditions.

$$\int_0^L \int_0^H \nabla \cdot \nabla u dy dx = \int_0^H \int_0^L u_{xx} dx dy + \int_0^L \int_0^H u_{yy} dy dx. \tag{4.2.35}$$

2. (a) Derive (4.2.19) and (4.2.20).
   (b) Adding (4.2.19) and (4.2.20) to obtain (4.2.21).

3. Solve the Laplace equation in a rectangle with the boundary conditions (a) to (d):

$$\Delta u = 0, \quad 0 < x < L,\ 0 < y < H.$$

If there is a solvability condition, state it and explain it physically.

(a) $u(0,y) = f_1(y),\ u(L,y) = 0,\ u(x,0) = 0,\ u(x,H) = 0$
(b) $u(0,y) = 0,\ u(L,y) = f_2(y),\ u(x,0) = 0,\ u_y(x,H) = 0$
(c) $u(0,y) = 0,\ u_x(L,y) = f_2(y),\ u_y(x,0) = 0,\ u_y(x,H) = 0$
(d) $u_x(0,y) = 0,\ u_x(L,y) = 0,\ u_y(x,0) = g_1(x),\ u_y(x,H) = 0$

4. Explain how you would decompose the problem into sub-problems and solve them.

$$\Delta u = 0, \quad 0 < x < L,\ 0 < y < H,$$

$$u(0,y) = 0,\ u(L,y) = 0,\ u(x,0) = g_1(x),\ u_y(x,H) = g_2(x).$$

5. Solve the Laplace equation inside the quarter circle of radius 1 with the following boundary conditions. The region is given by $\Omega = \{(r,\theta) \mid 0 \le r \le 1,\ 0 \le \theta \le \pi/2\}$.

(a) $u_\theta(r,0) = 0, \quad u_\theta(r,\frac{\pi}{2}) = 0, \quad u(1,\theta) = f(\theta)$
(b) $u(r,0) = 0, \quad u_\theta(r,\frac{\pi}{2}) = 0, \quad u(1,\theta) = f(\theta)$

6. Solve the Laplace equation inside a circular annulus with the following boundary conditions. The region is given by $\Omega = \{(r,\theta) \mid a \le r \le b,\ 0 \le \theta \le 2\pi\}$, where a and b are positive constants. If there is a solvability condition, state it and explain it physically.

(a) $u_r(a,\theta) = f(\theta), \quad u_r(b,\theta) = g(\theta)$
(b) $u(a,\theta) = f(\theta), \quad u_r(b,\theta) = g(\theta)$

4.3 Fundamental Solution

A fundamental solution is a special solution to $\Delta u = 0$ which depends only on the radius. In this section we derive the fundamental solution. We also derive the Green's identity using the divergence theorem. With the help of the Green's identity we find the solution representation for the Laplace or Poisson equation by the fundamental solution.

4.3.1 *Green's Identity*

The Green's identity is an application of the divergence theorem and in two dimension it is given by

$$\iint\limits_{\Omega} (v\Delta u - u\Delta v)d\mathbf{x} = \int_{\partial\Omega} (v\nabla u \cdot \mathbf{n} - u\nabla v \cdot \mathbf{n})ds = \int_{\partial\Omega} (v\frac{\partial u}{\partial n} - u\frac{\partial v}{\partial n})ds, \tag{4.3.1}$$

where $\partial u/\partial n$ is a simplified notation for the directional derivative $\nabla u \cdot \mathbf{n}$ and so is $\partial v/\partial n$. We use these notations interchangeably. To derive this formula we apply the divergence theorem to the functions $v\nabla u$ and $u\nabla v$. For $v\nabla u$ we have

$$\iint\limits_{\Omega} \nabla \cdot (v\nabla u)d\mathbf{x} = \int_{\partial\Omega} v\frac{\partial u}{\partial n}ds.$$

If we perform the differentiation on the left hand side, we obtain

$$\iint\limits_{\Omega} \nabla \cdot (v\nabla u)d\mathbf{x} = \iint\limits_{\Omega} v\Delta u d\mathbf{x} + \iint\limits_{\Omega} \nabla v \cdot \nabla u d\mathbf{x}.$$

Therefore,

$$\iint\limits_{\Omega} v\Delta u d\mathbf{x} + \iint\limits_{\Omega} \nabla v \cdot \nabla u d\mathbf{x} = \int_{\partial\Omega} v\frac{\partial u}{\partial n}ds. \tag{4.3.2}$$

Similarly, for $u\nabla v$ we have

$$\iint\limits_{\Omega} u\Delta v d\mathbf{x} + \iint\limits_{\Omega} \nabla v \cdot \nabla u d\mathbf{x} = \int_{\partial\Omega} u\frac{\partial v}{\partial n}ds. \tag{4.3.3}$$

Subtracting (4.3.3) from (4.3.2), we obtain the Green's identity (4.3.1). Note that from (4.3.2) we have

$$\iint\limits_{\Omega} v\Delta u d\mathbf{x} = \iint\limits_{\Omega} v\nabla \cdot \nabla u d\mathbf{x} = \int_{\partial\Omega} v\frac{\partial u}{\partial n}ds - \iint\limits_{\Omega} \nabla v \cdot \nabla u d\mathbf{x}.$$

This can be thought of as a multidimensional version of the integration by parts. In the three dimensional case the Green's identity is given by

$$\iiint\limits_{\Omega} (v\Delta u - u\Delta v)d\mathbf{x} = \iint\limits_{\partial\Omega} (v\frac{\partial u}{\partial n} - u\frac{\partial v}{\partial n})dS. \tag{4.3.4}$$

4.3.2 *Derivation of Fundamental Solution*

We look for a spherically symmetric solution for the Laplace equation in R^n having the form

$$u(\mathbf{x}) = v(r), \tag{4.3.5}$$

where $r = |\mathbf{x}| = \sqrt{x_1^2 + \cdots + x_n^2}$. Substituting (4.3.5) in $\Delta u = 0$, we see that v must satisfy

$$v''(r) + \frac{n-1}{r}v'(r) = 0. \tag{4.3.6}$$

This is a linear (or separable) equation in v'. Multiplying (4.3.6) by the integrating factor $\mu = \exp(\int \frac{n-1}{r}dr) = r^{n-1}$, we obtain

$$r^{n-1}v''(r) + \frac{n-1}{r}r^{n-1}v'(r) = [r^{n-1}v'(r)]' = 0.$$

Here, $\exp(a) = e^a$. Therefore,

$$v'(r) = Cr^{1-n},$$

$$v(r) = \begin{cases} C\ln r & n = 2, \\ \frac{Cr^{2-n}}{2-n} & n > 2. \end{cases}$$

We choose constants C to be the reciprocal of the surface area of unit ball in R^n and denote the fundamental solution with this choice of C as

$$K(\mathbf{x};\mathbf{x}_0) = \phi(|\mathbf{x}-\mathbf{x}_0|) = \begin{cases} \frac{\ln|\mathbf{x}-\mathbf{x}_0|}{\omega_n} & n = 2, \\ -\frac{1}{\omega_n|\mathbf{x}-\mathbf{x}_0|} & n = 3, \end{cases} \tag{4.3.7}$$

where $r = |\mathbf{x} - \mathbf{x}_0|$ and

$$\omega_n = \begin{cases} 2\pi, & n = 2, \\ 4\pi, & n = 3, \end{cases}$$

is the surface area of unit ball in R^n. They are called the fundamental solution of Laplace equation.

4.3.3 *Green's Identity and Fundamental Solution*

In this section we use the Green's identity to express a solution $u(\mathbf{x})$ in terms of the fundamental solution $K(\mathbf{x};\mathbf{x}_0)$, *i.e.*, to derive the formula (4.3.8) for the three dimensional case. The problem is that as we see in (4.3.7) $K(\mathbf{x};\mathbf{x}_0)$ has a singularity at $\mathbf{x} = \mathbf{x}_0$. To avoid the singularity, we first remove a ball $B(\mathbf{x}_0, \varepsilon)$ centered at $\mathbf{x}_0$ with radius ε from Ω and apply the Green's identity

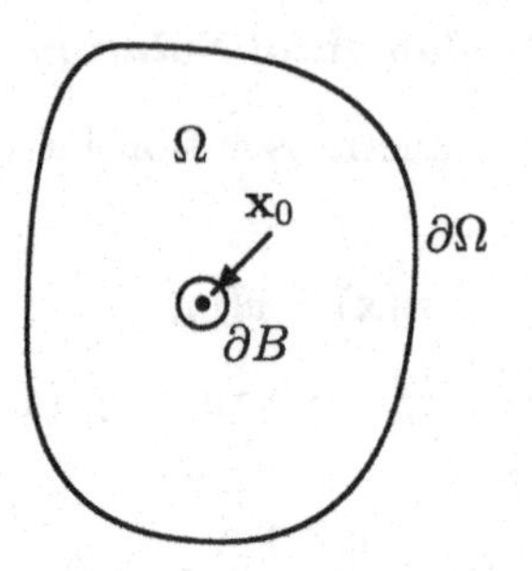

Fig. 4.2 The region Ω_ε.

to u in the remaining region $\Omega_\varepsilon = \Omega - B(\mathbf{x}_0, \varepsilon)$. Then, we let ε approach zero. Here, a ball $B(\mathbf{x}_0, \varepsilon)$ is defined as $B(\mathbf{x}_0, \varepsilon) = \{\mathbf{x} \mid |\mathbf{x} - \mathbf{x}_0| \leq \varepsilon\}$.

Since there are two boundaries for Ω_ε and $\Delta K = 0$ on Ω_ε, from (4.3.1) we have

$$\iiint\limits_{\Omega_\varepsilon} K\Delta u d\mathbf{x} = \iint\limits_{\partial\Omega} (K\frac{du}{dn} - u\frac{dK}{dn})dS + \iint\limits_{\partial B(\mathbf{x}_0,\varepsilon)} (K\frac{du}{dn} - u\frac{dK}{dn})dS. \quad (4.3.8)$$

We evaluate the second integral on the right hand side. We start from the first term. On the surface $\partial B(\mathbf{x}, \varepsilon)$, K is a constant. Using this and applying the divergence theorem, we have

$$\iint\limits_{\partial B(\mathbf{x}_0,\varepsilon)} K\frac{du}{dn}dS = K(\varepsilon) \iint\limits_{\partial B(\mathbf{x}_0,\varepsilon)} \frac{du}{dn}dS = -K(\varepsilon) \iiint\limits_{B(\mathbf{x}_0,\varepsilon)} \Delta u d\mathbf{x}.$$

For a function u for which Δu exists, the right hand side can be estimated as follows.

$$\max_{\mathbf{x}\in B(\mathbf{x}_0,\varepsilon)} \left| K(\varepsilon) \iiint\limits_{B(\mathbf{x}_0,\varepsilon)} \Delta u d\mathbf{x} \right| \leq (\max_{\mathbf{x}\in B(\mathbf{x}_0,\varepsilon)} |\Delta u|)|K(\varepsilon)| \iiint\limits_{B(\mathbf{x}_0,\varepsilon)} d\mathbf{x}$$

$$= (\max_{\mathbf{x}\in B(\mathbf{x}_0,\varepsilon)} |\Delta u|)\frac{1}{\omega_n \varepsilon}\frac{4\pi}{3}\varepsilon^3. \quad (4.3.9)$$

Therefore, as $\varepsilon \to 0$,

$$-K(\varepsilon) \iiint\limits_{B(\mathbf{x}_0,\varepsilon)} \Delta u dV \to 0.$$

Next, we discuss the second term. We use the fact that the exterior normal to Ω_ε on $\partial B(\mathbf{x}_0, \varepsilon)$ points toward $\mathbf{x}_0$. So, on $\partial B(\mathbf{x}_0, \varepsilon)$

$$\frac{dK}{dn} = \mathbf{n} \cdot \nabla K = -K'(\varepsilon). \tag{4.3.10}$$

Then, since $\partial K/\partial n$ is a constant on $\partial B(\mathbf{x}_0, \varepsilon)$,

$$\iint\limits_{\partial B(\mathbf{x}_0,\varepsilon)} u\frac{dK}{dn}dS = -\frac{1}{\omega_n \varepsilon^{n-1}} \iint\limits_{\partial B(\mathbf{x}_0,\varepsilon)} udS.$$

As the right hand side is the average of u on $\partial B(\mathbf{x}_0, \varepsilon)$, it approaches $-u(\mathbf{x}_0)$ as $\varepsilon \to 0$. Therefore, (4.3.8) implies that for any u for which Δu exists

$$u(\mathbf{x}_0) = \iiint\limits_{\Omega} K(\mathbf{x}; \mathbf{x}_0)\Delta u d\mathbf{x} - \iint\limits_{\partial\Omega} (K(\mathbf{x}; \mathbf{x}_0)\frac{du}{dn} - u\frac{dK}{dn}(\mathbf{x}; \mathbf{x}_0))dS. \tag{4.3.11}$$

In particular if u is a solution to the Laplace equation, $\Delta u = 0$ and the above formula reduces to

$$u(\mathbf{x}_0) = -\iint\limits_{\partial\Omega} (K\frac{du}{dn} - u\frac{dK}{dn})dS. \tag{4.3.12}$$

Another example of (4.3.11) is given in the following.

Example 4.3.1. A solution to the Poisson's equation

$$\Delta u(\mathbf{x}) = h, \quad \mathbf{x} \in R^n,$$

where $h \in C_c^2(R^n)$, is given by

$$u(\mathbf{x}_0) = \int_{R^n} K(\mathbf{x}; \mathbf{x}_0)h(\mathbf{x})d\mathbf{x}.$$

Exercises

1. Show that the direction of a gradient vector of u is the direction of the steepest ascent of u.
2. Suppose that a surface is given by $K(r) = C$, where C is a constant. In (4.3.10) knowing that the gradient vector is normal to the surface, show that

$$\frac{dK}{dn} = \mathbf{n} \cdot \nabla K = -K'(\varepsilon).$$

3. Find the estimate for $n = 2$ corresponding to (4.3.9).
4. Derive the formula for $n = 2$ corresponding to (4.3.12).

4.4 Green's Function

The Green's function is a special fundamental solution satisfying the corresponding homogeneous boundary conditions. If the geometry of boundaries is simple, it is possible to construct the explicit representations of solutions in terms of Green's functions. Examples of such solutions are given in the Subsections 4.4.2 and 4.4.3.

4.4.1 *Definition*

We now use the identity (4.3.11) to express the solution to the Laplace equation or the Poisson equation in a smooth region Ω. Specifically we consider the solution for

$$\Delta u(\mathbf{x}) = h(\mathbf{x}), \quad \mathbf{x} \in \Omega,$$

with the Dirichlet boundary condition

$$u(\mathbf{x}) = g(\mathbf{x}), \quad \mathbf{x} \in \partial\Omega.$$

The Neumann boundary condition is more complicated and will be treated in Section 12.1.

To find a solution to the Dirichlet boundary value problems, we need to eliminate $K\partial u/\partial n$ from the right hand side of (4.3.11). For this purpose, let $w(\mathbf{x};\mathbf{x}_0)$ be any solution of $\Delta w = 0$ of class $C^2(\bar{\Omega})$. Then,

$$G(\mathbf{x};\mathbf{x}_0) = K(\mathbf{x};\mathbf{x}_0) + w(\mathbf{x};\mathbf{x}_0) \tag{4.4.1}$$

is a fundamental solution of the Laplace equation. Therefore, we can replace K with G in (4.3.11) to have

$$u(\mathbf{x}_0) = \iiint_{\Omega} G(\mathbf{x};\mathbf{x}_0)\Delta u d\mathbf{x} - \iint_{\partial\Omega} (G(\mathbf{x};\mathbf{x}_0)\frac{du}{dn} - u\frac{dG}{dn}(\mathbf{x};\mathbf{x}_0))dS. \tag{4.4.2}$$

For the Dirichlet boundary conditions we know u but we do not know $\partial u/\partial n$ on the boundary. Therefore, we choose w appropriately so that $G = 0$ on the boundary $\partial\Omega$. Then, (4.4.2) reduces to

$$u(\mathbf{x}_0) = \iiint_{\Omega} G(\mathbf{x};\mathbf{x}_0)h(\mathbf{x})d\mathbf{x} + \iint_{\partial\Omega} \frac{dG}{dn}(\mathbf{x};\mathbf{x}_0)g(\mathbf{x})ds.$$

This motivates the definition for the Green's function.

Definition 4.4.1. The Green's function is the fundamental solution $G(\mathbf{x};\mathbf{x}_0)$ to the Laplace equation satisfying the corresponding homogeneous boundary condition, *i.e.*, $G = 0$ on the boundary for the Dirichlet boundary condition.

The idea is that we choose w to be another fundamental solution satisfying two properties.

1. The singularity is outside Ω so that $\Delta w = 0$ on Ω.
2. G satisfies the corresponding homogeneous boundary conditions.

If the boundary $\partial\Omega$ is complicated, this is in general difficult. However, if the geometry of Ω is simple, it is possible to find w so that $G = 0$ on the boundary. In the following, we study two examples of the Green's functions where the explicit form of G can be found.

4.4.2 *Green's Function for a Half Space*

We consider the solution to the Laplace equation with the Dirichlet boundary condition when Ω is a half space $R^3_+ = \{\mathbf{x} = (x_1, x_2, x_3) \in R^3 \mid x_3 > 0\}$, (or $R^2_+ = \{\mathbf{x} = (x_1, x_2) \in R^2 \mid x_2 > 0\}$).

$$\Delta u = 0, \quad \mathbf{x} \in R^n_+, \tag{4.4.3}$$

$$u(\mathbf{x}) = g(\mathbf{x}), \quad \mathbf{x} \in \partial R^n_+. \tag{4.4.4}$$

Definition 4.4.2. We define $\mathbf{x}^* = (x_1, \ldots, x_{n-1}, -x_n)$ to be the point conjugate to $\mathbf{x} \in R^n_+$ with respect to the surface $x_n = 0$. See Figure 4.3.

Theorem 4.4.3. *The Green's function for the Laplace equation in the half space with the Dirichlet boundary condition is given by*

$$G(\mathbf{x}; \mathbf{x}_0) = K(\mathbf{x}; \mathbf{x}_0) - K(\mathbf{x}; \mathbf{x}_0^*), \tag{4.4.5}$$

where

$$K(\mathbf{x}; \mathbf{x}_0) = \phi(|\mathbf{x} - \mathbf{x}_0|) = \begin{cases} \frac{\ln|\mathbf{x}-\mathbf{x}_0|}{\omega_n} & n = 2, \\ -\frac{1}{\omega_n|\mathbf{x}-\mathbf{x}_0|} & n = 3. \end{cases}$$

Furthermore, the solution to (4.4.3) and (4.4.4) is given by

$$u(\mathbf{x}_0) = \iint_{\partial\Omega} \frac{dG}{dn}(\mathbf{x}; \mathbf{x}_0) g(\mathbf{x}) dS = \frac{2x_n}{\omega_n} \iint_{\partial\Omega} \frac{g(\mathbf{x})}{|\mathbf{x} - \mathbf{x}_0|^n} ds, \tag{4.4.6}$$

where

$$\frac{dG}{dn}\Big|_{x_n=0} = \frac{2x_n}{\omega_n |\mathbf{x} - \mathbf{x}_0|^n}. \tag{4.4.7}$$

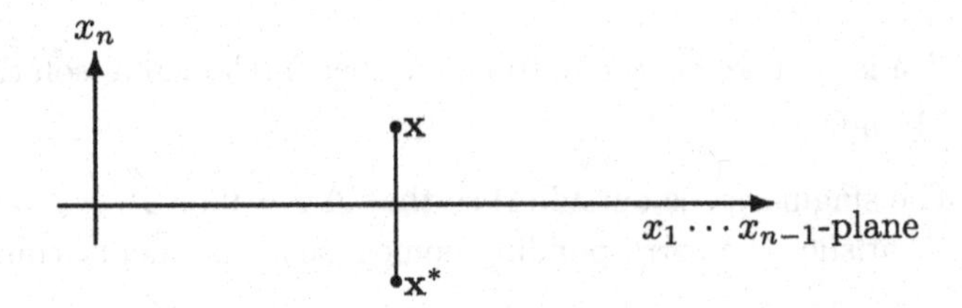

Fig. 4.3 Green's function for a half space.

Proof. From the way $\mathbf{x}_0^*$ is chosen (Figure 4.3), it is easy to see that G satisfies the Dirichlet boundary condition

$$G(\mathbf{x};\mathbf{x}_0)\mid_{x_n=0}=0.$$

Since on $\partial\Omega$ $\mathbf{n}=[0,\ldots,-1]$ and

$$\frac{\partial G}{\partial x_i}=\frac{1}{\omega_n|\mathbf{x}-\mathbf{x}_0|^n}\{(x_i-x_{0i})-(x_i-x_{0i}^*)\}=0,\ i\neq n,$$

$$\frac{\partial G}{\partial x_n}=-\frac{2x_{0n}}{\omega_n|\mathbf{x}-\mathbf{x}_0|^n},$$

therefore,

$$\frac{dG}{dn}\mid_{x_n=0}=\mathbf{n}\cdot\nabla G\mid_{x_n=0}=\frac{2x_{0n}}{\omega_n|\mathbf{x}-\mathbf{x}_0|^n}. \tag{4.4.8}$$

□

Example 4.4.4. Use (4.4.5) to verify that the Green's function for R_+^3 in the coordinate form is given by

$$\begin{aligned}G(\mathbf{x};\mathbf{x}_0)=&-\frac{1}{4\pi}[(x_1-x_{01})^2+(x_2-x_{02})^2+(x_3-x_{03})^2]^{-1/2}\\&+\frac{1}{4\pi}[(x_1-x_{01})^2+(x_2-x_{02})^2+(x_3+x_{03})^2]^{-1/2}.\end{aligned}$$

4.4.3 *Green's Function for a Ball*

As the second example of the Green's functions, we consider a solution to the Laplace equation with the Dirichlet boundary condition where Ω is a ball $B(\mathbf{0},a)$ with center $\mathbf{0}$ and radius a.

$$\Delta u(\mathbf{x})=0,\quad \mathbf{x}\in B(\mathbf{0},a), \tag{4.4.9}$$

$$u(\mathbf{x})=g(\mathbf{x}),\quad \mathbf{x}\in\partial B(\mathbf{0},a). \tag{4.4.10}$$

As in the previous subsection, we define a point conjugate to a given point $\mathbf{x}\in B(\mathbf{0},a)$.

Definition 4.4.5. We define $\mathbf{x}_0^*$ to be the point conjugate to $\mathbf{x}_0 \in B(\mathbf{0}, a)$ with respect to the surface of the ball $B(\mathbf{0}, a)$ if it is on the line connecting $\mathbf{0}$ and $\mathbf{x}$ and satisfying $|\mathbf{x}_0||\mathbf{x}_0^*| = a^2$. In other words,

$$\mathbf{x}_0^* = |\mathbf{x}_0^*|\frac{\mathbf{x}_0}{|\mathbf{x}_0|} = \frac{a^2}{|\mathbf{x}_0|^2}\mathbf{x}_0. \tag{4.4.11}$$

The relation between $\mathbf{x}_0$ and $\mathbf{x}_0^*$ is illustrated in Figure 4.4. If $\mathbf{x}$ is a point on $\partial B(\mathbf{0}, a)$, because of (4.4.11) the triangles $\triangle ox_0x$ and $\triangle oxx_0^*$ are similar and the ratio of $|\mathbf{x}_0 - \mathbf{x}|$ and $|\mathbf{x}_0^* - \mathbf{x}|$ is a constant. In the next lemma we show a useful relation between $\mathbf{x}$ and $\mathbf{x}_0$ when $\mathbf{x}$ is on $\partial B(\mathbf{0}, a)$.

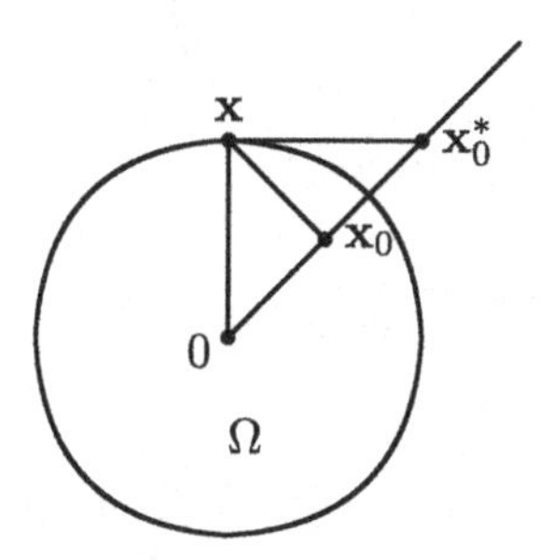

Fig. 4.4 Green's function on a ball.

Lemma 4.4.6. *For* $\mathbf{x} \in \partial B(\mathbf{0}, a)$, *we have*

$$|\mathbf{x}_0^* - \mathbf{x}| = \frac{a}{|\mathbf{x}_0|}|\mathbf{x}_0 - \mathbf{x}|. \tag{4.4.12}$$

Proof. Since the triangles $\triangle ox_0x$ and $\triangle oxx_0^*$ are similar,

$$\frac{|\mathbf{x}_0^* - \mathbf{x}|}{|\mathbf{x}_0 - \mathbf{x}|} = \frac{|\mathbf{x}|}{|\mathbf{x}_0|}.$$

If $\mathbf{x}$ is on the boundary of the ball $B(\mathbf{0}, a)$, $|\mathbf{x}| = a$ and therefore,

$$|\mathbf{x}_0^* - \mathbf{x}| = \frac{a}{|\mathbf{x}_0|}|\mathbf{x}_0 - \mathbf{x}|.$$

□

Theorem 4.4.7. *For $n = 3$, the Green's function for the boundary value problem (4.4.9) and (4.4.10) is given by*

$$G(\mathbf{x};\mathbf{x}_0) = K(\mathbf{x};\mathbf{x}_0) - \frac{a}{|\mathbf{x}_0|}K(\mathbf{x};\mathbf{x}_0^*)$$
$$= -\frac{1}{4\pi|\mathbf{x}_0 - \mathbf{x}|} + \frac{a}{|\mathbf{x}_0|}\frac{1}{4\pi|\mathbf{x}_0^* - \mathbf{x}|} \tag{4.4.13}$$

and for $n = 2$, it is given by

$$G(\mathbf{x};\mathbf{x}_0) = \frac{\ln|\mathbf{x}_0 - \mathbf{x}|}{2\pi} - \frac{\ln|\frac{\mathbf{x}_0}{a}||\mathbf{x}_0^* - \mathbf{x}|}{2\pi}. \tag{4.4.14}$$

The solution to (4.4.9) and (4.4.10) is given by

$$u(\mathbf{x}_0) = \iint\limits_{|\mathbf{x}|=a} \frac{\partial G(\mathbf{x};\mathbf{x}_0)}{\partial n} g(\mathbf{x})dS = \iint\limits_{|\mathbf{x}|=a} H(\mathbf{x};\mathbf{x}_0)g(\mathbf{x})dS, \tag{4.4.15}$$

where $\mathbf{n}$ is the outward unit normal vector to $\partial B(\mathbf{0}, a)$ and

$$H(\mathbf{x};\mathbf{x}_0) = \frac{1}{\omega_n|\mathbf{x} - \mathbf{x}_0|^n}\frac{(a^2 - |\mathbf{x}|^2)}{a} \tag{4.4.16}$$

is called the Poisson's kernel for ball $B(\mathbf{0}, a)$.

Proof. We show the proof for $n = 3$. The case $n = 2$ is left for exercise. We need to choose w so that G is zero on the surface of the ball $\partial B(\mathbf{0}, a)$. Choose $w = -C(\mathbf{x}_0)K(\mathbf{x};\mathbf{x}_0^*)$. Then, $\Delta w = 0$ for $\mathbf{x} \in B(\mathbf{0}, a)$. From Lemma 4.4.6, for $\mathbf{x} \in \partial B(\mathbf{0}, a)$, we have

$$K(\mathbf{x};\mathbf{x}_0^*) = -\frac{1}{4\pi|\mathbf{x}_0^* - \mathbf{x}|} = -\frac{1}{4\pi(\frac{a}{|\mathbf{x}_0|})|\mathbf{x}_0 - \mathbf{x}|}.$$

We choose $C(\mathbf{x}_0)$ so that for $\mathbf{x} \in \partial B(\mathbf{0}, a)$

$$G(\mathbf{x};\mathbf{x}_0) = K(\mathbf{x};\mathbf{x}_0) - C(\mathbf{x}_0)K(\mathbf{x};\mathbf{x}_0^*))$$
$$= -\frac{1}{4\pi|\mathbf{x} - \mathbf{x}_0|} + \frac{C(\mathbf{x}_0)}{4\pi(\frac{a}{|\mathbf{x}_0|})|\mathbf{x} - \mathbf{x}_0|} = 0.$$

This implies that $C(\mathbf{x}_0) = \frac{a}{|\mathbf{x}_0|}$. Then, $G(\mathbf{x};\mathbf{x}_0)$ is given by (4.4.13). Differentiating G in x_i, we have

$$\frac{\partial G}{\partial x_i} = \frac{1}{4\pi}\{\frac{x_i - x_{0i}}{|\mathbf{x}_0 - \mathbf{x}|^3} - \frac{a}{|\mathbf{x}_0|}\frac{x_i - x_{0i}^*}{|\mathbf{x}_0^* - \mathbf{x}|^3}\}.$$

By (4.4.11) and (4.4.12), we see

$$\frac{\partial G}{\partial x_i}|_{\mathbf{x}\in\partial B(\mathbf{0},a)} = \frac{1}{4\pi|\mathbf{x}_0 - \mathbf{x}|^3}\{1 - (\frac{|\mathbf{x}_0|}{a})^2\}x_i.$$

Then, using $\mathbf{n} = [x_1, \ldots, x_n]/|\mathbf{x}|$, for $\mathbf{x} \in \partial B(\mathbf{0}, a)$ we obtain

$$\begin{aligned}\frac{dG}{dn} &= \sum_{i=1}^{n} \frac{1}{4\pi|\mathbf{x}_0 - \mathbf{x}|^3}\{1 - (\frac{|\mathbf{x}_0|}{a})^2\}x_i\frac{x_i}{|\mathbf{x}|} \\ &= \frac{1}{4\pi|\mathbf{x}_0 - \mathbf{x}|^3}\frac{(a^2 - |\mathbf{x}_0|^2)}{a} = H(\mathbf{x}; \mathbf{x}_0).\end{aligned}$$

□

Remark 4.4.8. More comprehensive treatment of the Green's functions including the Green's functions for the heat and the wave equations will be discussed in Chapter 12.

4.4.4 *Symmetry of Green's Function*

We prove that the Green's function is symmetric, *i.e.*,

$$G(\mathbf{x}; \mathbf{x}_0) = G(\mathbf{x}_0; \mathbf{x}), \quad \mathbf{x} \neq \mathbf{x}_0. \tag{4.4.17}$$

We apply the Green's identity to two functions $G(\mathbf{x}; \mathbf{a})$ and $G(\mathbf{x}; \mathbf{b})$ and to the domain Ω_ε. Here, Ω_ε is the domain where the two small balls $B(\mathbf{a}, \varepsilon)$ and $B(\mathbf{b}, \varepsilon)$ contained in Ω are removed from Ω. Then, the divergence theorem is

$$\begin{aligned}&\iiint_{\Omega_\varepsilon} [G(\mathbf{x}; \mathbf{a})\Delta G(\mathbf{x}; \mathbf{b}) - G(\mathbf{x}; \mathbf{b})\Delta G(\mathbf{x}; a)]d\mathbf{x} \\ &= \iint_{\partial\Omega} [G(\mathbf{x}; \mathbf{a})\frac{\partial G}{\partial n}(\mathbf{x}; \mathbf{b}) - G(\mathbf{x}; \mathbf{b})\frac{\partial G}{\partial n}(\mathbf{x}; \mathbf{a})]dS + A + B,\end{aligned} \tag{4.4.18}$$

where

$$A = \iint_{\partial B(\mathbf{a},\varepsilon)} [G(\mathbf{x}; \mathbf{a})\frac{\partial G}{\partial n}(\mathbf{x}; \mathbf{b}) - G(\mathbf{x}; \mathbf{b})\frac{\partial G}{\partial n}(\mathbf{x}; \mathbf{a})]dS,$$

$$B = \iint_{\partial B(\mathbf{b},\varepsilon)} [G(\mathbf{x}; \mathbf{a})\frac{\partial G}{\partial n}(\mathbf{x}; \mathbf{b}) - G(\mathbf{x}; \mathbf{b})\frac{\partial G}{\partial n}(\mathbf{x}; \mathbf{a})]dS.$$

The left hand side of (4.4.18) is zero since $\Delta G(\mathbf{x}; \mathbf{b}) = \Delta G(\mathbf{x}; \mathbf{a}) = 0$ on Ω_ε. The first term on the right hand side of (4.4.18) is also zero. These imply $A + B = 0$ for each ε. We apply the same argument as in Subsection 4.3.3 to obtain

$$\lim_{\varepsilon\to 0} A = G(\mathbf{a}; \mathbf{b}), \quad \lim_{\varepsilon\to 0} B = -G(\mathbf{b}; \mathbf{a}).$$

Therefore,

$$0 = \lim_{\varepsilon\to 0}(A + B) = G(\mathbf{a}; \mathbf{b}) - G(\mathbf{b}; \mathbf{a}).$$

This shows the symmetry of the Green's function (4.4.17).

Exercises

1. For the Green's function for the half space, show the following.

   (a) On $\mathbf{x} \in \partial\Omega$ show that

$$|\mathbf{x}_0 - \mathbf{x}| = |\mathbf{x}_0^* - \mathbf{x}|, \quad \mathbf{n} = [0, \ldots, -1],$$

$$\frac{\partial G}{\partial x_i} = \frac{1}{\omega_n |\mathbf{x}_0 - \mathbf{x}|^n}\{(x_i - x_{0i}) - (x_i - x_{0i}^*)\} = 0,\ i \neq n,$$

$$\frac{\partial G}{\partial x_n} = -\frac{2x_n}{\omega_n |\mathbf{x}_0 - \mathbf{x}|^n}.$$

   (b) Using (a) show (4.4.8).

$$\frac{dG}{dn}\,|_{x_n=0} = \mathbf{n} \cdot \nabla G\,|_{x_n=0} = \frac{2x_{0n}}{\omega_n |\mathbf{x}_0 - \mathbf{x}|^n}.$$

2. Find $K(\mathbf{x}; \mathbf{x}_0^*)$, $G(\mathbf{x}; \mathbf{x}_0)$, and $H(\mathbf{x}; \mathbf{x}_0)$ for the Laplace equation on the circular disk with radius a ($n = 2$).
3. For the Green's function for the ball, show the following.

   (a) Show that $\mathbf{n} = [x_1, \ldots, x_n]/|\mathbf{x}|$.
   (b) Compute $\partial G/\partial x_i$ and $\partial G/\partial n$ for $n = 3$ and 2.

4. For $n = 2$ the following formula is called the Poisson's formula. This is a two dimensional version of (4.4.15) if the Polar coordinates are used. Set $\mathbf{x}_0 = (r, \theta)$ and $\mathbf{x} = (a, \phi)$ to derive the formula.

$$u(r, \theta) = \frac{(a^2 - r^2)}{2\pi} \int_0^{2\pi} \frac{g(a, \phi)}{a^2 - 2ar\cos(\theta - \phi) + r^2} d\phi.$$

5. Consider the Laplace equation on the first quadrant.

$$\Delta u = 0, \quad 0 < x,\ 0 < y,$$

$$u(x, 0) = f(x),\ u(0, y) = g(y).$$

   (a) Find the Green's function.
   (b) Use (a) to express the solution.

4.5 Properties of Harmonic Functions

There are various interesting properties for the harmonic functions. We list some of them. We assume that $\Omega \subset R^n$ is open and bounded.

4.5.1 *Mean Value Property*

The following is an application of (4.4.2) for a ball $B(\mathbf{x}_0, r) \in \Omega$.

Theorem 4.5.1. *Suppose that $\Delta u = 0$ on Ω. Then, for every ball $B(\mathbf{x}_0, r) \in \Omega$,*

$$u(\mathbf{x}_0) = \frac{1}{\omega_n r^{n-1}} \iint_{\partial B(\mathbf{x}_0, r)} u dS \tag{4.5.1}$$

i.e., if u is harmonic, u at $\mathbf{x}_0$ is the average of u on the surface of a ball centered at $\mathbf{x}_0$. Furthermore,

$$u(\mathbf{x}_0) = \frac{1}{\alpha_n r^n} \iiint_{B(\mathbf{x}_0, r)} d\mathbf{x}, \tag{4.5.2}$$

where α_n is the volume of unit ball in R^n given by

$$\alpha_n = \begin{cases} \pi & n = 2, \\ \frac{4\pi}{3} & n = 3. \end{cases}$$

Proof. We choose a ball $B(\mathbf{x}_0, r) \in \Omega$ and we use $w(\mathbf{x}; \mathbf{x}_0) = -\phi(r)$ in (4.4.1). Then,

$$G(\mathbf{x}; \mathbf{x}_0) = K(\mathbf{x}; \mathbf{x}_0) - \phi(r) = \phi(|\mathbf{x}_0 - \mathbf{x}|) - \phi(r).$$

Therefore, on $\partial B(\mathbf{x}_0, r)$

$$G|_{|\mathbf{x}_0 - \mathbf{x}| = r} = 0, \quad \frac{dG}{dn}|_{|\mathbf{x}_0 - \mathbf{x}| = r} = \phi'(r) = \frac{1}{\omega_n} r^{1-n},$$

and (4.4.2) becomes

$$u(\mathbf{x}_0) = \iiint_{B(\mathbf{x}_0, r)} G(\mathbf{x}; \mathbf{x}_0) \Delta u d\mathbf{x} + \frac{1}{\omega_n r^{n-1}} \iint_{\partial B(\mathbf{x}_0, r)} u dS. \tag{4.5.3}$$

In particular, if $\Delta u = 0$ on Ω, we obtain the first assertion (4.5.1). Note that $\omega_n r^{n-1}$ is the surface area of the sphere $|\mathbf{x}_0 - \mathbf{x}| = r$.

From (4.5.1) we see

$$\omega_n r^{n-1} u(\mathbf{x}_0) = \iint_{\partial B(\mathbf{x}_0, r)} u dS.$$

Integrating this in r from 0 to a (a positive constant), we have

$$\int_0^a \omega_n r^{n-1} u(\mathbf{x}_0) dr = u(\mathbf{x}_0) \frac{\omega_n}{n} a^n = \int_0^a \iint_{\partial B(\mathbf{x}_0, r)} u dS dr = \iiint_{B(\mathbf{x}_0, a)} u d\mathbf{x}.$$

Since $\alpha_n = \omega_n / n$, we obtain the second conclusion (4.5.2). □

The converse is true.

Theorem 4.5.2. *If $u \in C^2(\Omega)$ satisfies*

$$u(\mathbf{x}_0) = \frac{1}{\omega_n r^{n-1}} \iint\limits_{\partial B(\mathbf{x}_0, r)} u dS \tag{4.5.4}$$

for each ball $B(\mathbf{x}_0, r) \subset \Omega$, then u is harmonic.

Proof. The proof is by contraposition. Assume that u is not harmonic and show that there is a ball $B(\mathbf{x}_0, a) \subset \Omega$ in which (4.5.4) does not hold. Since we assume that u is not harmonic, there is a region $\Omega_1 \subset \Omega$ where, say, $\Delta u > 0$. We choose $\mathbf{x}_0 \in \Omega_1$, $B(\mathbf{x}_0, a) \subset \Omega_1$, $G(\mathbf{x}; \mathbf{x}_0) = K(\mathbf{x}; \mathbf{x}_0) - \phi(a) = \phi(|\mathbf{x}_0 - \mathbf{x}|) - \phi(a)$, and use (4.5.3). Since $G(\mathbf{x}; \mathbf{x}_0)$ is negative in the interior of $B(\mathbf{x}_0, a)$, from (4.5.3) we obtain

$$u(\mathbf{x}_0) < \frac{1}{\omega_n r^{n-1}} \iint\limits_{\partial B(\mathbf{x}_0, r)} u dS.$$

□

4.5.2 *The Maximum Principle and Uniqueness*

Theorem 4.5.3. *(The maximum principle) Suppose u is harmonic in Ω. Then,*

$$\max_{\bar{\Omega}} u = \max_{\partial\Omega} u. \tag{4.5.5}$$

Proof. The proof is easy if $\Delta u > 0$. If we assume that the maximum is in the interior, then $\Delta u \leq 0$ holds. This contradicts with $\Delta u > 0$. Therefore, the maximum is on the boundary. Now we treat the case where $\Delta u = 0$. Consider $v = u + \varepsilon|x|^2$, where $|x|^2 = \sum_{i=1}^n x_i^2$. Then, $\Delta v > 0$ and the maximum of v is on the boundary.

$$\max_{\bar{\Omega}}(u + \varepsilon|x|^2) = \max_{\partial\Omega}(u + \varepsilon|x|^2).$$

Then,

$$\max_{\bar{\Omega}} u + \varepsilon \min_{\bar{\Omega}} |x|^2 \leq \max_{\partial\Omega} u + \varepsilon \max_{\partial\Omega} |x|^2.$$

Since this is true for every $\varepsilon > 0$, we obtain (4.5.5). □

Remark 4.5.4. If u is harmonic in Ω, we apply the maximum principle to $-u$ and use $\min u = -\max(-u)$ to obtain

$$\min_{\bar{\Omega}} u = \min_{\partial\Omega} u. \tag{4.5.6}$$

Therefore, if $\Delta u = 0$ in Ω,

$$\max_{\bar{\Omega}} |u| = \max_{\partial\Omega} |u|. \tag{4.5.7}$$

Using the maximum principle, we can show the uniqueness of solutions to the Dirichlet problem for the Poisson equation.

Theorem 4.5.5. *There is at most one solution to*

$$\Delta u = h(\mathbf{x}) \quad \mathbf{x} \in \Omega, \tag{4.5.8}$$

$$u = g(\mathbf{x}) \quad \mathbf{x} \in \partial\Omega. \tag{4.5.9}$$

Proof. Suppose there are two solutions u_1 and u_2 satisfying (4.5.8) and (4.5.9). Consider $w = u_1 - u_2$. It is easy to show that w satisfies $\Delta w = 0$ in Ω and $w = 0$ on $\partial\Omega$. We apply the maximum principle (4.5.7) to show that

$$\max_{\bar{\Omega}} |w| = \max_{\partial\Omega} |w| = 0.$$

□

Exercises

1. Suppose that u is a harmonic function in the rectangular domain $\Omega = \{(x, y) \mid 0 < x < 1, 0 < y < 2\}$ and that $u(x, 0) = x$, $u(x, 2) = 2 - x$, $u(0, y) = y$, $u(2, y) = 1$ on the boundary. Find the maximum and minimum of u on $\bar{\Omega}$.
2. Suppose that u is a harmonic function on the circular disk $\Omega = \{r < 1\}$ and that $u(1, \theta) = \sin\theta$ on the boundary.
   (a) Find the maximum and minimum of u on $\bar{\Omega}$.
   (b) Find the value of u at the origin.
3. Suppose that u is a harmonic function on the ball $\Omega = \{r < 1\}$ and that $u(1, \phi, \theta) = \cos^2\theta \sin\phi$ on the boundary.
   (a) Find the maximum and minimum of u on $\bar{\Omega}$.
   (b) Find the value of u at the origin.
4. Show (4.5.6) and (4.5.7).

4.6 Well-posedness Issues

In Section 1.5 we discussed the well-posedness of the PDE problems. Among the three questions posed in that section, we have not seen the examples concerning the third question, *i.e.*, the continuous dependence of the solutions to the initial data. We also notice that we discussed the initial boundary value problems for the wave equation but we did not discuss the boundary value problems. The opposite is true for the Laplace equation. We discussed the boundary value problems but not the initial boundary value problems.

In this section using the separation of variables, we show that the boundary value problems for the wave equation or the initial boundary value problems for the Laplace equation have problems with the continuous dependence of the solutions to the initial data. The problem lies in the mismatch between the equation and the conditions we impose.

4.6.1 *Laplace Equation*

We consider the following initial boundary value problem to illustrate what happens for the solution of the Laplace equation.

$$u_{tt} + u_{xx} = 0, \quad 0 < x < L, \quad 0 < t, \tag{4.6.1}$$

$$u_x(0,t) = 0, \quad u_x(L,t) = 0. \tag{4.6.2}$$

$$u(x,0) = f(x), \quad u_t(x,0) = g(x), \tag{4.6.3}$$

Step 1: As we did before we assume

$$u(x,t) = X(x)T(t).$$

Then, by the separation of variables we obtain two ODE's

$$X'' + \lambda X = 0, \tag{4.6.4}$$

$$T'' - \lambda T = 0. \tag{4.6.5}$$

The boundary conditions for X are

$$u_x(0,t) = 0 \Rightarrow X'(0)T(t) = 0 \Rightarrow X'(0) = 0, \tag{4.6.6}$$

$$u_x(L,t) = 0 \Rightarrow X'(L)T(t) = 0 \Rightarrow X'(L) = 0. \tag{4.6.7}$$

Step 2: We solve (4.6.4). The eigenvalues are

$$\lambda_0 = 0, \quad \lambda_n = (\frac{n\pi}{L})^2, \quad n = 1, 2, \ldots$$

and the corresponding eigenfunctions are

$$X_0 = 1, \quad X_n = \cos\frac{n\pi x}{L}, \quad n = 1, 2, \ldots.$$

Now, we solve (4.6.5) for the eigenvalues of λ. If $\lambda = \lambda_0 = 0$,

$$T_0 = d_1 + d_2 t$$

and if $\lambda = \lambda_n = (n\pi/L)^2$,

$$T_n = d_1 \cosh\frac{n\pi ct}{L} + d_2 \sinh\frac{n\pi ct}{L}.$$

Step 3: Since we assume that $u = XT$, the solutions are

$$u_0 = X_0 T_0 = \frac{a_0}{2} + \frac{b_0}{2} t,$$

$$u_n = X_n T_n = (a_n \cosh\frac{n\pi ct}{L} + b_n \sinh\frac{n\pi ct}{L}) \cos\frac{n\pi x}{L}, \quad n = 1, 2, \ldots.$$

Using the superposition principle, we take

$$\begin{aligned} u(x,t) &= \sum_{n=0}^{\infty} u_n \\ &= \frac{a_0}{2} + \frac{b_0}{2} t + \sum_{n=1}^{\infty} (a_n \cosh\frac{n\pi ct}{L} + b_n \sinh\frac{n\pi ct}{L}) \cos\frac{n\pi x}{L} \end{aligned} \tag{4.6.8}$$

as the solution to the problem.

Our task now is to find a_n and b_n from the initial conditions. At $t = 0$,

$$u(x,0) = \frac{a_0}{2} + \sum_{n=1}^{\infty} a_n \cos\frac{n\pi x}{L} = f(x), \tag{4.6.9}$$

$$u_t(x,0) = \frac{b_0}{2} + \sum_{n=1}^{\infty} \frac{n\pi c}{L} b_n \cos\frac{n\pi x}{L} = g(x). \tag{4.6.10}$$

To find a_0 and b_0 we integrate over $0 < x < L$. Then,

$$a_0 = \frac{2}{L}\int_0^L f(x)dx, \quad b_0 = \frac{2}{L}\int_0^L g(x)dx. \tag{4.6.11}$$

To find a_m and b_m we multiply by $\cos(m\pi x/L)$ and integrate over $0 < x < L$. Then, we obtain

$$a_m = \frac{2}{L}\int_0^L f(x)\cos\frac{m\pi x}{L}dx,$$

$$b_m = \frac{L}{m\pi c}\frac{2}{L}\int_0^L g(x)\cos\frac{m\pi x}{L}dx,\ m = 1, 2, \ldots. \tag{4.6.12}$$

The problem is that if $a_n \neq 0$ or $b_n \neq 0$ for a large n, the solution grows exponentially fast. Since we can do this for any n we choose by adjusting the initial condition, we can make the solution as large as we wish in any positive time. This shows that a small error in the initial data will grow exponentially with arbitrarily large exponent. Therefore, the continuous dependence of solutions to the initial data may not hold. An example concerning this is discussed in Exercise 3.

4.6.2 *Wave Equation*

We use the following problem to show that the boundary value problems for the wave equation is not well-posed. Consider

$$u_{tt} - c^2 u_{xx} = 0, \quad 0 < x < L,\ 0 < t < \bar{T}, \tag{4.6.13}$$

$$u(0,t) = 0, \quad u(L,t) = 0, \tag{4.6.14}$$

$$u(x,0) = f(x), \tag{4.6.15}$$

$$u(x,\bar{T}) = 0. \tag{4.6.16}$$

Step 1: We assume

$$u(x,t) = X(x)T(t)$$

and obtain two ODE's

$$X'' + \lambda X = 0, \tag{4.6.17}$$

$$T'' + \lambda c^2 T = 0. \tag{4.6.18}$$

The boundary conditions for X are

$$X(0) = 0, \quad X(L) = 0. \tag{4.6.19}$$

Step 2: We solve (4.6.17) first. The eigenvalues are

$$\sqrt{\lambda}L = n\pi \Rightarrow \quad \lambda = (\frac{n\pi}{L})^2, \quad n = 1, 2, \ldots$$

and the corresponding eigenfunctions are

$$X_n = \sin\frac{n\pi x}{L}, \quad n = 1, 2, \ldots.$$

Now, we solve (4.6.18) for the eigenvalues of λ. If $\lambda = \lambda_n = (n\pi/L)^2$,

$$T_n = d_1 \cos \frac{n\pi ct}{L} + d_2 \sin \frac{n\pi ct}{L}.$$

Step 3: Since we assume that $u = XT$, the solutions are

$$u_n = X_n T_n = (a_n \cos \frac{n\pi ct}{L} + b_n \sin \frac{n\pi ct}{L}) \sin \frac{n\pi x}{L}, \quad n = 1, 2, \ldots.$$

Using the superposition principle, we take

$$\begin{aligned} u(x,t) &= \sum_{n=0}^{\infty} u_n \\ &= \sum_{n=1}^{\infty} (a_n \cos \frac{n\pi ct}{L} + b_n \sin \frac{n\pi ct}{L}) \sin \frac{n\pi x}{L} \end{aligned} \tag{4.6.20}$$

as the solution to the problem. We find a_n and b_n from the initial conditions. At $t = 0$,

$$u(x,0) = \sum_{n=1}^{\infty} a_n \sin \frac{n\pi x}{L} = f(x). \tag{4.6.21}$$

To find a_m we multiply by $\sin(m\pi x/L)$ and integrate over $0 < x < L$. Using the relation

$$\int_0^L \sin \frac{m\pi x}{L} \sin \frac{n\pi x}{L} dx = \begin{cases} \frac{L}{2} & m = n \\ 0 & m \neq n, \end{cases}$$

we obtain

$$a_m = \frac{2}{L} \int_0^L f(x) \sin \frac{m\pi x}{L} dx, \quad m = 1, 2, \ldots. \tag{4.6.22}$$

At $t = \bar{T}$,

$$\sum_{n=1}^{\infty} (a_n \cos \frac{n\pi c\bar{T}}{L} + b_n \sin \frac{n\pi c\bar{T}}{L}) \sin \frac{n\pi x}{L} = 0,$$

where a_n are given by (4.6.22). Multiply by $\sin(m\pi x/L)$ and integrate over $0 < x < L$. Then, b_m are given by

$$a_m \cos \frac{m\pi c\bar{T}}{L} + b_m \sin \frac{m\pi c\bar{T}}{L} = 0,$$

$$b_m = -a_m \frac{\cos \frac{m\pi c\bar{T}}{L}}{\sin \frac{m\pi c\bar{T}}{L}}.$$

So, for example when $\bar{T}$ is a multiple of L/c, the numerator becomes zero. As in the previous example, the continuous dependence of solutions to the initial data may not hold.

Exercises

1. Consider the boundary value problems for the wave equation. Show that it is not well posed.

$$u_{tt} - c^2 u_{xx} = 0, \quad 0 < x < L,\ 0 < t < \bar{T},$$

$$u(0,t) = 0, \quad u(L,t) = 0,$$

$$u(x,0) = 0,$$

$$u(x,\bar{T}) = f(x).$$

2. We consider the following initial boundary value problem to illustrate what happens for the solution of the Laplace equation.

$$u_{tt} + u_{xx} = 0, \quad 0 < x < L, \quad 0 < t,$$

$$u(0,t) = 0, \quad u(L,t) = 0,$$

$$u(x,0) = f(x), \quad u_t(x,0) = g(x).$$

3. Consider the initial boundary value problem (4.6.1) to (4.6.3). We show that we can construct the initial conditions so that for any large $\Delta > 0$, any small $\varepsilon > 0$, and any small $T > 0$, there are initial conditions

$$\max_{0 \le x \le L} \{|u(x,0)| + |u_t(x,0)|\} \le \varepsilon$$

such that

$$\max_{0 \le x \le L} |u(x,T)| \ge \Delta. \tag{4.6.23}$$

   (a) Find the solution if the initial conditions are given by

$$u(x,0) = f(x) = \varepsilon \cos \frac{n\pi x}{L}, \quad u_t(x,0) = 0.$$

   (b) Find the maximum value of $u(x,t)$ in (a) at each t.
   (c) Find the value of n for which u satisfies (4.6.23).

Chapter 5

First Order Equations Revisited

In this chapter we study the first order nonlinear equations. This is a continuation of Chapter 1 where we studied the first order linear equations. They are classified as semi-linear, quasilinear, and fully nonlinear. The definitions and examples are given in Sections 5.1 and 5.3. Roughly speaking, if a given equation is linear in the highest order derivatives, it is called the semi-linear or quasilinear equation and if it is nonlinear in the highest order derivatives, it is called the fully nonlinear equation.

We apply the method of characteristics studied in Chapter 1. We restrict our attention to the case where there are two independent variables. In Section 5.1 we study the quasilinear equations and in Section 5.3 we discuss the fully nonlinear equations. As an application of the quasilinear equations we study the scalar conservation law in Section 5.2 and in Section 5.4 the Hamilton-Jacobi-Bellman (HJB) equations are discussed as an application of fully nonlinear equations.

5.1 First Order Quasilinear Equations

In the linear equation (1.3.1), the coefficients and the non-homogeneous term g do not depend on u. In the first order quasilinear equations, the equation is linear in the highest order derivatives, but a, b, and c may depend on u. The general form of the first order quasilinear equation is given by

$$a(x,t,u)u_t + b(x,t,u)u_x = c(x,t,u), \tag{5.1.1}$$

where at least one of a or b depends on u. For example,

$$u_t + uu_x = -u$$

is a quasilinear equation where $a = 1$, $b = u$, and $c = -u$. If neither a nor b depend on u, it is called semi-linear. For example,

$$u_t + u_x = u^2$$

is a semi-linear equation. In both cases, the method of solution is similar to the linear case. In what follows we focus our attention to the quasilinear case. We use the method of characteristics to derive the set of ODE's as we did in the linear case. In (5.1.1) the ODE's for characteristics are

$$\frac{dt}{d\tau} = a(x,t,u), \quad \frac{dx}{d\tau} = b(x,t,u), \tag{5.1.2}$$

and along the characteristics we have

$$\frac{du}{d\tau} = u_t\frac{dt}{d\tau} + u_x\frac{dx}{d\tau} = a(x,t,u)u_t + b(x,t,u)u_x = c(x,t,u). \tag{5.1.3}$$

These equations correspond to (1.3.3) and (1.3.4) in Section 1.3 and they form a characteristic system. In the linear or semi-linear case, where a and b do not depend on u, we are able to solve (5.1.2) for x and t and use the result to solve (5.1.3) for u. In the nonlinear case, we may have to solve the simultaneous equations for three variables.

Example 5.1.1. Solve

$$u_t + u^3u_x = 0, \quad -\infty < x < \infty,\ 0 < t,$$

$$u(x,0) = \alpha x^{1/3},$$

where α is a constant.

Solution: Basic steps are similar to the linear case. However, the second step may be more involved than the linear problems.
Step 1: The first step is to parametrize the initial curve C_I as follows.

$$C_I:\ x_0 = s,\ t_0 = 0,\ u_0 = \alpha s^{1/3}.$$

Step 2: In the second step, we solve the system of ordinary differential equations (characteristic system) for each s

$$\frac{dx}{d\tau} = u^3,\ \frac{dt}{d\tau} = 1,\ \frac{du}{d\tau} = 0 \tag{5.1.4}$$

with the initial data

$$x(0) = x_0 = s,\ t(0) = t_0 = 0,\ u(0) = u_0 = \alpha s^{1/3}.$$

If we examine the three equations in (5.1.4), we notice that u is a constant along each characteristic and that we should solve u before solving x. Then,

$$t(s,\tau) = \tau + t_0 = \tau, \tag{5.1.5}$$

$$u(s,\tau) = u_0 = \alpha s^{1/3}. \tag{5.1.6}$$

Since u is a constant along each characteristic, using (5.1.6), we obtain

$$x(s,\tau) = u^3\tau + x_0 = u_0^3\tau + x_0 = \alpha^3 s\tau + s. \tag{5.1.7}$$

Step 3: The final step is to express u using (x,t). Solve (5.1.5) and (5.1.7) for s and τ and substitute them in (5.1.6), *i.e.*, putting $\tau = t$ in $x = \alpha^3 s\tau + s$, we have $s = x/(\alpha^3 t + 1)$. So,

$$u = u(x,t) = \alpha\left(\frac{x}{\alpha^3 t + 1}\right)^{1/3}.$$

If α is negative, u becomes infinite in finite time.

Next, consider the example where the initial data is similar to the one in Example 1.3.3 in Section 1.3. This will illustrate the difference between linear and quasilinear differential equations.

Example 5.1.2. Solve

$$u_t + uu_x = 0, \quad -\infty < x < \infty,\ 0 < t,$$

$$u(x,0) = h(x) = \begin{cases} 0, & x \le 0, \\ x, & 0 < x \le 1, \\ 1, & 1 < x. \end{cases}$$

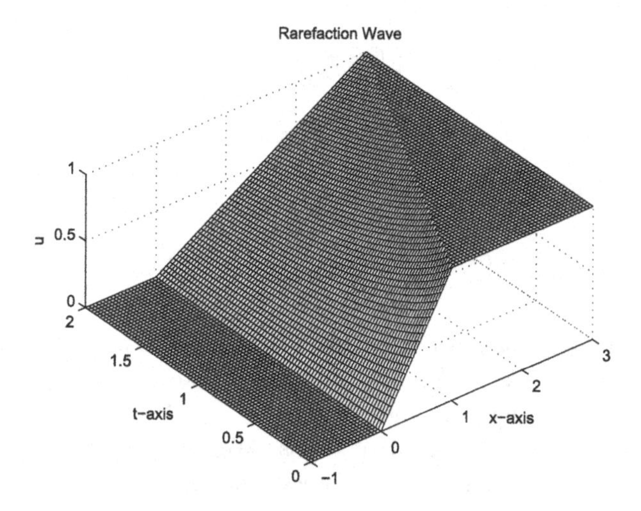

Fig. 5.1 Rarefaction wave.

Solution: Step 1: The initial condition is parametrized as

$$x = x_0 = s, \quad t = t_0 = 0, \quad u = u_0 = h(s) = \begin{cases} 0, & s \le 0, \\ s, & 0 < s \le 1, \\ 1, & 1 < s. \end{cases}$$

Steps 2 and 3: The characteristic system is

$$\frac{dx}{d\tau} = u, \quad \frac{dt}{d\tau} = 1, \quad \frac{du}{d\tau} = 0.$$

Solving the characteristic system, we have

$$u = u_0, \quad x = u_0\tau + x_0 = u_0\tau + s, \quad t = \tau + t_0 = \tau.$$

Eliminating τ from x and t, we see that the characteristics satisfy

$$x = u_0 t + s \quad \text{or} \quad s = x - u_0 t. \tag{5.1.8}$$

Since the initial condition for u is divided to three different functions, we consider each case separately.

1. For $s < 0$, $u = u_0 = 0$ and (5.1.8) is $x = s \le 0$. Therefore, for $x \le 0$ $u = 0$.
2. For $s > 1$, $u = u_0 = 1$ and (5.1.8) is $s = x - t > 1$. Therefore, for $t + 1 < x$ $u = 1$.
3. For $0 < s \le 1$, $u = u_0 = s$ and (5.1.8) is $x = u_0 t + s = st + s$. Therefore, $s = \frac{x}{1+t}$. Substituting this in $0 < s \le 1$ and $u = s$, we obtain

$$0 < \frac{x}{1+t} \le 1 \Rightarrow 0 < x \le 1 + t$$

$$u = \frac{x}{1+t}.$$

Summarizing the results, we have

$$u(x,t) = \begin{cases} 0, & x \le 0, \\ \frac{x}{1+t}, & 0 < x \le 1+t, \\ 1, & 1+t < x. \end{cases}$$

The solution is sketched in Figure 5.1. The characteristics (5.1.8) for three intervals of s are sketched in Figure 5.2. As we see in Figure 5.1 the characteristics starting from the interval $0 \le s \le 1$ expand as t increases and propagate toward the positive x-axis. We call this expanding wave a rarefaction wave. We should also observe that unlike the linear case the slopes of the characteristics depend on the solution.

The following is an example where the right hand side is not zero.

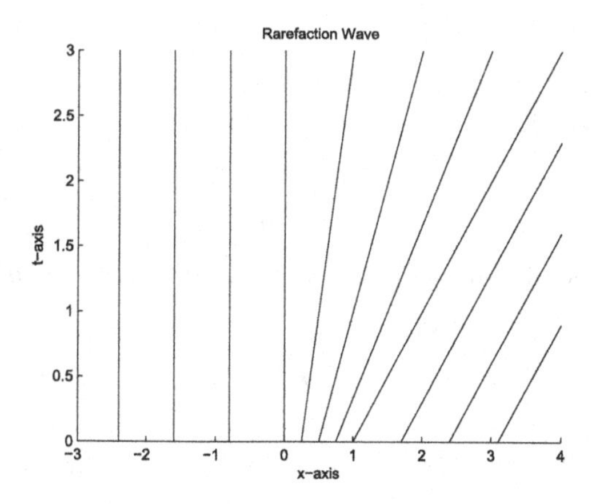

Fig. 5.2 Rarefaction wave.

Example 5.1.3. Solve

$$uu_t + u_x = -u, \tag{5.1.9}$$

$$u(0,t) = \alpha t, \tag{5.1.10}$$

where α is a constant.

Solution: Step 1: The initial data can be parametrized as follows.

$$x = x_0 = 0,\ t = t_0 = s,\ u = u_0 = \alpha s.$$

Step 2: Second, for each s, we apply the chain rule for multivariable calculus and solve the system of ordinary differential equations

$$\frac{dx}{d\tau} = 1,\ \frac{dt}{d\tau} = u,\ \frac{du}{d\tau} = -u$$

with the initial data

$$x(0) = x_0 = 0,\ t(0) = t_0 = s,\ u(0) = u_0 = \alpha s.$$

Solving $\frac{du}{d\tau} = -u$, we obtain

$$u = u_0 e^{-\tau} = \alpha e^{-\tau} s. \tag{5.1.11}$$

Next, solving $\frac{dt}{d\tau} = u$ with the above u, we obtain

$$t = t_0 + u_0(-e^{-\tau} + 1) = s(1 + \alpha - \alpha e^{-\tau}). \tag{5.1.12}$$

Also, from $\frac{dx}{d\tau} = 1$, we have

$$x = \tau + x_0 = \tau. \tag{5.1.13}$$

Step 3: Solving (5.1.12) and (5.1.13) for (s, τ) in terms of (x, t) and substituting the resulting equations in (5.1.11), we have

$$u = \alpha e^{-x} \frac{t}{(1 + \alpha - \alpha e^{-x})}. \tag{5.1.14}$$

It is interesting to observe that u becomes infinite at $x = -\ln(\frac{1+\alpha}{\alpha})$.

Exercises

1. Verify that (5.1.14) is a solution to (5.1.9) and (5.1.10).
2. Find the solution to the initial value problem.

$$u_t + uu_x = 0, \quad -\infty < x < \infty,\ 0 < t,$$

$$u(x,0) = x.$$

3. Consider the initial value problem. Here, α is a constant.

$$u_t + uu_x = -u, \quad -\infty < x < \infty,\ 0 < t,$$

$$u(x,0) = \alpha x.$$

   (a) Find the solution.
   (b) Find the values of α for which the solution becomes infinite in finite time t. Express the value of t in terms of α.

4. Consider the initial value problem

$$u_t + (\frac{u^3}{3})_x = 0, \quad -\infty < x < \infty,\ 0 < t,$$

$$u(x,0) = h(x) = \begin{cases} 0, & x \le 0, \\ x, & 0 < x \le 1, \\ 1, & 1 < x. \end{cases}$$

   (a) Sketch the (projection of) characteristics on the xt-plane.
   (b) Find the solution.

5. Consider the initial value problem where the initial data is given by a general function.

$$u_t + f(u)_x = u_t + a(u)u_x = 0, \quad -\infty < x < \infty,\ 0 < t,$$

$$u(x,0) = h(x),$$

   where $a(u) = f'(u)$. We assume that f is strictly convex ($a' > 0$). We follow the steps we took in this section to derive the solution implicitly.

   (a) Parametrize the initial condition by s.
   (b) Use the method of characteristics to express x, t, and u in terms of s and τ.
   (c) Show that the solution is given implicitly by

$$u = h(x - a(u)t). \tag{5.1.15}$$

   (d) Differentiate (5.1.15) in x to observe that if the slope of the initial is negative at one point, the slope of the solution becomes infinite in finite time.

5.2 An Application of Quasilinear Equations

5.2.1 *Scalar Conservation Law*

We derive a scalar conservation law using a traffic flow problem on a highway as an example. In the traffic flow problems we assume that the flow of cars can be regarded as a flow of continuum. We take the x-axis along the highway and assume that the traffic flows in the positive direction. Denote by $u = u(x,t)$ the number (or density) of cars per unit length at x and at time t. Also, denote by $f = f(x,t)$ the rate at which cars pass through x at t. This rate is commonly called flux and in this problem measured by the number of cars passing x per unit time. We now derive a partial differential equation governing the density and flux. The derivation is very similar to that of the heat equation. Consider a segment of highway $[x_1, x_2]$ $(x_1 < x_2)$. The rate of change of the number of cars in the segment is given by

$$\frac{d}{dt}\int_{x_1}^{x_2} u(x,t)dx = \int_{x_1}^{x_2} \frac{\partial u}{\partial t}(x,t)dx.$$

This should be the same as the rate at which the cars come in at x_1 minus the rate at which the cars leave at x_2 from the segment $[x_1, x_2]$. Therefore, we have

$$\int_{x_1}^{x_2} \frac{\partial u}{\partial t}(x,t)dx = f(x_1,t) - f(x_2,t) = -\int_{x_1}^{x_2} \frac{\partial f}{\partial x}(x,t)dx, \qquad (5.2.1)$$

$$\int_{x_1}^{x_2} [\frac{\partial u}{\partial t}(x,t) + \frac{\partial f}{\partial x}(x,t)]dx = 0.$$

Since x_1, x_2 are arbitrary, we obtain

$$\frac{\partial u}{\partial t} + \frac{\partial f}{\partial x} = 0. \qquad (5.2.2)$$

To derive (5.2.2) we used the following relation in (5.2.1).

$$\begin{bmatrix} \text{The rate of} \\ \text{change of number of} \\ \text{cars in the interval} \end{bmatrix} = [\text{Influx}] - [\text{Ouflux}], \qquad (5.2.3)$$

where the influx is the number of cars coming in to the interval per unit time and the outflux is the number of cars leaving the interval per unit time. This is called the conservation law and actually we already studied this in Section 2.1.

The difference lies in the flux. Unlike the heat equation the flux f depends on the density only. The relation

$$f = cu(1 - \frac{u}{u_1}) \tag{5.2.4}$$

is suggested from the experiment. Here, u_1 is called the saturation density at which cars can no longer move. This form of f makes sense because if the density is zero, there is no flux and if the density of cars reaches u_1, cars can no longer move. Therefore, there is no flux. Between $0 < u < u_1$ cars are moving and there is a flux. Substituting (5.2.4) in (5.2.3), we obtain

$$u_t + c(1 - 2\frac{u}{u_1})u_x = 0.$$

Setting $v = \frac{u}{u_1}$, we see

$$v_t + c(1 - 2v)v_x = 0. \tag{5.2.5}$$

In what follows we assume $c = 1$ and consider two initial value problems.

Example 5.2.1. Find the solution to (5.2.5) if $c = 1$ and the initial data is given by

$$v(x,0) = h(x) = \begin{cases} 1, & x \le 0, \\ 1 - x, & 0 < x \le 1, \\ 0, & 1 < x. \end{cases}$$

Solution: Step 1: The initial data can be parametrized as follows.

$$x = x_0 = s,\ t = t_0 = 0,\ v(x,0) = v_0 = h(s) = \begin{cases} 1, & s \le 0, \\ 1 - s, & 0 < s \le 1, \\ 0, & 1 < s. \end{cases}$$

Steps 2 and 3: The characteristic system is

$$\frac{dx}{d\tau} = (1 - 2v),\ \frac{dt}{d\tau} = 1,\ \frac{dv}{d\tau} = 0. \tag{5.2.6}$$

Examining the three equations in (5.2.6), we notice that v is a constant along each characteristic and that we should solve v before solving x. From the ODE's for v and t we have

$$t = \tau + t_0 = \tau,$$

$$v = v_0. \tag{5.2.7}$$

Substituting $v = v_0$ in $dx/d\tau = (1 - 2v)$ and solving it, we see that the characteristics satisfy

$$x = (1 - 2v_0)\tau + x_0 = (1 - 2v_0)t + s. \tag{5.2.8}$$

We consider the three intervals of the initial data separately.

1. For $s \le 0$, $v_0 = 1$ and (5.2.8) is $x = (1-2v_0)t + x_0 = -t + s$. So, for $s = x + t \le 0$, $v = 1$.
2. For $1 < s$, $v_0 = 0$ and (5.2.8) is $x = (1-2v_0)t + x_0 = t + s$. So, for $s = x - t > 1$, $v = 0$.
3. For $0 < s \le 1$, $v_0 = 1 - s$ and (5.2.8) is
$$x = (1-2v_0)t + x_0 = (-1+2s)t + s.$$
Therefore, $s = (x+t)/(2t+1)$ and for
$$0 < \frac{x+t}{2t+1} \le 1$$
the solution (5.2.7) is
$$v(x,t) = v_0 = 1 - \frac{x+t}{2t+1} = \frac{t-x+1}{2t+1}.$$

Summarizing the results, we have

$$v(x,t) = \begin{cases} 1, & x \le -t, \\ \frac{t-x+1}{2t+1}, & -t < x \le 1+t, \\ 0, & 1+t < x. \end{cases}$$

Example 5.2.2. Find the solution to (5.2.5) if $c = 1$ and the initial data is given by

$$v(x,0) = h(x) = \begin{cases} 0, & x \le 0, \\ x, & 0 < x \le 1, \\ 1, & 1 < x. \end{cases}$$

Solution: Step 1: The initial data can be parametrized as follows.

$$x = x_0 = s,\ t = t_0 = 0,\ v(x,0) = v_0 = h(s) = \begin{cases} 0, & s \le 0, \\ s, & 0 < s \le 1, \\ 1, & 1 < s. \end{cases}$$

Steps 2 and 3: The method of characteristics is

$$\frac{dx}{dt} = (1-2v),\ \frac{dt}{dt} = 1,\ \frac{dv}{dt} = 0.$$

We solve the ODE's from $dv/dt = 0$ and we have

$$v = v_0. \tag{5.2.9}$$

Substituting this in $dx/dt = (1-2v) = (1-2v_0)$, we see that the characteristics satisfy

$$x = (1-2v_0)t + x_0 = (1-2v_0)t + s. \tag{5.2.10}$$

We consider the three intervals of the initial data separately.

1. For $s \le 0$, $v_0 = 0$ and (5.2.10) is $x = (1-2v_0)t + x_0 = t + s$. So, for $s = x - t \le 0$, $v = 0$.
2. For $1 < s$, $v_0 = 1$ and (5.2.10) is $x = (1-2v_0)t + x_0 = -t + s$. So, for $s = x + t > 1$, $v = 1$.
3. For $0 < s \le 1$, $v_0 = s$ and (5.2.10) is

$$x = (1-2v_0)t + x_0 = (1-2s)t + s.$$

Therefore, $s = (x-t)/(1-2t)$ and the solution (5.2.9) is

$$v(x,t) = v_0 = \frac{x-t}{1-2t}.$$

Summarizing the results, we have

$$v(x,t) = \begin{cases} 0, & x \le t, \\ \frac{x-t}{1-2t}, & t < x \le 1-t, \\ 1, & 1-t < x. \end{cases}$$

The solution is valid for $0 \le t < 1/2$. To see what happens for $t \ge 1/2$, we draw the characteristics on the xt-plane. See Figure 5.3. Then, we see that there is a wedge shaped region where the three characteristics overlap. This means there are three values of densities u for each location in the wedge shaped region. This is not desirable. Extending the solution to $t \ge 1/2$ can be done after Example 5.2.3 in the next subsection and is left for exercises.

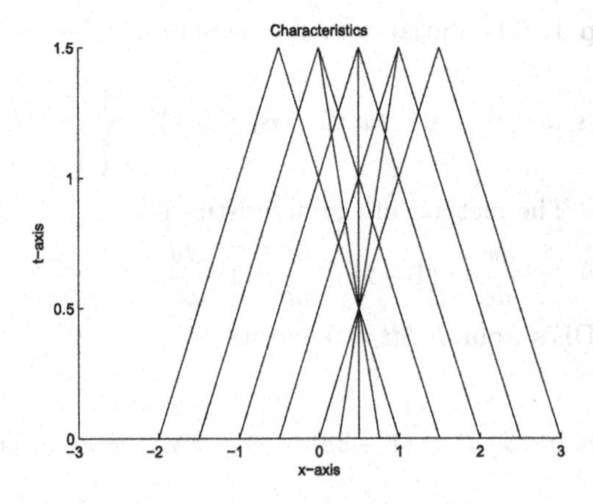

Fig. 5.3 Overlapping characteristics.

5.2.2 *Rankine-Hugoniot Condition*

In Example 5.2.2 we observed that the classical solution defined in Section 1.1 is inadequate for the definition of the solution. We need to extend or enlarge the definition of solutions so that we can also include discontinuous functions as solutions. We call such solutions weak solutions.

To motivate the definition of weak solutions, in this subsection we derive the condition called the Rankine-Hugoniot condition which should be satisfied across the discontinuity. This condition specifies the relation between the velocity of discontinuity and the states on the left and right of discontinuity. Suppose there is a discontinuity for u at $x(t)$ in an interval $[x_1, x_2]$ and u is smooth otherwise. Then, applying the conservation law for the intervals $[x_1, x(t)]$ and $[x(t), x_2]$, we obtain

$$\begin{aligned}
&\frac{d}{dt}\int_{x_1}^{x(t)} u(x,t)dx + \frac{d}{dt}\int_{x(t)}^{x_2} u(x,t)dx \\
&= u(x(t)_-, t)\frac{dx}{dt} + \int_{x_1}^{x(t)} u_t dx - u(x(t)_+, t)\frac{dx}{dt} + \int_{x(t)}^{x_2} u_t dx \\
&= [u_- - u_+)]\frac{dx}{dt} - \int_{x_1}^{x(t)} f_x dx - \int_{x(t)}^{x_2} f_x dx \\
&= [u_- - u_+]\frac{dx}{dt} - [f(u_-) - f(u(x_1,t))] \\
&\quad - [f(u(x_2,t)) - f(u_+)],
\end{aligned} \tag{5.2.11}$$

where

$$u_- = \lim_{x \to x(t)_-} u(x,t), \quad u_+ = \lim_{x \to x(t)_+} u(x,t).$$

In deriving the Rankine-Hugoniot relation the differentiation rule called the Leibniz rule in Appendix A.1 was used. Applying the conservation law, we also have,

$$\begin{aligned}
\frac{d}{dt}\int_{x_1}^{x(t)} udx + \frac{d}{dt}\int_{x(t)}^{x_2} udx &= \frac{d}{dt}\int_{x_1}^{x_2} u(x,t)dx \\
&= f(u(x_1,t)) - f(u(x_2,t)).
\end{aligned} \tag{5.2.12}$$

Combining (5.2.11) and (5.2.12), we obtain the Rankine-Hugoniot condition given by

$$[u_- - u_+]\frac{dx}{dt} = [f(u_-) - f(u_+)],$$

$$\frac{dx}{dt} = \frac{f(x(t)_+, t) - f(x(t)_-, t)}{u_+ - u_-}. \tag{5.2.13}$$

This condition relates the speed of discontinuity (or the speed of shock discontinuity) dx/dt with u on the left and right of discontinuity. Furthermore, if the characteristics on the left and right impinge on the discontinuity, it is called the shock discontinuity. If $a(u) = f'(u)$ is strictly increasing or decreasing (or f'' is positive or negative), this condition is given by

$$a(u_+) < \frac{dx}{dt} < a(u_-). \tag{5.2.14}$$

The case where a changes sign is not simple as (5.2.14). The condition (5.2.14) for discontinuities is called the Lax entropy condition and the importance of this condition is discussed in the next subsection.

The following example considers the case where the shock discontinuity develops from the smooth data.

Example 5.2.3. Solve

$$u_t + uu_x = 0, \quad -\infty < x < \infty,\ 0 < t,$$

$$u(x,0) = h(x) = \begin{cases} 1, & x \le 0, \\ 1-x, & 0 < x \le 1, \\ 0, & 1 < x. \end{cases}$$

Solution: Step 1: The initial condition is parametrized as

$$x = x_0 = s, \quad t = t_0 = 0, \quad u = u_0 = h(s) = \begin{cases} 1, & s \le 0, \\ 1-s, & 0 < s \le 1, \\ 0, & 1 < s. \end{cases}$$

Steps 2 and 3: The characteristic system is

$$\frac{dx}{d\tau} = u, \quad \frac{dt}{d\tau} = 1, \quad \frac{du}{d\tau} = 0.$$

We should solve u and t before solving x. Then,

$$u = u_0, \quad t = \tau + t_0 = \tau.$$

Since $u = u_0$ along the characteristics, solving x, we have

$$x = u_0\tau + x_0 = u_0\tau + s.$$

Eliminating τ from x and t, we see that the characteristics satisfy

$$x = u_0 t + s \quad \text{or} \quad s = x - u_0 t. \tag{5.2.15}$$

As the initial condition for u is divided to three different functions, we consider each case separately.

1. For $s \leq 0$, $u = u_0 = 1$ and (5.2.15) is $s = x - t$. Therefore, for $x \leq t$, $u = 1$.
2. For $s > 1$, $u = u_0 = 0$ and (5.2.15) is $s = x$. Therefore, for $1 < x$, $u = 0$.
3. For $0 < s \leq 1$, $u = u_0 = 1 - s$ and (5.2.15) is $x = u_0 t + s = (1-s)t + s$. Solving the characteristics for s, we have $s = \frac{x-t}{1-t}$. Substituting this s in $0 < s \leq 1$ and $u = 1 - s$, we obtain
$$0 < \frac{x-t}{1-t} \leq 1 \Rightarrow t < x \leq 1$$
$$u = 1 - \frac{x-t}{1-t} \Rightarrow u = \frac{1-x}{1-t}.$$

Summarizing the results, we have
$$u(x,t) = \begin{cases} 1, & x \leq t, \\ \frac{1-x}{1-t}, & t < x \leq 1, \\ 0, & 1 < x. \end{cases}$$
The solution we obtained is valid for $0 \leq t < 1$. At $t = 1$ the solution becomes discontinuous. As in Example 5.2.2 the characteristics overlap in the wedge shaped region for $t > 1$ as depicted in Figure 5.4 (Left).

One way to extend the solution to $t \geq 1$ is to introduce a line (or curve) of discontinuity originating at $(x,t) = (1,1)$ in the wedge shaped region. Then, as we observe in Figure 5.4 (Right), the characteristics impinge on the discontinuity. Along the characteristics impinging on the discontinuity from the left, the value of solution is $u = 1$ and along the characteristics impinging on the discontinuity from the right, the value of solution is $u = 0$. Therefore, $u_+ = 0$, $u_- = 1$, and the speed of discontinuity is
$$\frac{dx}{dt} = \frac{\frac{1}{2}u_+^2 - \frac{1}{2}u_-^2}{u_+ - u_-} = \frac{1}{2}.$$
Since the discontinuity starts at $(x,t) = (1,1)$, the location of the discontinuity can be found by integrating the above relation. Then,
$$x = \frac{1}{2}t + c \Rightarrow x = \frac{1}{2}t + \frac{1}{2}.$$
To find the constant c, we substituted $(x,t) = (1,1)$ in the equation. Therefore, the solution for $t \geq 1$ is given by
$$u(x,t) = \begin{cases} 1 & x < \frac{1}{2}t + \frac{1}{2}, \\ 0 & x > \frac{1}{2}t + \frac{1}{2}. \end{cases} \tag{5.2.16}$$
It is not difficult to see that (5.2.14) is satisfied for the solution (5.2.16). Figure 5.5 sketches the solution u for $0 \leq t \leq 2$. The slope of solution becomes steeper and steeper and the shock discontinuity develops at $t = 1$.

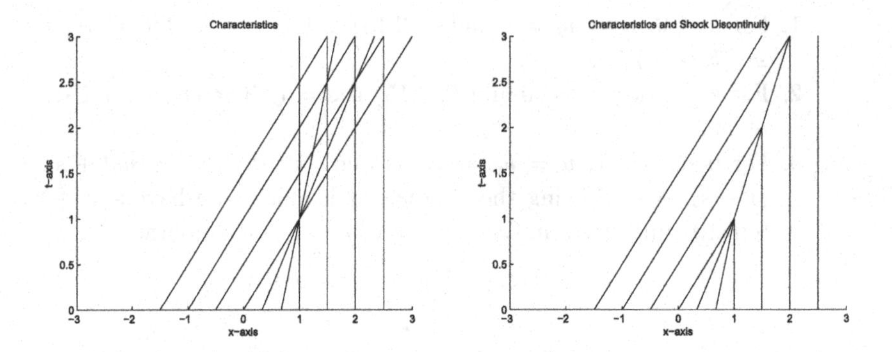

Fig. 5.4 Overlapping characteristics (Left) and Development of shock discontinuity (Right).

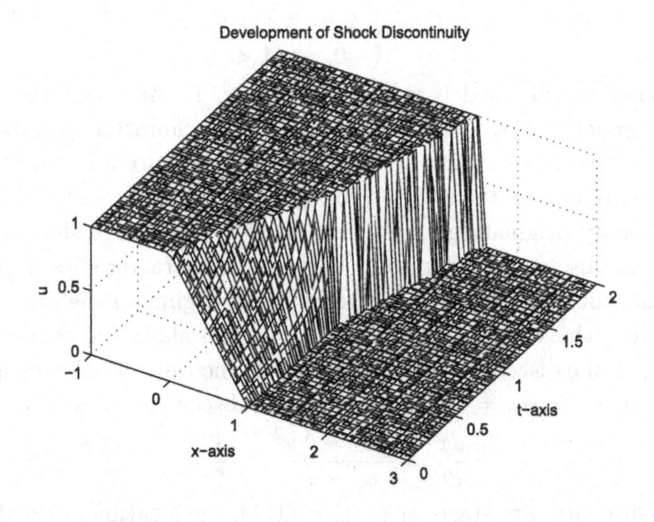

Fig. 5.5 Development of shock discontinuity.

5.2.3 *Weak Solutions*

We learned that the discontinuity develops for quasilinear equations of conservation laws of the form

$$u_t + f(u)_x = 0, \quad -\infty < x < \infty,\ 0 < t. \tag{5.2.17}$$

There have been efforts to include discontinuous solutions as solutions to (5.2.17). A definition accepted now is the following.

Definition 5.2.4. The weak solution to (5.2.17) is the function $u(x,t)$ satisfying

$$\int_0^\infty \int_{-\infty}^\infty [u\phi_t + f(u)\phi_x]dxdt = 0 \tag{5.2.18}$$

for every C^∞ test function ϕ with compact support.

This definition needs some explanation. A test function means a function ϕ with which we test if u satisfies (5.2.18) and the compact support means the closure of the set of (x,t) where ϕ is nonzero is bounded. In PDE's, roughly speaking, the closure of a region Ω is $\Omega \cup \partial\Omega$. With this definition u could be discontinuous. However, the question is whether the definition is good enough. We wish that the weak solutions would satisfy

1. if u is a classical solution to (5.2.17), it must be a weak solution. In other words, it satisfies (5.2.18).
2. u must satisfy the Rankine-Hugoniot condition across discontinuities.
3. the weak solution is unique.

As we will see the definition satisfies the first two requirements but fails the third one. To show the first requirement is left for exercises. Third requirement will be discussed in the next subsection. In the rest of the subsection we discuss the second requirement. For this purpose, assume that there is a discontinuity along $x = \sigma(t)$ and u is smooth on both sides of the discontinuity. We split the integral in (5.2.18) into two parts

$$\int_0^\infty \int_{-\infty}^{\sigma(t)} [u\phi_t + f(u)\phi_x]dxdt + \int_0^\infty \int_{\sigma(t)}^\infty [u\phi_t + f(u)\phi_x]dxdt = 0 \tag{5.2.19}$$

and apply the divergence theorem to the vector $[f(u)\phi, u\phi]$. This is illustrated in Figure 5.6 where Ω is the compact support for the test function ϕ. Since ϕ has a compact support, the line integral is performed only along $x = \sigma(t)$ where $\mathbf{n} = [1, -\sigma'(t)]$. Therefore, for the first integral in (5.2.19) the divergence theorem implies

$$\int_0^\infty \int_{-\infty}^{\sigma(t)} \nabla \cdot [f(u)\phi, u\phi]dxdt = \int_{x=\sigma(t)} [f(u)\phi, u\phi] \cdot [1, -\sigma'(t)]ds$$

and this leads to

$$\int_0^\infty \int_{-\infty}^{\sigma(t)} [u\phi_t + f(u)\phi_x]dxdt = -\int_0^\infty \int_{-\infty}^{\sigma(t)} [u_t + f(u)_x]\phi dxdt$$
$$+ \int_{x=\sigma(t)} [f(u_-) - \sigma'(t)u_-]\phi ds. \tag{5.2.20}$$

Similarly, for the second integral

$$\int_0^\infty \int_{\sigma(t)}^\infty \nabla \cdot [f(u)\phi, u\phi] dxdt = \int_{x=\sigma(t)} [f(u)\phi, u\phi] \cdot [-1, \sigma'(t)] ds$$

holds and this leads to

$$\int_0^\infty \int_{\sigma(t)}^\infty [u\phi_t + f(u)\phi_x] dxdt = -\int_0^\infty \int_{\sigma(t)}^\infty [u_t + f(u)_x]\phi dxdt$$
$$- \int_{x=\sigma(t)} [-f(u_+) + \sigma'(t)u_+]\phi ds. \qquad (5.2.21)$$

Adding (5.2.20) and (5.2.21) and using (5.2.19) and the equation (5.2.17), we have

$$\int_{x=\sigma(t)} \{[f(u_-) - f(u_+)] + \sigma'(t)[-u_- + u_+]\}\phi ds = 0.$$

Since this holds for every test function, we have

$$\sigma'(t) = \frac{f(u_+) - f(u_-)}{u_+ - u_-}.$$

This is precisely the Rankine-Hugoniot condition.

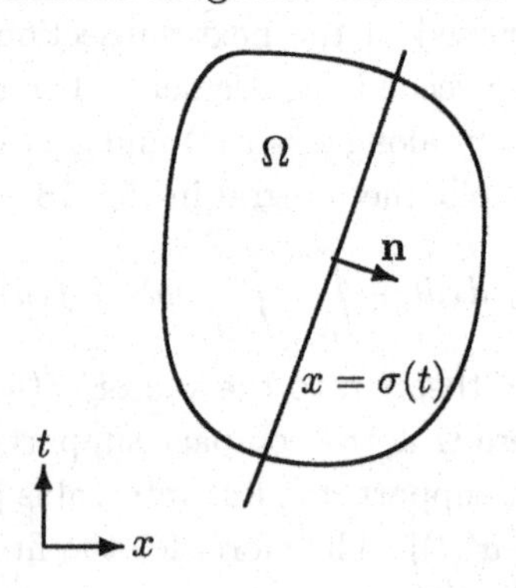

Fig. 5.6 Line of discontinuity.

5.2.4 *Entropy Condition and Admissibility Criterion*

As stated in the previous subsection, we examine the uniqueness of the weak solutions (the third item in the wish list for the weak solutions). For this purpose, we discuss special initial value problems called the Riemann problems where the initial data is given by two constant states u_l and u_r separated at $x = 0$.

$$u(x,0) = \begin{cases} u_l, & x < 0, \\ u_r, & 0 < x. \end{cases}$$

As a specific example, consider the following example.

Example 5.2.5. Solve

$$u_t + (\frac{1}{2}u^2)_x = 0, \quad -\infty < x < \infty,\ 0 < t,$$

$$u(x,0) = h(x) = \begin{cases} 2, & x < 0, \\ 0, & 0 < x. \end{cases}$$

Solution: Step 1: First we parametrize the initial data by s. Then,

$$x = x_0 = s,\ t = t_0 = 0,\ u = u(s,0) = h(s) = \begin{cases} 2, & s < 0, \\ 0, & 0 < s. \end{cases}$$

Step 2: We apply the method of characteristics. The characteristic system is

$$\frac{dx}{d\tau} = u, \quad \frac{dt}{d\tau} = 1, \quad \frac{du}{d\tau} = 0.$$

Solving the characteristic system, we see that the characteristics satisfy

$$u = u_0, \quad x = u_0\tau + x_0 = u_0\tau + s, \quad t = \tau + t_0 = \tau.$$

Step 3: Eliminating τ from x and t, we obtain

$$x = u_0 t + s \quad \text{or} \quad s = x - u_0 t.$$

Therefore, for $s < 0$ $u = u_0 = 2$ and the characteristics are $x = u_0 t + s = 2t + s$. For $s > 0$ $u = u_0 = 0$ and the characteristics are $x = u_0 t + s = s$.
Step 4: This is a new step where we fix the overlapping characteristics by the Rankine-Hugoniot condition. As we see in Figure 5.7 the characteristics overlap in the wedge shaped region originating at $(x,t) = (0,0)$. In this wedge shaped region u is multi-valued. We seek a weak solution where there is a shock discontinuity which divides the wedge shaped region into two so that u will be no longer multivalued. Since along the characteristics starting in $s < 0$ $u = 2$ and along the characteristics starting in $s > 0$ $u = 0$, $u_+ = 0$ and $u_- = 2$. By the Rankine-Hugoniot condition (5.2.13) the shock speed is given by

$$\frac{dx}{dt} = \frac{\frac{1}{2}u_+^2 - \frac{1}{2}u_-^2}{u_+ - u_-} = 1.$$

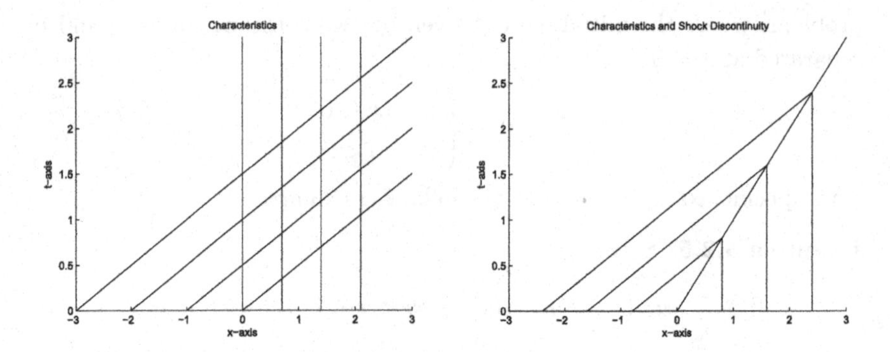

Fig. 5.7 Characteristics and shock discontinuity.

Therefore, the solution is given by

$$u(x,t) = \begin{cases} 2 & x < t, \\ 0 & x > t. \end{cases}$$

It is easy to check that the Lax entropy condition is satisfied for the discontinuity.

Another example is the case where the initial data is reversed. In this case we have a solution where the characteristics are expanding.

Example 5.2.6. Solve

$$u_t + (\frac{1}{2}u^2)_x = 0, \quad -\infty < x < \infty,\ 0 < t, \tag{5.2.22}$$

$$u(x,0) = h(x) = \begin{cases} 0, & x < 0, \\ 2, & 0 < x. \end{cases} \tag{5.2.23}$$

Solution: Step 1: First we parametrize the initial data by s. Then,

$$x = x_0 = s,\ t = t_0 = 0,\ u = u(s,0) = h(s) = \begin{cases} 0, & s < 0, \\ 2, & 0 < s. \end{cases}$$

Step 2: We apply the method of characteristics. The characteristic system is

$$\frac{dx}{d\tau} = u, \quad \frac{dt}{d\tau} = 1, \quad \frac{du}{d\tau} = 0.$$

Solving the characteristic system, we have

$$u = u_0, \quad x = u_0\tau + x_0 = u_0\tau + s, \quad t = \tau + t_0 = \tau.$$

Step 3: Eliminating τ from x and t, we obtain

$$x = u_0 t + s \quad \text{or} \quad s = x - u_0 t.$$

Therefore, for $s < 0$ $u = u_0 = 0$ and the characteristics are $x = u_0 t + s = s$. For $s > 0$ $u = u_0 = 2$ and the characteristics are $x = u_0 t + s = 2t + s$.

Step 4: As we see in Figure 5.8 there is a wedge shaped region originating at $(x, t) = (0, 0)$ where there is no characteristic. This is a new step where we fill the wedge shaped region with u. To do so we assume that u in the wedge is given by $u(x, t) = v(x/t) = v(\xi)$. Substituting this in (5.2.22), we obtain

$$-\frac{x}{t^2}v' + \frac{1}{t}vv' = 0 \Rightarrow v = \frac{x}{t}.$$

Therefore, $u = x/t$ in the wedge region. The left end of the wedge is $x = 0$ and $u = 0$ along $x = 0$. The right end of the wedge is $x = 2t$ and along this line $u = 2$. Hence, the solution is continuous along both ends of the wedge. This implies that the solution is continuous for all t and given by

$$u(x,t) = \begin{cases} 0, & x < 0, \\ \frac{x}{t}, & 0 \le x \le 2t, \\ 2, & 2t < x. \end{cases} \tag{5.2.24}$$

The solution in the wedge is called the centered rarefaction wave.

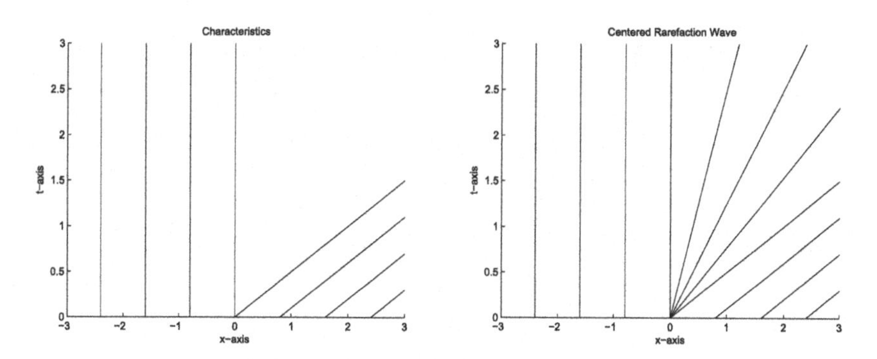

Fig. 5.8 Centered rarefaction wave.

When the initial data is given as in Example 5.2.6, we can also construct a weak solution where the discontinuity with speed one separates the constant states zero and two, *i.e.*,

$$u(x,t) = \begin{cases} 0 & x < t, \\ 2 & x > t \end{cases} \tag{5.2.25}$$

is a solution since the discontinuity satisfies the Rankine-Hugoniot condition. The problem with this solution is that the characteristics do not impinge on the discontinuity. Rather there are characteristics originating from the discontinuity. As a matter of fact, the Lax entropy condition (5.2.14) is violated across the discontinuity in the solution (5.2.25). Also, if we trace the characteristics backward in time, there are characteristics which do not reach the initial line. It is reasonable to expect the causality, that is, the present (the initial condition) affects the future (the solution). The fact that there are such characteristics violates the causality. Therefore, we select the solution (5.2.24) as a physically relevant solution and exclude (5.2.25) as an entropy violating solution. The Lax entropy condition provides a selection criterion for a physically admissible solution. There are other such criteria and they are referred as admissibility criteria. For example, if the flux f changes sign, the Lax entropy condition is no longer sufficient. Instead we use a stronger condition, the Oleinik entropy condition.

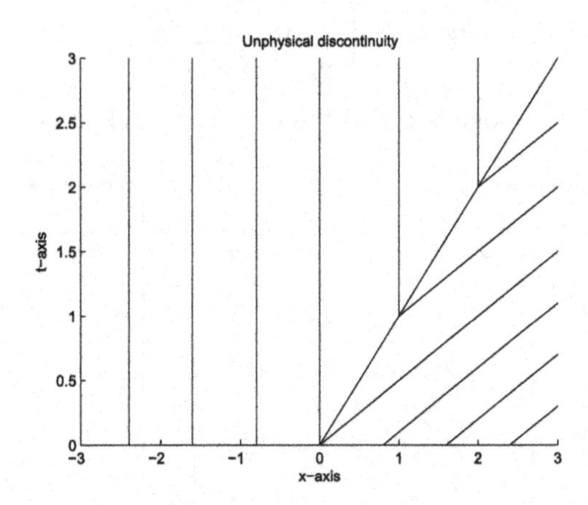

Fig. 5.9 Unphysical discontinuity.

5.2.5 *Traffic Flow Problem*

We reconsider the traffic flow problem. Remember that the traffic flow is in the positive x-direction. We discuss two Riemann problems. First, in Example 5.2.7 we consider the case where the traffic is saturated for $x > 0$ and not for $x < 0$. This corresponds to the case where the traffic is jammed for $x > 0$ or the signal at $x = 0$ turns red at $t = 0$. Second, in Example 5.2.8 we consider the case where there is no traffic on $x > 0$ and the traffic is saturated for $x < 0$. This corresponds to the case where the traffic light at $x = 0$ turns green at $t = 0$.

Example 5.2.7. (a) Solve

$$u_t + (u - u^2)_x = 0, \quad -\infty < x < \infty,\ 0 < t,$$

$$u(x,0) = \begin{cases} \frac{1}{2} & x < 0, \\ 1 & x > 0. \end{cases}$$

(b) Explain the solution.

Solution: (a) **Step 1:** First we parametrize the initial data by s. Then,

$$x = x_0 = s,\ t = t_0 = 0,\ u = u(s,0) = h(s) = \begin{cases} \frac{1}{2}, & s < 0, \\ 1, & 0 < s. \end{cases}$$

Step 2: We apply the method of characteristics. The characteristic system is

$$\frac{dx}{d\tau} = 1 - 2u, \quad \frac{dt}{d\tau} = 1, \quad \frac{du}{d\tau} = 0.$$

Solving the characteristic system, we have

$$u = u_0, \quad x = u_0\tau + x_0 = u_0\tau + s, \quad t = \tau + t_0 = \tau.$$

Step 3: Eliminating τ from x and t, we obtain

$$x = (1 - 2u_0)t + s \quad \text{or} \quad s = x - (1 - 2u_0)t.$$

Therefore, for $s < 0$ $u = u_0 = 1/2$ and the characteristics are $x = (1 - 2u_0)t + s = s$. For $s > 0$ $u = u_0 = 1$ and the characteristics are $x = (1 - 2u_0)t + s = -t + s$.

Step 4: As we see in Figure 5.10 (Left) the characteristics overlap in the wedge shaped region originating at $(x,t) = (0,0)$. In this wedge shaped region u is multivalued. We seek a weak solution where there is a shock discontinuity which divides the wedge shaped region into two so that u will be no longer multivalued. Since along the characteristics starting from $s < 0$ $u = 1/2$ and along the characteristics starting from $s > 0$ $u = 1$, $u_+ = 1$ and $u_- = 1/2$. The shock speed is given by

$$\frac{dx}{dt} = \frac{(u_+ - u_+^2) - (u_- - u_-^2)}{u_+ - u_-} = -\frac{1}{2}.$$

Therefore, the solution is given by

$$u(x,t) = \begin{cases} \frac{1}{2} & x < -\frac{1}{2}t, \\ 1 & x > -\frac{1}{2}t \end{cases}$$

and the relation between characteristics and shock discontinuity is drawn in Figure 5.10 (Right).

(b) If the traffic is jammed for $x > 0$ at $t = 0$, the line of shock discontinuity is the location where the jam starts. This line recedes into the negative x as t increases. For the red signal, it represents the end of queue.

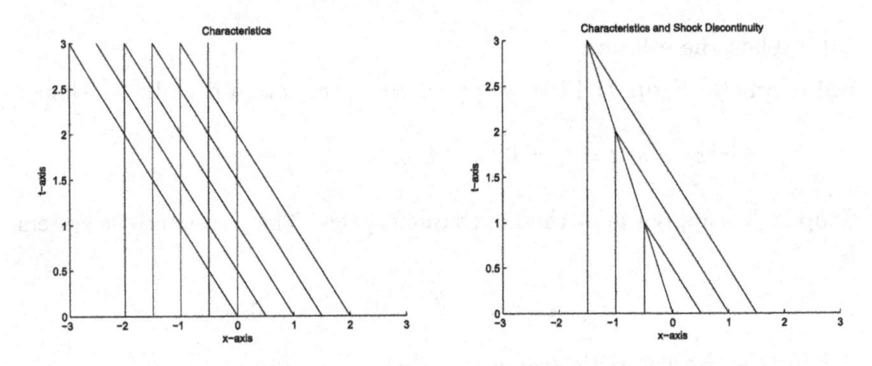

Fig. 5.10 Characteristics (Left), Characteristics and shock discontinuity (Right).

Example 5.2.8. (a) Solve

$$u_t + (u - u^2)_x = 0, \quad -\infty < x < \infty,\ 0 < t, \tag{5.2.26}$$

$$u(x,0) = \begin{cases} 1, & x < 0, \\ 0, & x > 0. \end{cases}$$

(b) Explain the solution.

Solution: (a) **Step 1:** First we parametrize the initial data by s. Then,

$$x = x_0 = s,\ t = t_0 = 0,\ u = u(s,0) = h(s) = \begin{cases} 1, & s < 0, \\ 0, & 0 < s. \end{cases}$$

Step 2: We apply the method of characteristics. The characteristic system is

$$\frac{dx}{d\tau} = 1 - 2u, \quad \frac{dt}{d\tau} = 1, \quad \frac{du}{d\tau} = 0.$$

Solving the characteristic system, we have

$$u = u_0, \quad x = u_0\tau + x_0 = u_0\tau + s, \quad t = \tau + t_0 = \tau.$$

Step 3: Eliminating τ from x and t, we obtain as the characteristics

$$x = (1 - 2u_0)t + s \quad \text{or} \quad s = x - (1 - 2u_0)t.$$

Therefore, for $s < 0$ $u = u_0 = 1$ and the characteristics are $x = (1-2u_0)t + s = -t + s$. So, for $s = x + t < 0$, $u = 1$. For $s > 0$ $u = u_0 = 0$ and the characteristics are $x = (1 - 2u_0)t + s = t + s$. So, for $s = x - t > 0$, $u = 0$.

Step 4: As we see in Figure 5.11 (Left) there is a wedge shaped region originating at $(x, t) = (0, 0)$ where there are no characteristics. We need to find the solution in the wedge shaped region. To do so we assume that u in the wedge is given by $u(x, t) = v(x/t) = v(\xi)$. Substituting this in (5.2.26), we obtain

$$-\frac{x}{t^2}v' + \frac{1}{t}(1 - 2v)v' = \frac{1}{t}(-\frac{x}{t} + 1 - 2v)v' = 0.$$

Therefore, $v = \text{const}$ or $v = (1 - x/t)/2$ in the wedge region. If we choose $u = v = \text{const}$ as a solution in the wedge, it is difficult to make connection continuous and if the connection is discontinuous, the Lax entropy condition may not be satisfied. On the other hand if we choose $v = (1 - x/t)/2$, the left end of the wedge is $x = -t$ and $u = 1$ along $x = -t$. The right end of the wedge is $x = t$ and along this line $u = 0$. Hence, the solution is continuous along both ends of the wedge. This suggests that if we choose $v = (1 - x/t)/2$, the solution is continuous for all t and given by

$$u(x,t) = \begin{cases} 1, & x < -t, \\ \frac{1}{2}(1 - \frac{x}{t}), & -t \le x \le t, \\ 0, & t < x. \end{cases} \tag{5.2.27}$$

The solution in the wedge is called the centered rarefaction wave. See Figure 5.11 (Right). The solution in three intervals are connected continuously across $x = -t$ and $x = t$.

(b) As the traffic light turns green, cars start to move. The boundary between the interval where cars move to the right and the interval where cars are still stopped is given by $x = -t$ and it recedes to the left.

Exercises

1. Follow Example 5.2.3 to find the solution for $t \ge 1/2$ in Example 5.2.2.

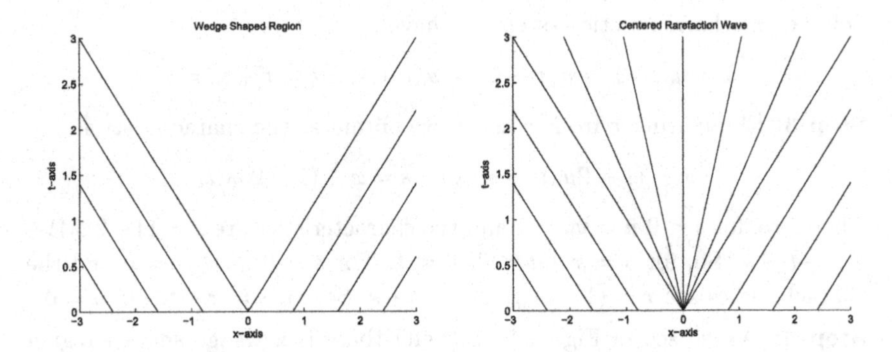

Fig. 5.11 Wedge (Left) and Centered rarefaction wave (Right).

2. Show that the classical solution to (5.2.17) satisfies (5.2.18).
3. Find the solutions to $u_t + (u^2)_x/2 = 0$ for the following initial conditions.

   (a) $u(x,0) = \begin{cases} 2 & x < 0 \\ 1 & 0 < x \end{cases}$

   (b) $u(x,0) = \begin{cases} 1 & x < 0 \\ 2 & 0 < x \end{cases}$

   (c) $u(x,0) = \begin{cases} 2 & x < 0 \\ 2 - x & 0 < x < 1 \\ 1 & 1 < x \end{cases}$

   (d) $u(x,0) = \begin{cases} 1 & x < 0 \\ 1 + x & 0 < x < 1 \\ 2 & 1 < x \end{cases}$

   (e) For (c) and (d) draw the profiles of solutions at $t = 1/2, 1, 2$.

4. Consider the initial value problem

$$u_t - (\frac{u^3}{3})_x = 0, \quad -\infty < x < \infty,\ 0 < t,$$

$$u(x,0) = h(x) = \begin{cases} 0, & x \le 0, \\ x, & 0 < x \le 1, \\ 1, & 1 < x. \end{cases}$$

   (a) Sketch the (projection of) characteristics on the xt-plane.
   (b) Find the solution.

5. Find the solutions to

$$v_t + (1 - 2v)v_x = 0$$

for the following initial data.

(a) $v(x,0) = \begin{cases} \frac{3}{4}, & x \leq 0, \\ \frac{3}{4} - \frac{1}{2}x, & 0 < x \leq 1, \\ \frac{1}{4}, & 1 < x. \end{cases}$

(b) $v(x,0) = \begin{cases} \frac{1}{4}, & x \leq 0, \\ \frac{1}{4} + \frac{1}{2}x, & 0 < x \leq 1, \\ \frac{3}{4}, & 1 < x. \end{cases}$

(c) For (a) and (b) draw the profiles of solutions at $t = 1/2, 1, 2$.

6. Consider an infinitely long cylindrical pipe along the x-axis. Suppose that a fluid is flowing in the positive x-direction. Denote by $\rho(x,t)$ and $f(x,t)$ the density (mass per unit length) and the flow rate (mass per unit time) of the fluid at x and t. Also suppose that fluid is leaking at the rate $h(x,t)$ (mass per unit length per unit time) from the wall.

   (a) Derive the equation
   $$\rho_t + f_x = -h.$$
   (b) If $f = \rho^2/2$, $h = a\rho$, and $\rho(x,0) = x$, find $\rho(x,t)$.

5.3 First Order Nonlinear Equations

If the equation is nonlinear in the highest order derivatives, it is called fully nonlinear. For example,

$$(u_x)^2 + (u_t)^2 = u, \tag{5.3.1}$$

is an example of fully nonlinear equation. The general nonlinear equation in two independent variables x and t can be written as

$$F(x,t,u,p,q) = 0,$$

where $p = u_x$ and $q = u_t$. Assume that the initial condition is given parametrically,

$$C_I : \ x = x_0(s), \ t = t_0(s) = 0, \ u = u_0(s).$$

If we take the similar approach and define

$$\frac{dx}{d\tau} = a, \ \frac{dt}{d\tau} = b, \tag{5.3.2}$$

where a and b are found later. Then, the differential equation for u along (5.3.2) is given by

$$\frac{du}{d\tau} = au_x + bu_t = ap + bq.$$

Unlike the linear or quasilinear equations, we cannot in general express the right hand side without p and q. This suggests that we need the differential equations for p and q, which are given by

$$\frac{dp}{d\tau} = u_{xx}\frac{dx}{d\tau} + u_{xt}\frac{dt}{d\tau} = u_{xx}a + u_{xt}b,$$

$$\frac{dq}{d\tau} = u_{tx}\frac{dt}{d\tau} + u_{tt}\frac{dt}{d\tau} = u_{xt}a + u_{tt}b.$$

Now, the problem is that we see new unknowns u_{xx}, u_{xt}, and u_{tt}. So, the question is what are good choices for a and b so that we do not have to deal with the new unknowns. Differentiating $F = 0$ with respect to x and y, we obtain

$$F_x + F_u u_x + F_p u_{xx} + F_q u_{xt} = 0,$$

$$F_t + F_u u_t + F_p u_{xt} + F_q u_{tt} = 0.$$

Comparing the above four expressions, we notice that if we choose $a = F_p$ and $b = F_q$, we can eliminate u_{xx}, u_{xt}, and u_{tt} from $dp/d\tau$ and $dq/d\tau$. Therefore, the set of differential equations are

$$\frac{dx}{d\tau} = F_p, \quad \frac{dt}{d\tau} = F_q,$$

$$\frac{du}{d\tau} = F_p p + F_q q,$$

$$\frac{dp}{d\tau} = -(F_x + F_u p), \quad \frac{dq}{d\tau} = -(F_t + F_u q).$$

We have five equations for five unknowns. To solve the initial value problem, we need the initial data for p and q. Denote them by $p_0(s)$ and $q_0(s)$. They are provided as follows. First, the equation $F(x, t, u, p, q) = 0$ holds. So,

$$F(x_0(s), t_0(s), u_0(s), p_0(s), q_0(s)) = 0. \tag{5.3.3}$$

Also, from

$$\frac{du}{ds} = u_x\frac{dx}{ds} + u_t\frac{dt}{ds},$$

along the initial curve C_I we have

$$u_0'(s) = p_0(s)x_0'(s) + q_0(s)t_0'(s). \tag{5.3.4}$$

From two equations (5.3.3) and (5.3.4), we find $p_0(s)$ and $q_0(s)$.

Example 5.3.1. Solve $(u_x)^2 + (u_t)^2 = 1$ with the initial data $u(x, 0) = 0$. This equation is called the Eikonal equation and well-known in geometric optics.

Solution: **Step 1:** First, we express the initial conditions in terms of s. For this problem, we set $F = p^2 + q^2 - 1 = 0$. The initial data are given by

$$x_0 = s, \ t_0 = 0, \ u_0 = 0.$$

For p_0 and q_0, from (5.3.3) and (5.3.4), we see that

$$p_0^2 + q_0^2 = 1,$$

$$u_0'(s) = p_0 = 0.$$

So, $q_0 = \pm 1$.

Step 2: For each s, we apply the method of characteristics and solve the system of ordinary differential equations

$$\frac{dx}{d\tau} = F_p = 2p, \quad \frac{dt}{d\tau} = F_q = 2q,$$

$$\frac{du}{d\tau} = F_p p + F_q q = 2,$$

$$\frac{dp}{d\tau} = -(F_x + F_u p) = 0, \quad \frac{dq}{d\tau} = -(F_t + F_u q) = 0$$

with the initial data

$$x(0) = x_0 = s, \ t(0) = t_0 = 0, \ u(0) = u_0 = 0,$$

$$p(0) = p_0 = 0, \ q(0) = q_0 = \pm 1.$$

The last three equations are easy to solve. We first solve $du/d\tau = 2$.

$$u = u_0 + 2\tau = 2\tau. \tag{5.3.5}$$

We solve $dp/d\tau = 0$, $dq/d\tau = 0$. Then,

$$p = p_0 = 0, \ q = q_0 = \pm 1. \tag{5.3.6}$$

Now, integrating $dx/d\tau = F_p = 2p$ and $dt/d\tau = F_q = 2q$, we have

$$x = x_0 = s, \ t = \pm 2\tau. \tag{5.3.7}$$

Step 3: Finally, we express u in terms of (x, t). This step is easy for this problem. Comparing (5.3.5) and (5.3.7), we have

$$u = \pm t.$$

Exercises

1. Solve the first order equation in R^2

$$u = xu_x + yu_y + \frac{1}{2}(u_x^2 + u_y^2), \quad -\infty < x < \infty,\ 0 < t,$$

$$u(x,0) = \frac{1}{2}(1 - x^2).$$

2. Solve

$$u_x^2 + u_y^2 = u^2, \quad -\infty < x < \infty,\ 0 < t,$$

with the following initial conditions.

(a) $x_0 = \cos s$, $y_0 = \sin s$, $u_0 = 1$.
(b) $x_0 = s$, $y_0 = 0$, $u_0 = 1$.

5.4 An Application of Nonlinear Equations - Optimal Control Problem

There are roughly two types of optimal control problems, stochastic and deterministic. We restrict our attention to the deterministic case. We consider an investor whose wealth $x(t)$ follows

$$dx = rxdt - c(t)dt, \tag{5.4.1}$$

where r is the interest rate and $c(t)$ is the rate of consumption. In Economics the wealth means the sum of his assets such as money and stock holdings. In the current setting x is money. Equation (5.4.1) says that the money grows at the rate r and the consumption reduces the money. His utility, a measure of satisfaction or happiness, is given by

$$[\int_0^T U(c(s))ds + G(x(T))], \tag{5.4.2}$$

where T is a future time, and U and G measure his utility of consumption during the time period $[0, T]$ and wealth at T (often called the bequest), respectively. In economics U and G are called utility functions and they are increasing and concave functions of the arguments. This reflects our human nature. Typical utility functions are a power utility function

$$U(c) = \frac{c^{1-\gamma}}{1-\gamma}, \tag{5.4.3}$$

where γ is a constant $0 < \gamma < 1$ and an exponential utility function

$$U(c) = -e^{-\beta c},$$

where β is a positive constant. In this section we use the power utility function given in (5.4.3). He tries to maximize his utility (5.4.2) by controlling his consumption. His optimal control problem is given as follows.

Definition 5.4.1. The control problem is defined as maximizing the value function

$$J(x,t,c) = [\int_t^T U(c(s))ds + G(x(T))]$$

with given dynamics

$$dx = (rx - c)dt.$$

The optimal value function is defined as

$$V(x,t) = \max_c J(x,t,c). \qquad (5.4.4)$$

We derive the differential equation for V. Since $V(x,t)$ is the result of optimal strategy, *i.e.*, the optimal choice of $c(t)$, for any choice of $c(t)$,

$$V(x,t) \geq [\int_t^{t+h} U(c(s))ds + V(x(t+h), t+h)]. \qquad (5.4.5)$$

Expanding $V(x(t+h), t+h)$ in Taylor series, we have

$$V(x(t+h), t+h) = V(t,x) + \frac{\partial V}{\partial t}h + \frac{\partial V}{\partial x}(rx - c)h + O(h^2).$$

Substituting this in (5.4.5) and dividing the inequality by h, and taking the limit as $h \to 0_+$, we have

$$\frac{\partial V}{\partial t} + \frac{\partial V}{\partial x}(rx - c) + U(c) \leq 0.$$

If c is the optimal strategy, the equality holds in (5.4.5). Therefore, we obtain

$$\frac{\partial V}{\partial t} + rx\frac{\partial V}{\partial x} + \max_c[-c\frac{\partial V}{\partial x} + U(c)] = 0. \qquad (5.4.6)$$

Since we use the power utility function (5.4.3), the above maximization is

$$-\frac{\partial V}{\partial x} + U'(c) = -\frac{\partial V}{\partial x} + c^{-\gamma} = 0 \Rightarrow c = (\frac{\partial V}{\partial x})^{-\frac{1}{\gamma}}.$$

Substituting the above c in (5.4.6), we see that the optimal value function V satisfies

$$\frac{\partial V}{\partial t} + rx\frac{\partial V}{\partial x} + \frac{\gamma}{1-\gamma}(\frac{\partial V}{\partial x})^{1-\frac{1}{\gamma}} = 0 \qquad (5.4.7)$$

with the terminal condition

$$V(x,T) = G(x).$$

After introducing the change of variable $t = T - \eta$, the above terminal problem becomes an initial value problem

$$\frac{\partial V}{\partial \eta} - rx\frac{\partial V}{\partial x} - \frac{\gamma}{1-\gamma}(\frac{\partial V}{\partial x})^{1-\frac{1}{\gamma}} = 0$$

$$V(x,0) = G(x).$$

Equations based on the above derivation are called the Hamilton-Jacobi-Bellman (HJB) equations and are widely used in the control theory, economics, and finance.

Since the case where $r > 0$ is rather involved, we consider the case where $r = 0$ as an example. This is a reasonable assumption if we deposit all money in a bank account which does not yields any interest. The case where $r > 0$ is treated in Exercise 3.

Example 5.4.2. Solve the HJB equation

$$\frac{\partial V}{\partial \eta} - \frac{\gamma}{1-\gamma}(\frac{\partial V}{\partial x})^{1-\frac{1}{\gamma}} = 0, \quad -\infty < x < \infty,\ 0 < \eta,$$

with the initial data

$$V(x,0) = \frac{x^{1-\gamma}}{1-\gamma}.$$

Solution: For this problem,

$$F(x,\eta,V,p,q) = q - \frac{\gamma}{1-\gamma}p^{1-\frac{1}{\gamma}},$$

where $p = \partial V/\partial x$ and $q = \partial V/\partial \eta$.

Step 1: First, parametrize the initial data. Then,

$$x = x_0 = s,\ \eta = \eta_0 = 0,\ V = V_0 = \frac{s^{1-\gamma}}{1-\gamma}.$$

Also, we need the initial data for p and q. From (5.3.4) and (5.3.3),

$$\frac{dV_0}{ds} = V_{0x}x_0'(s) + V_{0t}\eta_0'(s) = s^{-\gamma} \Rightarrow V_{0x} = p_0 = s^{-\gamma},$$

$$F\,|_{\eta=0} = 0 \Rightarrow q_0 = \frac{\gamma}{1-\gamma}(\frac{\partial V_0}{\partial x})^{1-\frac{1}{\gamma}} = \frac{\gamma}{1-\gamma}s^{1-\gamma}.$$

Step 2: Now, we use the method of characteristics. We need to find the solutions to five ODE's.

$$\frac{dx}{d\tau} = F_p = p^{-\frac{1}{\gamma}}, \quad \frac{d\eta}{d\tau} = F_q = 1,$$

$$\frac{dV}{d\tau} = F_p p + F_q q = p^{1-\frac{1}{\gamma}} + q,$$

$$\frac{dp}{d\tau} = -(F_x + F_V p) = 0, \quad \frac{dq}{d\tau} = -(F_\eta + F_V q) = 0.$$

By inspection we start from the last two ODE's and use them to solve the rest. Then,

$$\frac{dp}{d\tau} = 0 \Rightarrow p = p_0 = s^{-\gamma},$$

$$\frac{dq}{d\tau} = 0 \Rightarrow q = q_0 = \frac{\gamma}{1-\gamma} s^{1-\gamma},$$

$$\frac{dx}{d\tau} = p^{-\frac{1}{\gamma}} = s \Rightarrow x = s\tau + s, \tag{5.4.8}$$

$$\frac{d\eta}{d\tau} = 1 \Rightarrow \eta = \tau + \eta_0 = \tau, \tag{5.4.9}$$

$$\frac{dV}{d\tau} = p^{1-\frac{1}{\gamma}} + q = s^{1-\gamma} + \frac{\gamma}{1-\gamma} s^{1-\gamma} = \frac{1}{1-\gamma} s^{1-\gamma},$$

$$V = \frac{1}{1-\gamma} s^{1-\gamma}\tau + V_0 = \frac{1}{1-\gamma} s^{1-\gamma}(\tau + 1).$$

Step 3: From (5.4.8) and (5.4.9), we have

$$\tau = \eta, \quad s = \frac{x}{1+\eta}.$$

Therefore,

$$V(x,\eta) = \frac{1}{1-\gamma} x^{1-\gamma}(1+\eta)^{\gamma} = \frac{1}{1-\gamma}[x(1+\eta)^{\frac{\gamma}{1-\gamma}}]^{1-\gamma}.$$

Changing back to $t = T - \eta$, we have

$$V(x,t) = \frac{1}{1-\gamma}[x(1+T-t)^{\frac{\gamma}{1-\gamma}}]^{1-\gamma}.$$

What this says is that if he needs the utility $x^{1-\gamma}/(1-\gamma)$ at $t = T$ from his wealth x, his wealth at $t = 0$ should be $x(1+T)^{\frac{\gamma}{1-\gamma}}$.

Exercises

1. Solve

$$u_t + (u_x)^2 = 0, \quad -\infty < x < \infty,\ 0 < t$$

with the following initial conditions.

(a) $u(x,0) = x$.
(b) $u(x,0) = -x$.

2. Find the solution to

$$u_t - (u_x)^3 = 0, \quad -\infty < x < \infty,\ 0 < t,$$

$$u(x,0) = 2x^{3/2}.$$

3. This is the HJB equation when $r > 0$. We solve it in a different way.

$$\frac{\partial V}{\partial \eta} - rx\frac{\partial V}{\partial x} - \frac{\gamma}{1-\gamma}\left(\frac{\partial V}{\partial x}\right)^{1-\frac{1}{\gamma}} = 0,$$

$$V(x,0) = x^{1-\gamma}.$$

(a) Assume that $V(x,\eta) = a(\eta)x^b$ and find the constant b and the ODE satisfied by $a(\eta)$.
(b) Solve the ODE for $a(\eta)$ to find the solution $V(x,\eta)$.

5.5 Systems of First Order Equations

5.5.1 2×2 *System*

To motivate this section, consider the initial value problem for the wave equation

$$u_{tt} - c^2 u_{xx} = 0, \quad -\infty < x < \infty,\ 0 < t,$$

$$u(x,0) = f(x), \quad u_t(x,0) = g(x).$$

Setting $v = u_x$ and $w = u_t$, we have

$$v_t = w_x,$$
$$w_t = c^2 v_x,$$

$$\begin{bmatrix} v \\ w \end{bmatrix}(x,0) = \begin{bmatrix} f'(x) \\ g(x) \end{bmatrix}.$$

Writing the equation in a matrix form, we see

$$\begin{bmatrix} v \\ w \end{bmatrix}_t + \begin{bmatrix} 0 & -1 \\ -c^2 & 0 \end{bmatrix} \begin{bmatrix} v \\ w \end{bmatrix}_x \doteq \mathbf{u}_t + A\mathbf{u}_x = 0. \tag{5.5.1}$$

We denote a system with 2 unknowns by 2×2 system. One thing we can try is to diagonalize A. For this purpose, let's compute the eigenvalues and eigenvectors of A. The eigenvalues are

$$\begin{vmatrix} -\lambda & -1 \\ -c^2 & -\lambda \end{vmatrix} = \lambda^2 - c^2 = 0 \Rightarrow \lambda = \pm c. \tag{5.5.2}$$

So, the eigenvector for $\lambda = c$ is

$$\begin{bmatrix} -c & -1 \\ -c^2 & -c \end{bmatrix} \begin{bmatrix} v \\ w \end{bmatrix} = 0 \Rightarrow \begin{bmatrix} v \\ w \end{bmatrix} = \begin{bmatrix} 1 \\ -c \end{bmatrix} \tag{5.5.3}$$

and the eigenvector for $\lambda = -c$ is

$$\begin{bmatrix} c & -1 \\ -c^2 & c \end{bmatrix} \begin{bmatrix} v \\ w \end{bmatrix} = 0 \Rightarrow \begin{bmatrix} v \\ w \end{bmatrix} = \begin{bmatrix} 1 \\ c \end{bmatrix}. \tag{5.5.4}$$

We use the change of variable given by

$$\begin{bmatrix} v \\ w \end{bmatrix} = \begin{bmatrix} 1 & 1 \\ -c & c \end{bmatrix} \begin{bmatrix} y \\ z \end{bmatrix} \tag{5.5.5}$$

and substitute (5.5.5) in (5.5.1). After multiplying the resulting system by the inverse matrix

$$\begin{bmatrix} 1 & 1 \\ -c & c \end{bmatrix}^{-1} = \frac{1}{2c} \begin{bmatrix} c & -1 \\ c & 1 \end{bmatrix},$$

we see

$$\begin{bmatrix} y \\ z \end{bmatrix}_t + \begin{bmatrix} 1 & 1 \\ -c & c \end{bmatrix}^{-1} \begin{bmatrix} 0 & -1 \\ -c^2 & 0 \end{bmatrix} \begin{bmatrix} 1 & 1 \\ -c & c \end{bmatrix} \begin{bmatrix} y \\ z \end{bmatrix}_x = 0.$$

This yields

$$\begin{bmatrix} y \\ z \end{bmatrix}_t + \begin{bmatrix} c & 0 \\ 0 & -c \end{bmatrix} \begin{bmatrix} y \\ z \end{bmatrix}_x = 0.$$

From (5.5.5) the initial conditions for $[y, z]^T$ are

$$\begin{bmatrix} y \\ z \end{bmatrix}(x, 0) = \begin{bmatrix} 1 & 1 \\ -c & c \end{bmatrix}^{-1} \begin{bmatrix} f'(x) \\ g(x) \end{bmatrix} = \frac{1}{2c} \begin{bmatrix} cf'(x) - g(x) \\ cf'(x) + g(x) \end{bmatrix}. \tag{5.5.6}$$

Since $[y, z]^T$ satisfies the first order constant coefficient linear equations, the result in Example 1.3.2 implies

$$\begin{bmatrix} y \\ z \end{bmatrix}(x, t) = \begin{bmatrix} \frac{1}{2c}[cf'(x - ct) - g(x - ct)] \\ \frac{1}{2c}[cf'(x + ct) + g(x + ct)] \end{bmatrix}. \tag{5.5.7}$$

So,

$$\begin{bmatrix} v \\ w \end{bmatrix} = \begin{bmatrix} 1 & 1 \\ -c & c \end{bmatrix} \begin{bmatrix} y \\ z \end{bmatrix}$$

$$= \begin{bmatrix} \frac{1}{2}[f'(x-ct) + f'(x+ct)] + \frac{1}{2c}[g(x+ct) - g(x-ct)] \\ \frac{c}{2}[f'(x+ct) - f'(x-ct)] + \frac{1}{2}[g(x+ct) + g(x-ct)] \end{bmatrix}. \quad (5.5.8)$$

Since $w = u_t$,

$$u(x,t) - u(x,0)$$
$$= \int_0^t \frac{c}{2}[f'(x+cs) - f'(x-cs)] + \frac{1}{2}[g(x+cs) + g(x-cs)]ds$$
$$= \frac{1}{2}[f(x+ct) + f(x-ct)] - f(x) + \frac{1}{2}\int_x^{x+ct} g(\tau)\frac{d\tau}{c} - \frac{1}{2}\int_x^{x-ct} g(\xi)\frac{d\tau}{\xi},$$

where $\tau = x + cs$ and $\xi = x - cs$. Arranging the terms, we have

$$u(x,t) = \frac{1}{2}[f(x+ct) + f(x-ct)] + \frac{1}{2c}\int_{x-ct}^{x+ct} g(s)ds.$$

Example 5.5.1. Find the solution to

$$\begin{bmatrix} v \\ w \end{bmatrix}_t + \begin{bmatrix} 0 & -1 \\ -c^2 & 0 \end{bmatrix} \begin{bmatrix} v \\ w \end{bmatrix}_x = 0, \quad -\infty < x < \infty,\ 0 < t,$$

$$\begin{bmatrix} v \\ w \end{bmatrix}(x,0) = \begin{bmatrix} \sin 2x \\ \frac{1}{1+x^2} \end{bmatrix}.$$

Solution: The eigenvalues and eigenvectors are already found. From the transform (5.5.5) we have

$$\begin{bmatrix} y \\ z \end{bmatrix}(x,0) = \frac{1}{2c}\begin{bmatrix} c & -1 \\ c & 1 \end{bmatrix} \begin{bmatrix} v \\ w \end{bmatrix}(x,0)$$
$$= \frac{1}{2c}\begin{bmatrix} c\sin 2x - \frac{1}{1+x^2} \\ c\sin 2x + \frac{1}{1+x^2} \end{bmatrix}.$$

Then, (5.5.7) and (5.5.8) imply

$$\begin{bmatrix} y \\ z \end{bmatrix}(x,t) = \frac{1}{2c}\begin{bmatrix} c\sin 2(x-ct) - \frac{1}{1+(x-ct)^2} \\ c\sin 2(x+ct) + \frac{1}{1+(x+ct)^2} \end{bmatrix}$$

and

$$\begin{bmatrix} v \\ w \end{bmatrix} = \begin{bmatrix} 1 & 1 \\ -c & c \end{bmatrix} \begin{bmatrix} y \\ z \end{bmatrix},$$
$$= \begin{bmatrix} \frac{1}{2}[\sin 2(x-ct) + \sin 2(x+ct)] + \frac{1}{2c}[\frac{1}{1+(x+ct)^2} - \frac{1}{1+(x-ct)^2}] \\ \frac{c}{2}[\sin 2(x+ct) - \sin 2(x-ct)] + \frac{1}{2}[\frac{1}{1+(x+ct)^2} + \frac{1}{1+(x-ct)^2}] \end{bmatrix}.$$

5.5.2 $n \times n$ *System*

As in 2×2 case we denote a system with n unknowns by $n \times n$ system. An initial value problem for a general $n \times n$ system of constant coefficient linear first order equations is given by

$$\mathbf{u}_t + A\mathbf{u}_x = 0,$$

$$\mathbf{u}(x,0) = \mathbf{f}(x) = \begin{bmatrix} f_1(x) \\ f_2(x) \\ \vdots \\ f_n(x) \end{bmatrix},$$

where $\mathbf{u} = [u_1, \ldots, u_n]^T(x,t)$ and A is an $n \times n$ constant matrix. We say that the system is strictly hyperbolic if the eigenvalues of A are real and distinct. Suppose that λ_1, λ_2, ..., and λ_n are the eigenvalues of A with

$$\lambda_1 < \lambda_2 < \cdots < \lambda_n$$

and the corresponding eigenvectors are $\mathbf{p}_1$, $\mathbf{p}_2$, ..., and $\mathbf{p}_n$, respectively. Then, setting

$$\mathbf{u} = P\mathbf{v}, \quad P = [\mathbf{p}_1, \mathbf{p}_2, \ldots, \mathbf{p}_n],$$

we see that since

$$AP = P\Lambda, \quad \Lambda = \begin{bmatrix} \lambda_1 & & & \\ & \lambda_2 & & \\ & & \ddots & \\ & & & \lambda_n \end{bmatrix},$$

$\mathbf{v}$ satisfies

$$\mathbf{v}_t + \Lambda\mathbf{v}_x = 0,$$

$$\mathbf{v}(x,0) = P^{-1}\mathbf{f}(x).$$

This is a diagonal system and the solution is given by

$$\mathbf{v}(x,t) = \begin{bmatrix} \tilde{\mathbf{p}}_1\mathbf{f}(x-\lambda_1 t) \\ \tilde{\mathbf{p}}_2\mathbf{f}(x-\lambda_2 t) \\ \vdots \\ \tilde{\mathbf{p}}_n\mathbf{f}(x-\lambda_n t) \end{bmatrix} = \begin{bmatrix} \sum_{j=1}^n \tilde{p}_{1j} f_j(x-\lambda_1 t) \\ \sum_{j=1}^n \tilde{p}_{2j} f_j(x-\lambda_2 t) \\ \vdots \\ \sum_{j=1}^n \tilde{p}_{nj} f_j(x-\lambda_n t) \end{bmatrix},$$

where $\tilde{\mathbf{p}}_i$ are the row vectors of P^{-1}. Then,

$$\mathbf{u}(x,t) = P\mathbf{v}(x,t).$$

Exercises

1. Find the solution to the initial value problem

$$\begin{bmatrix} u \\ v \end{bmatrix}_t - \begin{bmatrix} 1 & 1 \\ 4 & -2 \end{bmatrix} \begin{bmatrix} u \\ v \end{bmatrix}_x = \begin{bmatrix} 0 \\ 0 \end{bmatrix}, \quad -\infty < x < \infty,\ 0 < t,$$

$$[u, v]^T(x, 0) = [\cos x, \sin x], \quad -\infty < x < \infty.$$

2. Find the solution to the initial value problem

$$\begin{bmatrix} u \\ v \\ w \end{bmatrix}_t - \begin{bmatrix} 0 & 1 & 0 \\ 4 & 0 & 1 \\ 0 & 0 & 0 \end{bmatrix} \begin{bmatrix} u \\ v \\ w \end{bmatrix}_x = \begin{bmatrix} 0 \\ 0 \\ 0 \end{bmatrix}, \quad -\infty < x < \infty,\ 0 < t,$$

$$[u, v, w]^T(x, 0) = [\cos x, \sin x, 1], \quad -\infty < x < \infty.$$

Chapter 6

Fourier Series and Eigenvalue Problems

When we discussed the separation of variables, we encountered occasions where we expanded a given function in sine or cosine series. It was mentioned that the Fourier sine and cosine series are the special cases of the Fourier series. In this chapter we study the Fourier series in detail and discuss the relations among these series. In Section 6.2 we introduce the Fourier series on the interval $[-L, L]$ and then using this to derive the Fourier sine and cosine series for the interval $[0, L]$. We will see that we have the Fourier sine or cosine series depending on a function being odd or even. In Section 6.3 we discuss the convergence of Fourier series. Specifically, we study three types of convergence, the mean-square (or L^2), the pointwise, and the uniform convergence. The relation between the term by term differentiation of the Fourier series of f and the Fourier series of f' is discussed in Section 6.4. This gives justification for the eigenfunction expansions that we study in Chapter 7. In Section 6.5 we discuss, a related topic, the eigenvalue problems. In one-dimension it is called the Sturm-Liouville Problems.

6.1 Even, Odd, and Periodic Functions

6.1.1 *Even and Odd Functions*

Definition 6.1.1. If a given function $y = f(x)$ satisfies

$$f(-x) = f(x), \tag{6.1.1}$$

it is an even function. If f satisfies

$$f(-x) = -f(x), \tag{6.1.2}$$

it is an odd function.

Example 6.1.2. Check if the following functions are even, odd, or neither.

(a) $f(x) = \cos x$, (b) $f(x) = \sin x$, (c) $f(x) = e^{2x}$.

Solution: We need to see if (6.1.1) or (6.1.2) holds.
(a) $f(-x) = \cos(-x) = \cos x = f(x)$. So, even.
(b) $f(-x) = \sin(-x) = -\sin x = -f(x)$. So, odd.
(c) $f(-x) = e^{-2x}$. This is not equal to either $f(x)$ or $-f(x)$.
So, neither.

There are several important properties of even or odd functions.

1. An even function is symmetric about the y-axis and an odd function is symmetric about the origin.
2. Similar to numbers, the following relations hold.

$$\begin{aligned}(\text{even})\times(\text{even}) &= (\text{even})\\ (\text{odd})\times(\text{odd}) &= (\text{even})\\ (\text{even})\times(\text{odd}) &= (\text{odd})\end{aligned} \tag{6.1.3}$$

   That is, if we multiply an even function and an even function, we get an even function, *etc.*
3. If f is even,

$$\int_{-L}^{L} f(x)dx = 2\int_{0}^{L} f(x)dx.$$

4. If f is odd,

$$\int_{-L}^{L} f(x)dx = 0.$$

Example 6.1.3. Compute the following integrals.
(a) $\int_{-2}^{2}(1 - x + x^2 - x^{99})dx$. (b) $\int_{-L}^{L} \cos\frac{m\pi x}{L} \sin\frac{n\pi x}{L} dx$.
Solution: (a) 1 and x^2 are even and x and x^{99} are odd. Since the interval of integration is symmetric about $x = 0$, we see

$$\begin{aligned}\int_{-2}^{2}(1 - x + x^2 - x^{99})dx &= 2\int_{0}^{2}(1 + x^2)dx\\ &= 2[x + \frac{x^3}{3}]_0^2 = \frac{28}{3}.\end{aligned}$$

(b) We apply (even)×(odd) = (odd) for the integrand. Therefore, the integral is zero.

6.1.2 *Periodic Functions*

Periodic functions are another important class of functions which are useful in the Fourier series. A function is said to be periodic with period $T > 0$ if

$$f(x+T) = f(x) \tag{6.1.4}$$

for every value of x. From the above definition, we can easily see that if T is a period of f, then $2T$ is also a period and consequently any integer multiple of T is a period of f. The smallest value of T for which (6.1.4) holds is called the fundamental period or simply the period.

Example 6.1.4. Find the fundamental period of $f(x) = \sin(n\pi x/L)$.

Solution: We know that 2π is the fundamental period of $\sin x$, the fundamental period T satisfies

$$\frac{n\pi(x+T)}{L} = \frac{n\pi x}{L} + 2\pi.$$

Therefore, $T = 2L/n$.

Example 6.1.5. (a) Draw the periodic extension of $f(x) = x$ given on $(-1, 1]$ for three periods.
(b) Draw the even and odd periodic extension of $f(x) = 1 - x$ given on $(0, 1)$ to $(-3, 3)$ for three periods.

Solution: (a) The solution is given in Figure 6.1.

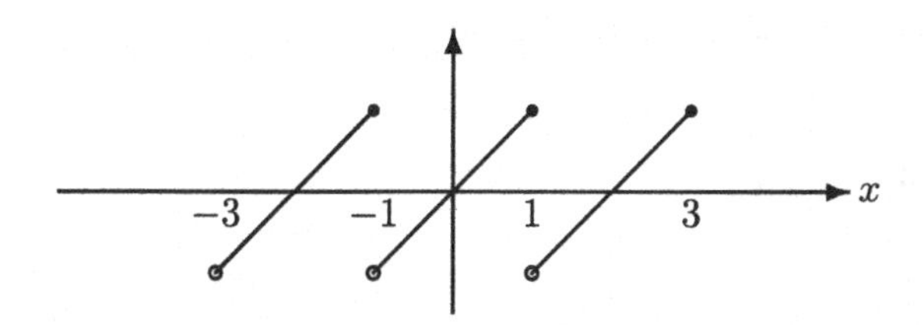

Fig. 6.1 Periodic extension of $f(x) = x$ on $(-1, 1]$.

(b) The solution is given in Figure 6.2.

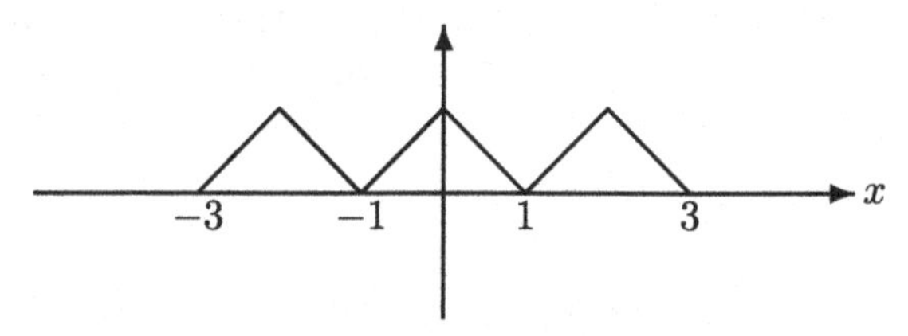

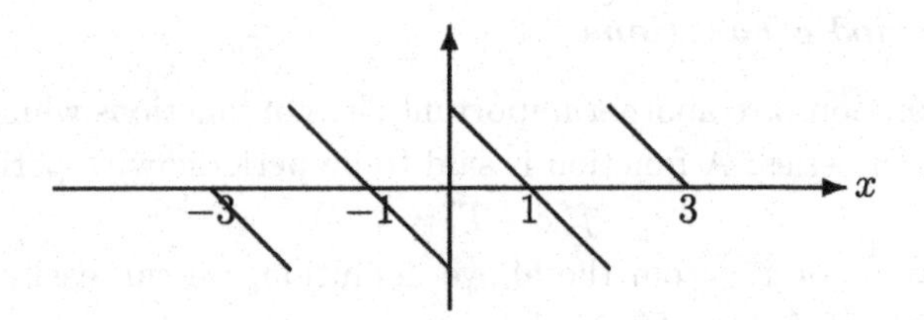

Fig. 6.2 Even and odd extensions of $f(x) = 1 - x$ on $(0, 1)$.

Exercises

1. State if f is even, odd, or neither.
   (a) $f(x) = \cos 2x \sin 3x$. (b) $f(x) = e^x \cos x$.
   (c) $f(x) = \cos x + \sin^2 3x$.
2. Compute the integrals.
   (a) $\int_{-3}^{3} \cos^2 2x \sin 3x dx$. (b) $\int_{-L}^{L} (x + x^3)^3 dx$.
   (c) $\int_{-1}^{1} [1 + x^{57}] dx$.
3. Determine if the following functions are periodic. If so, find the fundamental period.
   (a) $f(x) = \tan \frac{n\pi x}{L}$. (b) $f(x) = \tanh 5x$.
   (c) $f(x) = \sin x \cos x$.

6.2 Fourier Series

We study three types of Fourier series. They are the Fourier series, Fourier sine, and Fourier cosine series. We learn the relations between them and the relation with the even, odd, and periodic functions.

6.2.1 *Fourier Series*

To introduce the Fourier series, we usually restrict the interval to $[-L, L]$, where L is a positive constant. The Fourier series of $f(x)$ specified on $-L \leq x \leq L$ is given by

$$f(x) = \frac{a_0}{2} + \sum_{n=1}^{\infty} [a_n \cos \frac{n\pi x}{L} + b_n \sin \frac{n\pi x}{L}], \tag{6.2.1}$$

where

$$a_n = \frac{1}{L}\int_{-L}^{L} f(x)\cos\frac{n\pi x}{L}dx, \tag{6.2.2}$$

$$b_n = \frac{1}{L}\int_{-L}^{L} f(x)\sin\frac{n\pi x}{L}dx. \tag{6.2.3}$$

The coefficients a_n and b_n are called the Fourier coefficients and to obtain them, we use the following relations about the trigonometric functions.

$$\int_{-L}^{L}\cos\frac{m\pi x}{L}\cos\frac{n\pi x}{L}dx = \begin{cases} L & m = n \\ 0 & m \neq n \end{cases},$$

$$\int_{-L}^{L}\sin\frac{m\pi x}{L}\sin\frac{n\pi x}{L}dx = \begin{cases} L & m = n \\ 0 & m \neq n \end{cases},$$

$$\int_{-L}^{L}\cos\frac{m\pi x}{L}\sin\frac{n\pi x}{L}dx = 0.$$

As defined in Section 2.3 these relations are called the orthogonality relations and the sequence of functions

$$\{1, \cos\frac{\pi x}{L}, \sin\frac{\pi x}{L}, \ldots, \cos\frac{m\pi x}{L}, \sin\frac{m\pi x}{L}, \ldots\}$$

satisfying the orthogonality relations among each function is called mutually orthogonal. To find a_0 integrate (6.2.1) over $-L \leq x \leq L$. Then, using the orthogonality relations with $m = 0$

$$\begin{aligned}
&\int_{-L}^{L} f(x)dx \\
&= \int_{-L}^{L}\frac{a_0}{2}dx + \sum_{n=1}^{\infty}[a_n\int_{-L}^{L}\cos\frac{n\pi x}{L}dx + b_n\int_{-L}^{L}\sin\frac{n\pi x}{L}dx] \\
&= La_0.
\end{aligned}$$

Therefore,

$$a_0 = \frac{1}{L}\int_{-L}^{L} f(x)dx.$$

To find a_m we multiply (6.2.1) by $\cos\frac{m\pi x}{L}$ and integrate over $-L \leq x \leq L$. Then,

$$\begin{aligned}
&\int_{-L}^{L} f(x)\cos\frac{m\pi x}{L}dx \\
&= \int_{-L}^{L}\frac{a_0}{2}\cos\frac{m\pi x}{L}dx + \sum_{n=1}^{\infty}\int_{-L}^{L} a_n\cos\frac{m\pi x}{L}\cos\frac{n\pi x}{L}dx \\
&\quad + \sum_{n=1}^{\infty}\int_{-L}^{L} b_n\cos\frac{m\pi x}{L}\sin\frac{n\pi x}{L}dx.
\end{aligned}$$

Using the orthogonality relations, we see

$$a_m = \frac{1}{L}\int_{-L}^{L} f(x)\cos\frac{m\pi x}{L}dx.$$

Similarly, we obtain

$$b_m = \frac{1}{L}\int_{-L}^{L} f(x)\sin\frac{m\pi x}{L}dx.$$

Let's work out a couple of examples so that we see how to find the Fourier series of a given function. In the first example we treat the case where f is odd.

Example 6.2.1. Find the Fourier series of

$$f(x) = x, \quad -L \leq x \leq L. \tag{6.2.4}$$

Solution: To find a_0 we integrate (6.2.4) with $f(x) = x$ on $[-L, L]$. Then,

$$a_0 = \frac{1}{L}\int_{-L}^{L} xdx = 0.$$

To find a_m $(m = 1, 2, \ldots)$, multiply (6.2.4) by $\cos(m\pi x/L)$ and integrate. Then, we notice that the integrand is an odd function. Therefore,

$$a_m = \frac{1}{L}\int_{-L}^{L} x\cos\frac{m\pi x}{L}dx = 0.$$

To find b_m $(m = 1, 2, \ldots)$, multiply (6.2.4) by $\sin(m\pi x/L)$ and we use the fact that the integrand is even. Then,

$$\begin{aligned} b_m &= \frac{1}{L}\int_{-L}^{L} x\sin\frac{m\pi x}{L}dx = \frac{2}{L}\int_{0}^{L} x\sin\frac{m\pi x}{L}dx \\ &= \frac{2}{L}[x(-\frac{L}{m\pi})\cos\frac{m\pi x}{L}]_0^L + \frac{2}{L}\frac{L}{m\pi}\int_0^L \cos\frac{m\pi x}{L}dx \\ &= (-1)^{m+1}\frac{2L}{m\pi}. \end{aligned} \tag{6.2.5}$$

The solution is

$$f(x) = x = \sum_{n=1}^{\infty}(-1)^{n+1}\frac{2L}{n\pi}\sin\frac{n\pi x}{L}. \tag{6.2.6}$$

In the next example we treat the case where f is even.

Example 6.2.2. Find the Fourier series of

$$f(x) = |x|, \quad -L \le x \le L. \tag{6.2.7}$$

Solution: To find a_0 we integrate (6.2.7) with $f(x) = |x|$ on $[-L, L]$. Then, since $f(x) = |x|$ is even,

$$a_0 = \frac{1}{L}\int_{-L}^{L} |x| dx = \frac{2}{L}\int_0^L x dx = L.$$

To find a_m $(m = 1, 2, \ldots)$, multiply (6.2.7) by $\cos(m\pi x/L)$ and integrate. Since the integrand is an even function, we obtain

$$\begin{aligned} a_m &= \frac{1}{L}\int_{-L}^{L} |x| \cos\frac{m\pi x}{L} dx = \frac{2}{L}\int_0^L x \cos\frac{m\pi x}{L} dx \\ &= \frac{2}{L}[x\frac{L}{m\pi}\sin\frac{m\pi x}{L}]_0^L - \frac{2}{L}\int_0^L \frac{L}{m\pi}\sin\frac{m\pi x}{L} dx \\ &= \frac{2}{L}[(\frac{L}{m\pi})^2\cos\frac{m\pi x}{L}]_0^L = \frac{2L}{(m\pi)^2}((-1)^m - 1). \end{aligned} \tag{6.2.8}$$

To find b_m $(m = 1, 2, \ldots)$, multiply (6.2.7) by $\sin(m\pi x/L)$ and use the fact that the integrand is now odd. Then,

$$b_m = \frac{1}{L}\int_{-L}^{L} |x| \sin\frac{m\pi x}{L} dx = 0.$$

Therefore, the solution is

$$f(x) = |x| = \frac{L}{2} + \sum_{n=1}^{\infty} \frac{2L}{(n\pi)^2}((-1)^n - 1)\cos\frac{n\pi x}{L}. \tag{6.2.9}$$

If f is neither odd nor even, we see both a_m and b_m.

6.2.2 *Fourier Sine and Cosine Series*

These series apply to functions specified on $[0, L]$. We can apply either series to a function f given on $[0, L]$. Roughly speaking, the difference is that if we use the Fourier sine series, the series converges to the odd periodic extension of f, and if we use the Fourier cosine series, the series converges to the even periodic extension of f.

Fourier Sine Series

To derive the Fourier sine series, we compute the Fourier coefficients of the odd extension of f. Then, since $f(x)\cos(n\pi x/L)$ is an odd function and

$f(x)\sin(n\pi x/L)$ is an even function, from the properties of odd and even functions, we see

$$a_n = \frac{1}{L}\int_{-L}^{L} f(x)\cos\frac{n\pi x}{L}dx = 0,$$

$$b_n = \frac{1}{L}\int_{-L}^{L} f(x)\sin\frac{n\pi x}{L}dx = \frac{2}{L}\int_{0}^{L} f(x)\sin\frac{n\pi x}{L}dx. \tag{6.2.10}$$

Therefore, the Fourier series is simplified to

$$f(x) = \sum_{n=1}^{\infty} b_n \sin\frac{n\pi x}{L}, \tag{6.2.11}$$

where b_n is computed from (6.2.10).

Example 6.2.3. Find the Fourier sine series of $f(x) = x$ given on $0 \leq x < L$.

Solution: This is the same as (6.2.6).

Fourier Cosine Series

We compute the Fourier coefficients of the even extension of f. Then, since $f(x)\cos(n\pi x/L)$ is an even function and $f(x)\sin(n\pi x/L)$ is an odd function, from the properties of even and odd functions, we see

$$a_n = \frac{1}{L}\int_{-L}^{L} f(x)\cos\frac{n\pi x}{L}dx = \frac{2}{L}\int_{0}^{L} f(x)\cos\frac{n\pi x}{L}dx, \tag{6.2.12}$$

$$b_n = \frac{1}{L}\int_{-L}^{L} f(x)\sin\frac{n\pi x}{L}dx = 0.$$

Therefore, the Fourier series is simplified to

$$f(x) = \frac{a_0}{2} + \sum_{n=1}^{\infty} a_n \cos\frac{n\pi x}{L}, \tag{6.2.13}$$

where a_n is computed from (6.2.12). The above series is called the Fourier cosine series.

Example 6.2.4. Find the Fourier cosine series of $f(x) = x$ given on $0 \leq x < L$.

Solution: This is the same as (6.2.9).

Remarks about the Fourier Sine and Cosine Series

After studying the Fourier sine and cosine series one may wonder which series we use if f is given on $[0, L]$. Either one is fine. If we use the Fourier sine series, the series converges to the odd periodic extension of f, and if we use the Fourier cosine series, the series converges to the even periodic extension of f. More details of convergence is discussed in the next section.

In the case of initial boundary value problems, the boundary conditions dictate which Fourier series we use. In the Dirichlet problems we use the Fourier sine series and in the Neumann problems we use the Fourier cosine series. However, there are cases where the boundary conditions are neither the Dirichlet nor Neumann. As we saw in Chapters 2, 3, and 4, if one end is fixed and the other end is insulated or free, the series we have is different from the Fourier sine and cosine series. However, they are also eigenfunction expansions and they must share the similar properties. In Section 6.5, we study the properties of eigenfunctions in more details.

Exercises

1. Find the Fourier series of the following functions.

(a) $f(x) = \begin{cases} x+1 & -1 < x < 0, \\ 1-x & 0 < x < 1. \end{cases}$ (b) $f(x) = -x, \quad -L < x < L.$

(c) $f(x) = \begin{cases} -x-2 & -2 < x < 0, \\ 2-x & 0 < x < 2. \end{cases}$ (d) $f(x) = \begin{cases} 1 & -\pi < x < 0, \\ 0 & 0 < x < \pi. \end{cases}$

2. Find the Fourier cosine and sin series of the following functions.

(a) $f(x) = 1 - x, \quad 0 \le x \le 1.$ (b) $f(x) = \begin{cases} 1 & 0 < x < \pi, \\ 0 & \pi < x < 2\pi. \end{cases}$

6.3 Fourier Convergence Theorems

In this section we discuss the convergence of Fourier series. We discuss three types of convergence related to the Fourier series. They are the mean square (or L^2), pointwise, and uniform convergences. We study the conditions for convergence and in the pointwise convergence we also study to which function the Fourier series converges.

6.3.1 *Mean-square Convergence*

We start from the mean-square convergence or the convergence in the mean of order two. This convergence is also called the L^2-convergence. We use the integral of the square of the difference. Therefore, the difference could be large at particular points such as points of jump discontinuity.

Definition 6.3.1. Suppose $f(x)$ and $g(x)$ are defined on (a, b). Then, the inner product of f and g in (a, b) is defined as

$$(f, g) = \int_a^b f(x)g(x)dx.$$

Furthermore, if $(f, g) = 0$, we say that f and g are orthogonal.

Definition 6.3.2. We say that the series $\sum_{n=1}^{N} a_n\phi_n(x)$ converges in the mean-square sense to $f(x)$ on (a, b) if

$$\int_a^b \left| f(x) - \sum_{n=1}^{N} a_n\phi_n(x) \right|^2 dx \to 0 \quad \text{as } N \to \infty.$$

As discussed in Subsection 2.7.2 the set of functions f satisfying $\int_a^b |f(x)|^2 dx < \infty$ is called L^2-functions. We use the notation $\|f\| = \{\int_a^b |f(x)|^2 dx\}^{1/2}$ and use $L^2(a, b)$ to denote the set of L^2-functions on the interval (a, b). We examine the finite-term approximation of the Fourier series.

Theorem 6.3.3. *Let $\{\phi_n\}$ be a mutually orthogonal set of functions, $\|f\| < \infty$, and N be a fixed positive integer. The coefficients $\{a_n\}$ which minimize*

$$\int_a^b \left| f(x) - \sum_{n=1}^{N} a_n\phi_n(x) \right|^2 dx$$

are given by

$$a_n = \frac{(f, \phi_n)}{(\phi_n, \phi_n)} = \frac{\int_a^b f(x)\phi_n(x)dx}{\int_a^b \phi_n(x)\phi_n(x)dx}. \tag{6.3.1}$$

Remark 6.3.4. In the context of Chapters 2 to 4 $\{\phi_n\}$ and a_n represent the eigenfunctions and the Fourier coefficients, respectively. The form (6.3.1) is typical for the Fourier series and eigenfunction expansions.

Proof. Set the error of the approximation by

$$E_N = \int_a^b \left| f(x) - \sum_{n=1}^{N} a_n \phi_n(x) \right|^2 dx.$$

Expanding E_N, we obtain

$$\begin{aligned} E_N = &\int_a^b |f(x)|^2 dx - 2\sum_{n=1}^{N} a_n \int_a^b f(x)\phi_n(x)dx \\ &+ \sum_{n=1}^{N}\sum_{m=1}^{N} a_n a_m \int_a^b \phi_n(x)\phi_m(x)dx. \end{aligned}$$

Since $\{\phi_n\}$ is mutually orthogonal,

$$\begin{aligned} E_N &= \int_a^b |f(x)|^2 dx - 2\sum_{n=1}^{N} a_n \int_a^b f(x)\phi_n(x)dx + \sum_{n=1}^{N} a_n^2 \int_a^b \phi_n^2(x)dx \\ &= \|f\|^2 - 2\sum_{n=1}^{N} a_n (f, \phi_n) + \sum_{n=1}^{N} a_n^2 \|\phi_n\|^2 \\ &= \sum_{n=1}^{N} \|\phi_n\|^2 \left[a_n - \frac{(f, \phi_n)}{\|\phi_n\|^2} \right]^2 + \|f\|^2 - \sum_{n=1}^{N} \frac{(f, \phi_n)^2}{\|\phi_n\|^2}. \end{aligned} \tag{6.3.2}$$

We can minimize E_N if we choose a_n as in (6.3.1). □

If we choose a_n in (6.3.2) as (6.3.1), we see

$$0 \le E_N = \|f\|^2 - \sum_{n=1}^{N} a_n^2 \|\phi_n\|^2. \tag{6.3.3}$$

Since E_N is nonnegative, we obtain an inequality called Bessel's inequality

$$\sum_{n=1}^{N} a_n^2 \|\phi_n\|^2 \le \|f\|^2.$$

From (6.3.3) we have the Parseval's equality.

Theorem 6.3.5. *The Fourier series $\sum_{n=1}^{\infty} a_n \phi_n$ of $f(x)$ converges to $f(x)$ in the mean-square sense if and only if the following Parseval's equality holds.*

$$\sum_{n=1}^{\infty} a_n^2 \|\phi_n\|^2 = \|f\|^2.$$

Proof. The mean-square convergence means $E_N \to 0$ as $N \to \infty$. This implies that in (6.3.3)

$$\lim_{N\to\infty} \sum_{n=1}^{N} a_n^2 \|\phi_n\|^2 = \|f\|^2 .$$

□

The following is a useful consequence of the Bessel's inequality. This is called the Riemann-Lebesgue Lemma and will be used to show the pointwise convergence of the Fourier series.

Lemma 6.3.6. *Let $f \in L^2(a,b)$. Then, the Fourier coefficient a_n approaches zero as n approaches infinity.*

Remark 6.3.7. Consider the case where $(a,b) = (-L, L)$. What happens is that as n approaches infinity $\sin(n\pi x/L)$ or $\cos(n\pi x/L)$ oscillates so rapidly that the integrands in the Fourier coefficients such as

$$\frac{1}{L}\int_{-L}^{L} f(x)\cos\frac{m\pi x}{L}dx \text{ or } \frac{2}{L}\int_{0}^{L} f(x)\sin\frac{n\pi x}{L}dx$$

oscillate around zero so many times and cancel out to be zero.

Theorem 6.3.8. *The Fourier series converges in the mean-square sense to $f(x)$ in (a,b) provided that*

$$\int_a^b |f(x)|^2\, dx < \infty.$$

The proof requires the Lebesgue integral theory and unfortunately it is beyond the scope of the book. Instead we use the Parseval's equality to compute the sum of infinite series.

Example 6.3.9. (a) Find the Fourier series of $f(x) = x$ on $[-1,1]$

(b) Apply the Parseval's equality to show that

$$\sum_{n=1}^{\infty} \frac{1}{n^2} = \frac{\pi^2}{6}. \tag{6.3.4}$$

Solution: (a) If

$$x = \frac{a_0}{2} + \sum_{n=1}^{\infty}[a_n \cos n\pi x + b_n \sin n\pi x],$$

then

$$a_n = \int_{-1}^{1} x\cos n\pi x dx = 0, \ n = 0, 1, \ldots,$$

$$b_n = \int_{-1}^{1} x \sin n\pi x dx = 2[-\frac{1}{n\pi} x \cos n\pi x]_0^1 + 2\int_0^1 \frac{1}{n\pi} \cos n\pi x dx$$
$$= \frac{2}{n\pi}(-1)^{n+1},\ n = 1, 2, \ldots.$$

(b) The Parseval's equality implies

$$\sum b_n^2 \int_{-1}^{1} \sin^2 n\pi x dx = \int_{-1}^{1} x^2 dx.$$

The left hand side is

$$\sum b_n^2 \int_{-1}^{1} \sin^2 n\pi x dx = \sum (\frac{2}{n\pi})^2 \int_{-1}^{1} \sin^2 n\pi x dx = \sum (\frac{2}{n\pi})^2. \tag{6.3.5}$$

The right hand side is

$$\int_{-1}^{1} x^2 dx = 2\int_0^1 x^2 dx = \frac{2}{3}. \tag{6.3.6}$$

Therefore, setting (6.3.5) and (6.3.6) equal, we obtain (6.3.4).

6.3.2 *Pointwise Convergence*

To understand what we do let's take a partial sum of the Fourier series

$$f_N(x) = \frac{a_0}{2} + \sum_{n=1}^{N} [a_n \cos \frac{n\pi x}{L} + b_n \sin \frac{n\pi x}{L}]$$

and fix an x, say x_0. Then, we have

$$f_N(x_0) = \frac{a_0}{2} + \sum_{n=1}^{N} [a_n \cos \frac{n\pi x_0}{L} + b_n \sin \frac{n\pi x_0}{L}].$$

This is a partial sum of numbers. What we want to look at is what happens to the number $f_N(x_0)$ as N goes to infinity. We call this type of convergence pointwise convergence. The definition is given as follows.

Definition 6.3.10. An infinite series $\sum \phi_n(x)$ converges to $f(x)$ pointwise in (a, b) if it converges to $f(x)$ for each $x \in (a, b)$. In other words for each $x \in (a, b)$ we have

$$\left| f(x) - \sum_{n=1}^{N} \phi_n(x) \right| \to 0 \quad \text{as } N \to \infty.$$

The more precise definition is given as follows. For every $\varepsilon > 0$ and for every $x \in (a, b)$ there exists N_0 such that for every $N \geq N_0$

$$\left| f(x) - \sum_{n=1}^{N} \phi_n(x) \right| < \varepsilon.$$

Note that N_0 depends on x and ε. We need one more definition before we proceed to Theorem 6.3.13 where the pointwise convergence is stated.

Definition 6.3.11. A function f is said to be piecewise continuous on an interval $a \leq x \leq b$ if the interval can be partitioned to sub-intervals $a = x_0 < x_1 < \cdots < x_n = b$ so that
1. f is continuous on each open sub-interval $x_{i-1} < x < x_i$.
2. Both the left limit $\lim_{x\to x_{i-1}} f(x)$ and the right $\lim_{x\to x_i} f(x)$ exist. This means that the discontinuity at x_i is a finite jump discontinuity.

Example 6.3.12. Are the following functions piecewise continuous?

(a) $f(x) = \begin{cases} -x-2 & -2 < x < 0 \\ 2-x & 0 < x < 2 \end{cases}$ on $(-2, 2)$

(b) $f(x) = \begin{cases} 1/x & -1 < x < 0,\ 0 < x < 1 \\ 0 & x = 0 \end{cases}$ on $(-1, 1)$

Solution: (a) Yes. Both left and right limit exist at $x = 0$. (b) No. We have infinite discontinuity at $x = 0$.

Theorem 6.3.13. *Suppose that f and f' are piecewise continuous on the interval $-L \leq x \leq L$. Also, suppose that f is defined outside the interval $-L \leq x \leq L$ so that it is periodic with period $2L$. Then, f has a Fourier series*

$$f(x) = \frac{a_0}{2} + \sum_{n=1}^{\infty}[a_n \cos\frac{n\pi x}{L} + b_n \sin\frac{n\pi x}{L}],$$

whose coefficients are given by (6.2.2) and (6.2.3). The Fourier series converges to $f(x)$ at each point x where f is continuous and to $[f(x_+) + f(x_-)]/2$ at points x where f is discontinuous.

Proof. We denote the partial sum of the Fourier series by

$$f_N(x) = \frac{a_0}{2} + \sum_{n=1}^{N}[a_n \cos\frac{n\pi x}{L} + b_n \sin\frac{n\pi x}{L}],$$

where

$$a_n = \frac{1}{L}\int_{-L}^{L} f(y) \cos\frac{n\pi y}{L} dy,$$

$$b_n = \frac{1}{L}\int_{-L}^{L} f(y) \sin\frac{n\pi y}{L} dy.$$

Substituting a_n and b_n in f_N, we have

$$f_N(x) = \frac{1}{2L}\int_{-L}^{L}\{1+2\sum_{n=1}^{N}[\cos\frac{n\pi y}{L}\cos\frac{n\pi x}{L} + \sin\frac{n\pi y}{L}\sin\frac{n\pi x}{L}]\}f(y)dy$$

$$= \frac{1}{2L}\int_{-L}^{L}\{1+2\sum_{n=1}^{N}\cos\frac{n\pi(y-x)}{L}\}f(y)dy.$$

The identity

$$\frac{1}{2L}\int_{-L}^{L}\{1+2\sum_{n=1}^{N}\cos\frac{n\pi(y-x)}{L}\}dy = 1 \tag{6.3.7}$$

implies that the difference $f_N(x) - f(x)$ can be expressed as

$$f_N(x) - f(x) = \frac{1}{2L}\int_{-L}^{L}\{1+2\sum_{n=1}^{N}\cos\frac{n\pi(y-x)}{L}\}(f(y)-f(x))dy.$$

To sum the series in the integrand, we use the formula $\cos n\theta = \frac{1}{2}(e^{in\theta} + e^{-in\theta})$ with $\theta = \frac{\pi(y-x)}{L}$. Then,

$$1+2\sum_{n=1}^{N}\cos n\theta = \sum_{n=-N}^{N} e^{in\theta} = \frac{e^{-iN\theta} - e^{i(N+1)\theta}}{1-e^{i\theta}}$$

$$= \frac{e^{i(N+\frac{1}{2})\theta} - e^{-i(N+\frac{1}{2})\theta}}{e^{\frac{1}{2}i\theta} - e^{-\frac{1}{2}i\theta}} = \frac{\sin(N+\frac{1}{2})\theta}{\sin\frac{1}{2}\theta} = D_N(\theta).$$

The above sum is called the Dirichlet kernel and denoted as $D_N(\theta)$. The graphs of $D_N(\theta)$ for several values of N are sketched in Figure 6.3. Since $D_N(\theta)$ is periodic with period 2π, by (6.3.7)

$$\frac{1}{2\pi}\int_{-\pi-\frac{\pi x}{L}}^{\pi-\frac{\pi x}{L}} D_N(\theta)d\theta = \frac{1}{2\pi}\int_{-\pi}^{\pi} D_N(\theta)d\theta = 1. \tag{6.3.8}$$

Since $D_N(\theta)f(x+L\theta/\pi)$ is also periodic with period 2π and $D_N(\theta)$ is an even function in θ,

$$f_N(x) = \frac{1}{2\pi}\int_{-\pi-\frac{\pi x}{L}}^{\pi-\frac{\pi x}{L}} D_N(\theta)f(x+\frac{L\theta}{\pi})d\theta = \frac{1}{2\pi}\int_{-\pi}^{\pi} D_N(\theta)f(x+\frac{L\theta}{\pi})d\theta$$

$$= \frac{1}{2\pi}\int_{0}^{\pi} D_N(\theta)f(x+\frac{L\theta}{\pi})d\theta + \frac{1}{2\pi}\int_{0}^{\pi} D_N(\theta)f(x-\frac{L\theta}{\pi})d\theta. \tag{6.3.9}$$

We show that

$$\lim_{N\to\infty}\frac{1}{2\pi}\int_{0}^{\pi} D_N(\theta)f(x\pm\frac{L\theta}{\pi})d\theta = \frac{1}{2}f(x_\pm). \tag{6.3.10}$$

We consider + sign. The case of − sign is treated similarly. Since (6.3.8) holds, this is equivalent to showing that

$$\begin{aligned}&\lim_{N\to\infty}\frac{1}{2\pi}\int_0^{\pi}[f(x+\frac{L\theta}{\pi})-f(x_+)]D_N(\theta)d\theta\\&=\lim_{N\to\infty}\frac{1}{2\pi}\int_0^{\pi}\frac{[f(x+\frac{L\theta}{\pi})-f(x_+)]}{\frac{L\theta}{\pi}}\frac{\frac{L\theta}{\pi}}{\sin\frac{1}{2}\theta}\sin(N+\frac{1}{2})\theta d\theta\\&=0.\end{aligned}\tag{6.3.11}$$

As f' is piecewise continuous, by the l'Hôpital's rule we have

$$\lim_{\theta\to 0}\frac{[f(x+\frac{L\theta}{\pi})-f(x_+)]}{\frac{L\theta}{\pi}}=f'(x_+).$$

Therefore,

$$\frac{[f(x+\frac{L\theta}{\pi})-f(x_+)]}{\frac{L\theta}{\pi}}\frac{\frac{L\theta}{\pi}}{\sin\frac{1}{2}\theta}$$

is piecewise continuous. The Riemann-Lebesgue Lemma implies that (6.3.11) holds. □

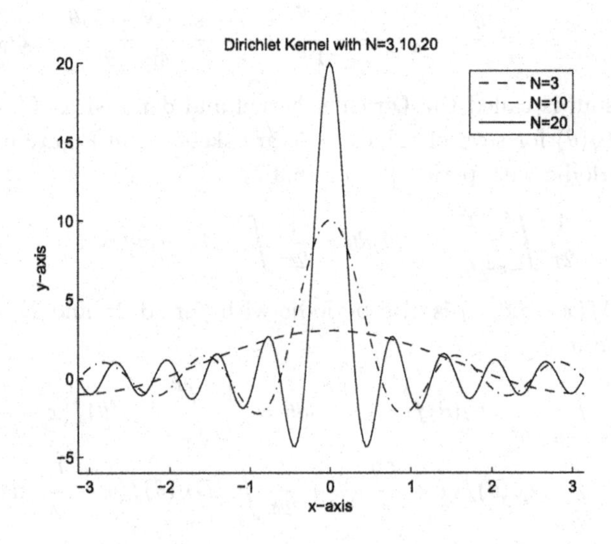

Fig. 6.3 Dirichlet kernel with $N = 3,\ 10,\ 20$.

Remark 6.3.14. The above theorem uses the pointwise convergence. The problem with the pointwise convergence is that how the series converges depends on the point we choose. As we see in Figure 2.4 the convergence looks good away from $x = 1$ but near $x = 1$, where there is a discontinuity, the approximation is not good. Suppose x_1 and x_2 are two different points where $f(x)$ is continuous. In the pointwise convergence, it could happen that at some N_1 the partial sum $f_{N_1}(x_1)$ of $f(x_1)$ is much better than that of $f(x_2)$ and at another N_2 the approximation of $f(x_2)$ is much better than that of $f(x_1)$ even though the partial sums $f_N(x_1)$ and $f_N(x_2)$ eventually converge in the pointwise sense to $f(x_1)$ and $f(x_2)$, respectively. The reason why this could happen is that N_0 in Definition 6.3.10 depends on x and ε.

In the following examples we practice what the theorem says.

Example 6.3.15. In Example 6.2.1, the Fourier series of

$$f(x) = x, \quad -L \leq x \leq L$$

is obtained. Find the graph of the function to which the Fourier series of f converges on the interval $[-3L, 3L]$.

Solution: First, we extend f as a periodic function to the interval $[-3L, 3L]$. After this is done, the graph of f extended to the interval $[-3L, 3L]$ looks like Figure 6.4. Then, we use the last sentence of the Theorem 6.3.13, *i.e.*, the Fourier series converges to $f(x)$ at each point where f is continuous and to $[f(x_+) + f(x_-)]/2$ at points where f is discontinuous. We do nothing where the graph in Figure 6.4 is connected and put a solid dot in the middle where the graph has a jump. In Figure 6.4 there are jump discontinuities at $x = -3L, -L, L, 3L$. So, we put solid dots at the average height $[f(x_+) + f(x_-)]/2$ at these x. The function to which the Fourier series of f converges is given in Figure 6.5.

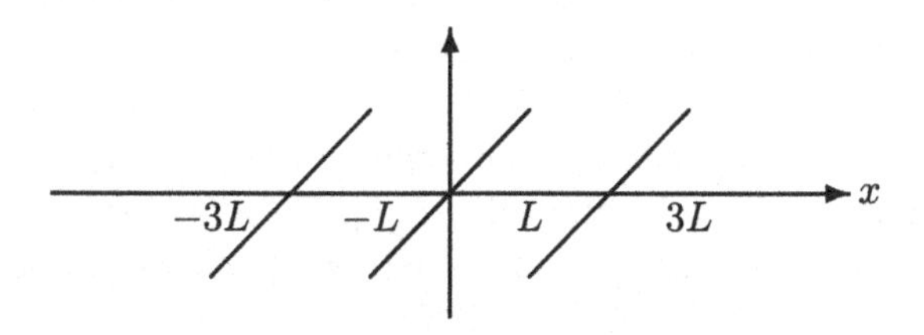

Fig. 6.4 The periodic extension of f.

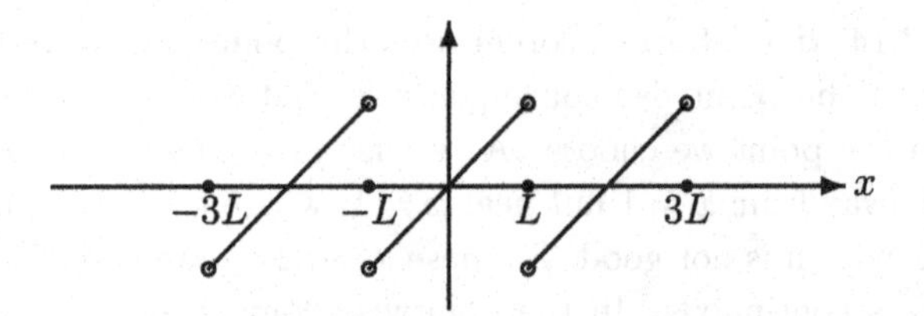

Fig. 6.5 The function to which the Fourier series of f converges.

Next we consider the Fourier sine and cosine series.

Example 6.3.16. In Examples 6.2.3 and 6.2.4, the Fourier sine and cosine series of

$$f(x) = x, \quad -L \leq x \leq L$$

are obtained. Find the graph of the function to which the Fourier sine and cosine series of f converge on the interval $[-3L, 3L]$.

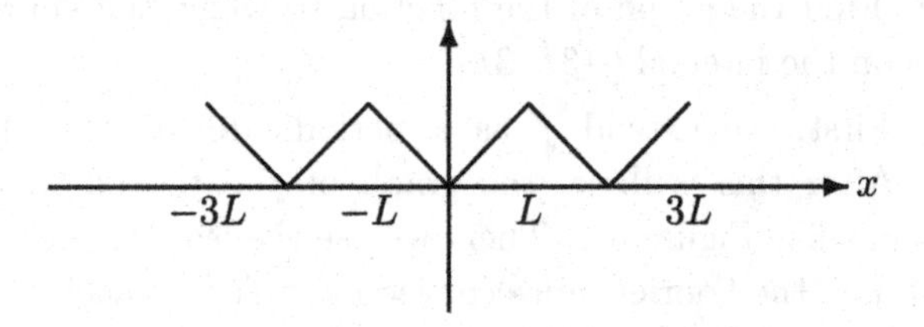

Fig. 6.6 The even periodic extension of f.

Solution: We consider the Fourier sine series. First, we extend f as an odd function to the interval $[-L, L]$. Then, we extend the resulting function as a periodic function to the interval $[-3L, 3L]$. The graph of f extended to the interval $[-3L, 3L]$ is the same as Figure 6.4. Therefore, the function to which the Fourier sine series of f converges is the same as given in Figure 6.5.

Next, consider the Fourier cosine series. For the cosine series we extend f as an even function to the interval $[-L, L]$. Then, we extend the resulting function as a periodic function to the interval $[-3L, 3L]$. The graph of f extended to the interval $[-3L, 3L]$ is Figure 6.6. Since there is no jump discontinuity in the graph, the modification based the last sentence of the

Theorem 6.3.13 is not necessary. Therefore, the final graph is the same as given in Figure 6.6. The graphs of partial sums of Fourier sine and cosine series of $f(x) = x$ with $L = 1$ are shown in Figure 6.7 for the sake of comparison.

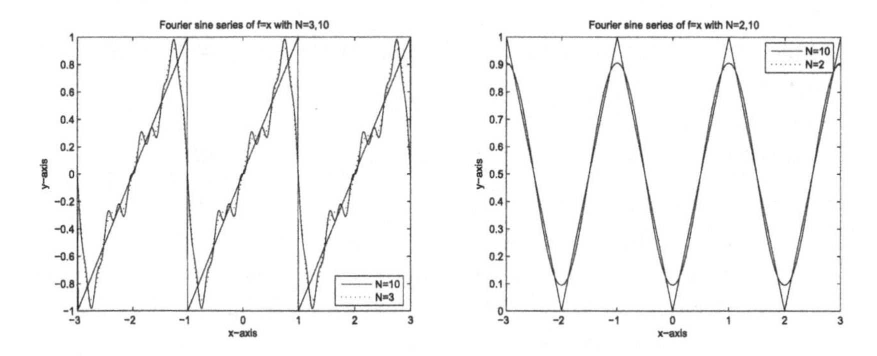

Fig. 6.7 Fourier sine (Left) and cosine (Right) series.

6.3.3 *Uniform Convergence*

In the pointwise convergence, it may happen that for a given N the difference $|f(x) - \sum_{n=1}^{N} \phi_n(x)|$ is less than ε at one point and greater than ε at another point. This may be inconvenient in approximation. If we prefer to have an N for which the difference is less than ε for every x in the domain, we need another type of convergence called the uniform convergence. In this subsection we introduce this and discuss a sufficient condition for the uniform convergence.

We say that the series $\sum \phi_n(x)$ converges uniformly to $f(x)$ in $[a, b]$ if

$$\max_{a \le x \le b} \left| f(x) - \sum_{n=1}^{N} \phi_n(x) \right| \to 0 \quad \text{as } N \to \infty.$$

The more precise definition is given as follows. For every $\varepsilon > 0$ there exists N_0 such that for every $N \ge N_0$ and for every $x \in [a, b]$

$$\left| f(x) - \sum_{n=1}^{N} \phi_n(x) \right| < \varepsilon.$$

Comparing the definitions of pointwise and uniform convergence, we notice that the order of N_0 and x are reversed. From the definitions of uniform and pointwise convergence, it is clear that the uniform convergence implies the pointwise convergence and for the bounded domain the mean square convergence, but not the other way. The following example illustrates the difference of these convergence types.

Example 6.3.17. Consider $\phi_n(x) = nx/[1+n^2x^2]-(n-1)x/[1+(n-1)^2x^2]$ and set the partial sum to be $S_N(x) = \sum_{n=1}^{N} \phi_n(x)$.
(a) Find the function $f(x)$ to which the partial sum $S_N(x) = Nx/(1+N^2x^2)$ converges pointwise on $[0, 1]$.
(b) Does $S_N(x)$ converge to $f(x)$ uniformly?
(c) Does $S_N(x)$ converge to $f(x)$ in the mean square sense?

Solution: (a) Choose a point x in $[0, 1]$. Then, it is clear that $f(x) = \lim_{N\to\infty} S_N(x) = 0$ for each x in $[0, 1]$.
(b) To apply the definition of the uniform convergence, we compute the maximum of $S_N(x)$ in $[0, 1]$. Then, from $S'_N(x) = 0$, the maximum is attained at $x = 1/N$ and the maximum is $S_N(1/N) = 1/2$. Therefore,

$$\max_{0\le x\le 1}\left|f(x) - \sum_{n=1}^{N}\phi_n(x)\right| = \frac{1}{2} \nrightarrow 0 \quad \text{as } N\to\infty.$$

So, $S_N(x)$ does not converge to $f(x) = 0$ uniformly.
(c) We compute the following to see if the integral approaches zero as $N \to \infty$.

$$\int_0^1 [f(x) - S_N(x)]^2 dx = \int_0^1 \frac{N^2x^2}{(1+N^2x^2)^2}dx.$$

Setting $u = Nx$, we have

$$\int_0^1 \frac{N^2x^2}{(1+N^2x^2)^2}dx = \int_0^N \frac{u^2}{(1+u^2)^2}\frac{du}{N} = \frac{1}{N}[-\frac{1}{2(1+u^2)}u]_0^N$$
$$+\frac{1}{2N}\int_0^N \frac{du}{(1+u^2)} = -\frac{1}{2(1+N^2)} + \frac{\pi}{4N}.$$

This approaches zero as $N \to \infty$. Therefore, $S_N(x)$ converges to $f(x) = 0$ in the mean square sense.

The following inequality is known as the Cauchy-Schwarz inequality and will be used in Theorem 6.3.19.

Lemma 6.3.18. ***(Cauchy-Schwarz inequality)*** *For infinite series we have*

$$\sum_{n=1}^{\infty}|a_nb_n| \le (\sum_{n=1}^{\infty}|a_n|^2)^{1/2}(\sum_{n=1}^{\infty}|b_n|^2)^{1/2}. \tag{6.3.12}$$

Proof. First we show that the inequality

$$\sum_{n=1}^{N} |a_n b_n| \leq (\sum_{n=1}^{N} |a_n|^2)^{1/2} (\sum_{n=1}^{N} |b_n|^2)^{1/2} \tag{6.3.13}$$

holds for every N. We have for every t

$$0 \leq \sum_{n=1}^{N} |a_n + t b_n|^2 \leq \sum_{n=1}^{N} a_n^2 + 2t \sum_{n=1}^{N} |a_n b_n| + t^2 \sum_{n=1}^{N} b_n^2. \tag{6.3.14}$$

The minimum of the right hand side is attained at $t = -(\sum_{n=1}^{N} |a_n b_n|)/(\sum_{n=1}^{N} b_n^2)$ and it is nonnegative. Therefore, substituting this t in (6.3.14), we obtain (6.3.13). Now we take the limit as N goes to infinity. □

We are now ready to prove a sufficient condition for the uniform convergence. Suffice to say that we need more restrictions for f to obtain the uniform convergence.

Theorem 6.3.19. *The Fourier series of $f(x)$ converges uniformly to $f(x)$ on $[-L, L]$ provided that*

(i) $f(-L) = f(L)$.

(ii) $f(x)$ is continuous and $f'(x)$ is piecewise continuous on $[-L, L]$.

Proof. We use the Weierstrass M-test in Appendix A.2 to show that Fourier series converges to $f(x)$ uniformly and absolutely. Let a_n and b_n be the Fourier coefficients of $f(x)$ and let A_n and B_n be the Fourier coefficients of $f'(x)$. The integration by parts leads to

$$\begin{aligned} a_n &= \frac{1}{L} \int_{-L}^{L} f(x) \cos \frac{n\pi x}{L} dx \\ &= [\frac{1}{n\pi} f(x) \sin \frac{n\pi x}{L}]_{-L}^{L} - \frac{L}{n\pi} \frac{1}{L} \int_{-L}^{L} f'(x) \sin \frac{n\pi x}{L} dx. \end{aligned}$$

This implies

$$a_n = -\frac{L}{n\pi} B_n, \quad n \neq 0.$$

Similarly,

$$b_n = \frac{L}{n\pi} A_n.$$

By the Bessel's inequality for $f'(x)$

$$\sum_{n=1}^{\infty} (|A_n|^2 + |B_n|^2) < \infty.$$

Therefore, by the Cauchy-Schwarz inequality we have

$$\begin{aligned}
&\sum_{n=1}^{\infty}(|a_n \cos\frac{n\pi x}{L}| + |b_n \sin\frac{n\pi x}{L}|) \\
&\leq \sum_{n=1}^{\infty}(|a_n| + |b_n|) \\
&= \sum_{n=1}^{\infty}\frac{L}{n\pi}(|A_n| + |B_n|) \\
&\leq \frac{L}{\pi}(\sum_{n=1}^{\infty}\frac{1}{n^2})^{1/2}[2\sum_{n=1}^{\infty}(|A_n|^2 + |B_n|^2)]^{1/2} < \infty.
\end{aligned}$$

This show that the Fourier series converges absolutely and uniformly.

Since $f(x)$ is continuous, the pointwise convergence theorem applies. Therefore, $S_N(x)$ approaches $f(x)$ at each x. This implies that the difference between $f(x)$ and $S_N(x)$ is bounded by the remainder of the Fourier series.

$$\begin{aligned}
\max_{-L\leq x\leq L}|f(x) - S_N(x)| &\leq \max_{-L\leq x\leq L}\sum_{n=N+1}^{\infty}|a_n \cos\frac{n\pi x}{L} + b_n \sin\frac{n\pi x}{L}| \\
&\leq \sum_{n=N+1}^{\infty}(|a_n| + |b_n|) \qquad (6.3.15)
\end{aligned}$$

As $\sum_{n=1}^{\infty}(|a_n| + |b_n|)$ is finite, the right hand side of (6.3.15) approaches zero as $N \to \infty$. This shows that $S_N(x)$ converges uniformly to 0. □

Exercises

1. For the following partial sums, consider the same questions as in Example 6.3.17.
   (a) $S_N(x) = x^N$ on the interval $0 \leq x \leq 1$.
   (b) $S_N(x) = 2Nxe^{-Nx^2}$ on the interval $0 \leq x \leq 1$.
   (c) $S_N(x) = Nx(1-x^2)^N$ on the interval $0 \leq x \leq 1$.
2. Use the observation that $\|f + sg\|^2 \geq 0$ holds for any real number s to show the Cauchy-Schwarz inequality for integrals

$$|(f, g)| \leq \|f\| \, \|g\| \, .$$

3. (a) Find the Fourier series of $f(x) = -x$, $-L \leq x \leq L$ and on the interval $-3L \leq x \leq 3L$ sketch the graph of the Fourier series (more precisely, the function to which the Fourier series converges).

(b) Find the Fourier cosine series of $f(x) = -x$, $0 \le x \le L$ and on the interval $-3L \le x \le 3L$ sketch the graph of the Fourier cosine series.

(c) Find the Fourier sine series of $f(x) = -x$, $0 \le x \le L$ and on the interval $-3L \le x \le 3L$ sketch the graph of the Fourier sine series.

4. For (a) and (b), do (i) and (ii).

(i) Find the Fourier series.

(ii) Sketch the graph of the function to which the Fourier series of f converges for three periods.

(a) $f(x) = \begin{cases} 1 + x, & -1 < x < 0, \\ 1 - x, & 0 < x < 1. \end{cases}$

(b) $f(x) = \begin{cases} -1 - x, & -1 < x < 0, \\ 1 - x, & 0 < x < 1. \end{cases}$

5. Suppose $f(x) = 1 - x$ is given on $0 < x < 1$.

(a) Find the Fourier cosine series of $f(x)$ and sketch the graph of the Fourier cosine series (more precisely, the function to which the Fourier cosine series converges) for three periods.

(b) Find the Fourier sine series of $f(x)$ and sketch the graph of the Fourier sine series (more precisely, the function to which the Fourier sine series converges) for three periods.

6. (a) Find the Fourier sine series of $f(x) = 1$ on $(0, \pi)$.

(b) Use the Parseval's equality to show that

$$\sum_{n=1}^{\infty} \frac{1}{(2n-1)^2} = \frac{\pi^2}{8}.$$

7. (a) Find the Fourier cosine series of $f(x) = x$ on $(0, \pi)$.

(b) Use the Parseval's equality to show that

$$\sum_{n=1}^{\infty} \frac{1}{(2n-1)^4} = \frac{\pi^4}{96}.$$

6.4 Derivatives of Fourier Series

Suppose that the Fourier series of f on $[-L, L]$ is given by

$$f(x) = \frac{a_0}{2} + \sum_{n=1}^{\infty} a_n \cos\frac{n\pi x}{L} + \sum_{n=1}^{\infty} b_n \sin\frac{n\pi x}{L}, \tag{6.4.1}$$

where

$$a_n = \frac{1}{L}\int_{-L}^{L} f(x)\cos\frac{n\pi x}{L}dx, \quad n = 0, 1, \ldots,$$

$$b_n = \frac{1}{L}\int_{-L}^{L} f(x)\sin\frac{n\pi x}{L}dx, \quad n = 1, 2, \ldots.$$

If we differentiate the right hand side of (6.4.1), we have

$$-\sum_{n=1}^{\infty} a_n \frac{n\pi}{L}\sin\frac{n\pi x}{L} + \sum_{n=1}^{\infty} b_n \frac{n\pi}{L}\cos\frac{n\pi x}{L}. \tag{6.4.2}$$

This is called the term by term differentiation of the right hand side of (6.4.1).

The question is if and when this is the same as the Fourier series of $f'(x)$. To answer this question we assume that the periodic extension of f is continuous and f' is piecewise continuous. Under these assumptions f' has a Fourier series denoted as

$$f'(x) = \frac{A_0}{2} + \sum_{n=1}^{\infty} A_n \cos\frac{n\pi x}{L} + \sum_{n=1}^{\infty} B_n \sin\frac{n\pi x}{L}, \tag{6.4.3}$$

$$A_n = \frac{1}{L}\int_{-L}^{L} f'(x)\cos\frac{n\pi x}{L}dx, \quad n = 0, 1, \ldots,$$

$$B_n = \frac{1}{L}\int_{-L}^{L} f'(x)\sin\frac{n\pi x}{L}dx, \quad n = 1, 2, \ldots.$$

Since we assume that the periodic extension of f is continuous, $f(-L) = f(L)$. Therefore, performing the integration by parts, we have

$$A_0 = \frac{1}{L}\int_{-L}^{L} f'(x)dx = \frac{1}{L}[f(L) - f(-L)] = 0,$$

$$\begin{aligned} A_n &= \frac{1}{L}\int_{-L}^{L} f'(x)\cos\frac{n\pi x}{L}dx \\ &= \frac{1}{L}[f(x)\cos\frac{n\pi x}{L}]_{-L}^{L} + \frac{1}{L}\int_{-L}^{L} f(x)\frac{n\pi}{L}\sin\frac{n\pi x}{L}dx \\ &= \frac{1}{L}\int_{-L}^{L} f(x)\frac{n\pi}{L}\sin\frac{n\pi x}{L}dx = \frac{n\pi}{L}b_n, \end{aligned} \tag{6.4.4}$$

$$\begin{aligned}B_n &= \frac{1}{L}\int_{-L}^{L} f'(x)\sin\frac{n\pi x}{L}dx \\ &= \frac{1}{L}[f(x)\sin\frac{n\pi x}{L}]_{-L}^{L} - \frac{1}{L}\int_{-L}^{L} f(x)\frac{n\pi}{L}\cos\frac{n\pi x}{L}dx \\ &= -\frac{1}{L}\int_{-L}^{L} f(x)\frac{n\pi}{L}\cos\frac{n\pi x}{L}dx = -\frac{n\pi}{L}a_n. \qquad (6.4.5)\end{aligned}$$

So, (6.4.4) and (6.4.5) imply that (6.4.2) and the right hand side of (6.4.3) are the same. We summarize this result as

Theorem 6.4.1. *Suppose that the periodic extension of f is continuous and f' is piecewise continuous. Then, the term by term differentiation of the Fourier series of $f(x)$ agrees with the Fourier series of $f'(x)$.*

Example 6.4.2. Are the periodic extensions of the following functions continuous?

(a) $f(x) = x$ on $[-1, 1]$. (b) $f(x) = 4 - x^2$ on $[-2, 2]$.

Solution: (a) No, because $f(-1) \neq f(1)$. The periodic extension is give in Figure 6.4. (b) Yes, because $f(-2) = f(2)$ and $f(x)$ is continuous on $[-2, 2]$.

In the next two examples we examine Theorem 6.4.1. The first example is the case where the assumptions of the theorem is satisfied. In the second example we treat the case where the assumptions are not satisfied. We verify that when there is a discontinuity in the Fourier series, the term-by-term differentiation of the Fourier series of $f(x)$ and the Fourier series of $f'(x)$ do not necessarily agree.

Example 6.4.3. (a) Find the Fourier cosine series of $f(x) = x$ on $[0, 1]$.

(b) Draw a graph of function to which the Fourier cosine series converges on $[-3, 3]$.

(c) Find the Fourier sine series of $g(x) = 1$ on $[0, 1]$.

(d) Differentiate the series in (a) term-by-term and compare the result with the series found in (c).

Solution: (a) The Fourier cosine series is given by (6.2.13) where

$$a_0 = \int_0^1 x dx = \frac{1}{2},$$

$$a_n = 2\int_0^1 x\cos n\pi x dx = 2[x\frac{1}{n\pi}\sin n\pi x]_0^1 - 2\int_0^1 \frac{1}{n\pi}\sin n\pi x dx$$
$$= 2[(\frac{1}{n\pi})^2\cos n\pi x]_0^1 = 2(\frac{1}{n\pi})^2[(-1)^n - 1].$$

Therefore, the Fourier cosine series of x is

$$x = \frac{1}{4} + \sum_{n=1}^{\infty} 2(\frac{1}{n\pi})^2[(-1)^n - 1]\cos n\pi x. \qquad (6.4.6)$$

(b) See Figure 6.6 with $L = 1$.

(c)

$$b_n = 2\int_0^1 \sin n\pi x dx = 2[-\frac{1}{n\pi}\cos n\pi x]_0^1 = -2(\frac{1}{n\pi})[(-1)^n - 1].$$

Therefore, the Fourier sine series of 1 is

$$1 = -\sum_{n=1}^{\infty} 2(\frac{1}{n\pi})[(-1)^n - 1]\sin n\pi x. \qquad (6.4.7)$$

(d) Differentiating (6.4.6) term-by-term, we obtain (6.4.7).

Example 6.4.4. (a) Find the Fourier sine series of $f(x) = x$ on $[0, 1]$.
(b) Draw a graph of function to which the Fourier sine series converges on $[-3, 3]$.
(c) Find the Fourier cosine series of $g(x) = 1$ on $[0, 1]$.
(d) Differentiate the series in (a) term by term differentiation and compare the result with the series found in (b).
Solution: (a)

$$b_n = 2\int_0^1 x\sin n\pi x dx = 2[-x\frac{1}{n\pi}\cos n\pi x]_0^1 + 2\int_0^1 \frac{1}{n\pi}\cos n\pi x dx$$
$$= -2(\frac{1}{n\pi})(-1)^n.$$

Therefore, we have

$$x = \sum_{n=1}^{\infty} 2(\frac{1}{n\pi})(-1)^{n+1}\sin n\pi x. \qquad (6.4.8)$$

(b) See Figure 6.5 with $L = 1$.
(c) $a_0 = 2$ and $a_n = 0$ for $n > 0$. So, the Fourier series of 1 is 1.
(d) The term by term differentiation of (6.4.8) is $\sum_{n=1}^{\infty} 2(-1)^{n+1}\cos n\pi x$. The series in (c) and (d) are not the same.

Exercises

1. Are the periodic extensions of the following functions continuous?

   (a) $f(x) = \begin{cases} x+1 & -1 \le x < 0 \\ 1-x & 0 \le x < 1 \end{cases}$ on $[-1, 1]$.

   (b) $f(x) = x^3 - x^2$ on $[-1, 1]$.

2. Suppose $f(x) = x - 1$ is given on $0 \le x \le 1$.

   (a) Find the Fourier cosine series of $f(x)$ and sketch the graph of the Fourier cosine series for three periods.

   (b) Find the Fourier sine series of $f(x)$ and sketch the graph of the Fourier sine series for three periods.

   (c) Suppose we differentiate the series in (a) term-by-term. Is it the same as the Fourier sine series of $f'(x)$? Explain the reason for your answer. You do not have to compute the Fourier sine series of $f'(x)$.

   (d) Suppose we differentiate the series in (b) term-by-term. Is it the same as the Fourier cosine series of $f'(x)$? Explain the reason for your answer. You do not have to compute the Fourier cosine series of $f'(x)$.

3. In this problem we use the term by term integration to find the Fourier series. The justification is that if $f(x)$ is piecewise continuous on $[a, b]$, the indefinite integral $F(x) = \int_a^x f(s)ds$ $(a < x < b)$ is continuous and Theorem 6.3.13 applies to $F(x)$.

   (a) Show that the Fourier sine series of x on $[0, L]$ is

   $$x = \sum_{n=1}^{\infty} \frac{2L}{n\pi}(-1)^{n+1} \sin \frac{n\pi x}{L}. \tag{6.4.9}$$

   (b) Integrate the both sides over $[0, x]$ $(0 < x < L)$ to show that the Fourier series of x^2 is given by

   $$x^2 = \sum_{n=1}^{\infty} 4(\frac{L}{n\pi})^2(-1)^{n+1} - \sum_{n=1}^{\infty} 4(\frac{L}{n\pi})^2(-1)^{n+1} \cos \frac{n\pi x}{L}. \tag{6.4.10}$$

   (c) Show the following relation.

   $$\sum_{n=1}^{\infty} 4(\frac{L}{n\pi})^2(-1)^{n+1} = \frac{1}{L}\int_0^L x^2 dx = \frac{1}{3}L^2. \tag{6.4.11}$$

(d) Integrate x^2 in (6.4.10) with (6.4.11) on $[0, x]$ and use (6.4.9) show that the Fourier series of x^3 is given by

$$x^3 = \sum_{n=1}^{\infty} \frac{2L^3}{n\pi}(-1)^{n+1} \sin \frac{n\pi x}{L} - \sum_{n=1}^{\infty} 12(\frac{L}{n\pi})^3(-1)^{n+1} \sin \frac{n\pi x}{L}.$$

6.5 Eigenvalue Problems

We summarize the eigenvalue problems studied in earlier chapters and prepare for the next chapter where we study the separation of variables in the multi-dimensional case. In Section 6.3 we studied the convergence of Fourier series. Consequently we know that the Fourier sine and cosine series converge. However, we did not discuss convergence issue for other series consisting of $\sin(n - 1/2)\pi x$ or $\cos(n - 1/2)\pi x$ that we encounter in Chapters 2, 3, and 4. We take up this issue in this section. In one dimension this is called the Sturm-Liouville problem.

6.5.1 *The Sturm-Liouville Problems*

The Sturm-Liouville problems are the eigenvalue problems where the equation is given by

$$-\frac{d}{dx}(p\frac{du}{dx}) + qu = \lambda hu, \quad a < x < b, \tag{6.5.1}$$

and the boundary conditions are the first order homogeneous boundary conditions given by

$$\alpha_1 u(a) + \beta_1 \frac{du}{dx}(a) = 0, \tag{6.5.2}$$

$$\alpha_2 u(b) + \beta_2 \frac{du}{dx}(b) = 0. \tag{6.5.3}$$

If $p = 1$, $q = 0$, and $h = 1$, the problem reduces to what we did in Chapters 2, 3, and 4. The Sturm-Liouville problems are generalization of the eigenvalue problems encountered in the separation of variables. The form of (6.5.1) look unfamiliar, but in the derivation of heat equation we studied in Section 2.1 we obtain the equation of the form if k in (2.1.7) depends on x.

The coefficients α_i and β_i are real constants, and for p, q, and h we assume that $p(x)$ is C^1, and $q(x)$ and $h(x)$ are continuous. We also assume that $p > 0$ and $h > 0$. Depending on p, q, h, and the domain the Sturm-Liouville problem is classified into two types. If one of the following holds, the problem is called the singular Sturm-Liouville problem.

1. The coefficient $p(x)$ becomes zero at one or both end points $x = a$ or $x = b$.
2. At least one of the coefficients p, q, or h becomes infinite at a or b.
3. One of the endpoints is infinite.

Otherwise it is called the regular Sturm-Liouville problem. The eigenvalue problems we have encountered so far are regular. The following are examples of the singular Sturm-Liouville problems.

Example 6.5.1. Bessel's Equation: This is given by

$$-\frac{d}{dr}(r\frac{dR}{dr}) + \frac{\mu}{r}R = \lambda r R, \quad 0 < r < b.$$

In this example, $p = r$, $q = \mu/r$, and $h = r$. This equation will appear in Sections 8.1 and 8.2 where we study the separation of variables for circular or cylindrical domains.

Example 6.5.2. Legendre Equation: This is given by

$$-(\sin\phi\Phi_\phi)_\phi + \frac{n^2}{\sin\phi}\Phi = \gamma\sin\phi\Phi, \quad 0 < \phi < \pi.$$

In this example, $p = \sin\phi$, $q = n^2/\sin\phi$, $h = \sin\phi$, and $\lambda = \gamma$. This equation will appear in Section 8.3 where we study the separation of variables in the spherical coordinates.

The properties of the eigenvalues and the eigenfunctions for the regular Sturm-Liouville problem are summarized in Theorem 6.5.3.

Theorem 6.5.3. *The eigenvalues and eigenfunctions of the regular Sturm-Liouville problem have the following properties.*

1. The eigenvalues λ are real.
2. There are infinite number of eigenvalues. We denote them by

   $$\lambda_1 < \lambda_2 < \cdots < \lambda_n < \cdots.$$

   There is a smallest eigenvalue but not a largest eigenvalue and $\lambda_n \to \infty$ as $n \to \infty$.
3. Corresponding to each eigenvalue, there is a unique eigenfunction ϕ_n and $\phi_n(x)$ has exactly $n - 1$ zeros on (a, b).

4. Eigenfunctions belonging to different eigenvalues are orthogonal with respect to the weight function h, *i.e.*,
$$\int_a^b h(x)\phi_m(x)\phi_n(x)dx = 0 \quad \text{if } \lambda_n \neq \lambda_m.$$
5. The eigenfunctions form a complete set. This means that every piecewise continuous function $f(x)$ can be expressed as
$$f(x) = \sum_{n=1}^{\infty} a_n\phi_n(x).$$
The partial sum $\sum_{n=1}^{N} a_n\phi_n(x)$ converges to $[f(x_+) + f(x_-)]$ as $N \to \infty$ on (a, b). Also, if
$$\int_a^b |f(x)|^2\, dx < \infty.$$
The partial sum converges in the mean-square sense to $f(x)$ in (a, b).

The proofs of 1, 3, and 4 are carried out in Subsection 6.5.2. The proof of 5 is similar to the proof in Section 6.3 in spirit. The proofs of 2 and 5 are available in more advanced textbook such as [Evans (1998)].

Remark 6.5.4. The periodic boundary conditions
$$u(a) = u(b), \quad u'(a) = u'(b)$$
we studied in Section 4.2 are not in the form of the Sturm-Liouville problem. Therefore, Theorem 6.5.3 does not have to apply. As a matter of fact, the eigenfunctions for each eigenvalue are not unique.

6.5.2 *Proofs*

We prove Properties 1, 3 and 4 stated in Theorem 6.5.3.

Proof of 1. Suppose λ_n is an eigenvalue and $\phi_n(x)$ is the corresponding eigenfunction for (6.5.1) to (6.5.3). Then, we have
$$-\frac{d}{dx}(p\frac{d\phi_n}{dx}) + q\phi_n = \lambda_n h\phi_n, \quad a < x < b, \tag{6.5.4}$$
$$\alpha_1\phi_n(a) + \beta_1\frac{d\phi_n}{dx}(a) = 0, \tag{6.5.5}$$
$$\alpha_2\phi_n(b) + \beta_2\frac{d\phi_n}{dx}(b) = 0. \tag{6.5.6}$$

Multiplying (6.5.4) by $\bar{\phi}_n$, the complex conjugate of ϕ_n, and integrating it by parts over (a, b), we have

$$\int_a^b (p|\frac{d\phi_n}{dx}|^2 + q|\phi_n|^2)dx - [p\frac{d\phi_n}{dx}\bar{\phi}_n]_a^b = \lambda_n \int_a^b h|\phi_n|^2 dx. \qquad (6.5.7)$$

Next, taking the complex conjugate of (6.5.4), multiplying it by ϕ_n, and integrating it by parts over (a, b), we obtain

$$\int_a^b (p|\frac{d\phi_n}{dx}|^2 + q|\phi_n|^2)dx - [p\frac{d\bar{\phi}_n}{dx}\phi_n]_a^b = \bar{\lambda}_n \int_a^b h|\phi_n|^2 dx. \qquad (6.5.8)$$

Subtracting (6.5.8) from (6.5.7) and using the boundary conditions (6.5.5) and (6.5.6), we see that

$$(\lambda_n - \bar{\lambda}_n) \int_a^b h|\phi_n|^2 dx = [p\frac{d\bar{\phi}_n}{dx}\phi_n]_a^b - [p\frac{d\phi_n}{dx}\bar{\phi}_n]_a^b = 0.$$

This implies that $\lambda_n = \bar{\lambda}_n$ and λ_n is real.

Proof of 3. Suppose ϕ_1 and ϕ_2 are different eigenfunctions for an eigenvalue λ. Then,

$$-\frac{d}{dx}(p\frac{d\phi_1}{dx}) + q\phi_1 = \lambda h\phi_1, \qquad (6.5.9)$$

$$-\frac{d}{dx}(p\frac{d\phi_2}{dx}) + q\phi_2 = \lambda h\phi_2. \qquad (6.5.10)$$

Computing (6.5.9)$\times\phi_2-$(6.5.10)$\times\phi_1$, we have

$$\phi_1\frac{d}{dx}(p\frac{d\phi_2}{dx}) - \phi_2\frac{d}{dx}(p\frac{d\phi_1}{dx}) = 0. \qquad (6.5.11)$$

The identity

$$\phi_1\frac{d}{dx}(p\frac{d\phi_2}{dx}) - \phi_2\frac{d}{dx}(p\frac{d\phi_1}{dx}) = \frac{d}{dx}[p(\phi_1\frac{d\phi_2}{dx} - \phi_2\frac{d\phi_1}{dx})]$$

and (6.5.11) yield

$$p(\phi_1\frac{d\phi_2}{dx} - \phi_2\frac{d\phi_1}{dx}) = \text{constant}. \qquad (6.5.12)$$

By the boundary conditions (6.5.2) or (6.5.3), the above constant is zero and hence dividing (6.5.12) by $p\phi_1\phi_2$, we have

$$\frac{\phi_1'}{\phi_1} = \frac{\phi_2'}{\phi_2}. \qquad (6.5.13)$$

Integrating (6.5.13) yields

$$\ln|\phi_1| = \ln|\phi_2| + c \Rightarrow \phi_1 = C\phi_2,$$

where c is a constant and $C = \pm e^c$. Therefore, ϕ_1 and ϕ_2 are linearly dependent.

Proof of 4. Suppose λ_m is an eigenvalue different from λ_n and ϕ_m is the corresponding eigenfunction. Multiply (6.5.4) by ϕ_m and perform the integration by parts. Then,

$$\int_a^b (p\frac{d\phi_n}{dx}\frac{d\phi_m}{dx} + q\phi_n\phi_m)dx - [p\frac{d\phi_n}{dx}\phi_m]_a^b = \lambda_n \int_a^b h\phi_n\phi_m dx. \quad (6.5.14)$$

Also we note that λ_m and ϕ_m satisfy

$$-\frac{d}{dx}(p\frac{d\phi_m}{dx}) + q\phi_m = \lambda_m h\phi_m, \quad a < x < b. \quad (6.5.15)$$

If we multiply (6.5.15) by ϕ_n and integrate over (a, b), then

$$\int_a^b (p\frac{d\phi_n}{dx}\frac{d\phi_m}{dx} + q\phi_n\phi_m)dx - [p\frac{d\phi_m}{dx}\phi_n]_a^b = \lambda_m \int_a^b h\phi_n\phi_m dx. \quad (6.5.16)$$

Subtracting (6.5.16) from (6.5.14) and using the boundary conditions, we have

$$(\lambda_n - \lambda_m)\int_a^b h\phi_n\phi_m dx = [p\frac{d\phi_m}{dx}\phi_n]_a^b - [p\frac{d\phi_n}{dx}\phi_m]_a^b = 0.$$

Since $\lambda_n \neq \lambda_m$, we have $\int_a^b h\phi_n\phi_m dx = 0$.

Exercises

1. Consider

$$u'' + a(x)u' + (\lambda b(x) + c(x))u = 0.$$

Multiply the equation by $\mu(x)$. Find $\mu(x)$ such that the resulting equation can be reduced to the Sturm-Liouville form

$$\frac{d}{dx}(p(x)\frac{du}{dx}) + (\lambda h(x) + q(x))u = 0.$$

2. Show that the eigenvalues λ are real for the periodic boundary value problem

$$-\frac{d}{dx}(p\frac{du}{dx}) + qu = \lambda hu, \quad a < x < b,$$

$$u(a) = u(b), \quad u'(a) = u'(b).$$

3. Consider the convection diffusion equation

$$u_t = a^2 u_{xx} - cu_x,$$

where a and c are positive constants.

(a) Use the separation of variables to obtain two ODE's.
(b) Rewrite the ODE for the spacial variable to a Sturm-Liouville form.
(c) Solve the initial boundary value problem

$$u(0,t)=0, \quad u(L,t)=0,$$

$$u(x,0)=f(x).$$

4. Show that the eigenfunctions for different eigenvalues are orthogonal.

$$\frac{d}{dx}(p(x)\frac{du}{dx})+(\lambda h(x)+q(x))u=0, \quad a<x<b$$

$$u(a)-u_x(a)=0, \quad u(b)=0.$$

5. Consider the eigenvalue problem

$$x^2\frac{d^2u}{dx^2}+x\frac{du}{dx}+\lambda u=0, \quad u(1)=0, \quad u(b)=0.$$

(a) Put the equation in the Sturm-Liouville form.
(b) Show that $\lambda \geq 0$.
(c) Determine all eigenvalues. Hint: it might be better if you set $y=\ln x$ and change the independent variable from x to y. Also you could assume that the eigenfunctions have the form $\phi = x^r$.
(d) With what weight the eigenvalues are orthogonal? Verify the orthogonality.
(e) Show that nth eigenfunction has $n-1$ zeros.

Chapter 7

Separation of Variables in Higher Dimensions

In this chapter we study the separation of variables for the multidimensional problems. We discuss the heat and wave equations in two or three dimensions and the Laplace equation in three dimensions. The key issue in this chapter is how the separation of variables we studied in one dimensional problems is extended to the multidimensional problems. One thing we have to cope with is the geometry of the domains. Typical geometries discussed in the textbooks are rectangles, circles, circular cylinders, and spheres. In this chapter we mainly confine our attention to the rectangular coordinates. We learn that the eigenfunctions are composed of the eigenfunctions for one dimensional problems. We repeatedly solve the one dimensional problems. We also discuss the eigenvalue problems for higher dimensions. In Section 7.2 we briefly explain the eigenvalue problems and then the Rayleigh quotient, which relates the eigenvalue problems and the minimization problems, will be discussed. In Section 7.3 we discuss an extension of separation of variables, the eigenfunction expansion, to more complex problems including the non-homogeneous boundary conditions and the source terms.

7.1 Rectangular Domains

We use the wave and the Laplace equations to illustrate how the separation of variables is carried out in the rectangular coordinates. The heat equation is similar to the wave equation and given in Exercises.

Example 7.1.1. Solve

$$u_{tt} = c^2 \triangle u = c^2(u_{xx} + u_{yy}), \quad 0 < x < L,\ 0 < y < H,\ 0 < t, \tag{7.1.1}$$

$$u_x(0, y, t) = 0, \quad u_x(L, y, t) = 0, \quad 0 < y < H, \tag{7.1.2}$$

$$u(x,0,t) = 0, \quad u(x,H,t) = 0, \quad 0 < x < L, \tag{7.1.3}$$

$$u(x,y,0) = f(x,y), \quad u_t(x,y,0) = g(x,y). \tag{7.1.4}$$

Solution: As we saw in the one dimensional case, the steps in the separation of variables consist of separating the variables, solving the resulting ODE's, and assembling the solutions by the superposition principle.

Step 1 (Separating Variables): In this step we separate the variables. First, assume that the solution is given as

$$u = v(x,y)T(t). \tag{7.1.5}$$

Then,

$$vT'' = c^2(v_{xx} + v_{yy})T.$$

Dividing by c^2vT, we have

$$\frac{T''}{c^2T} = \frac{v_{xx} + v_{yy}}{v} = -\lambda.$$

As we did in the one dimensional problems, we fix x and y, and change t. Then, we see that $T''/(c^2T)$ and λ are constants. Similarly, $(v_{xx} + v_{yy})/v$ is also a constant. Therefore, we obtain

$$v_{xx} + v_{yy} + \lambda v = 0, \tag{7.1.6}$$

$$T'' + \lambda c^2 T = 0. \tag{7.1.7}$$

For (7.1.6) we assume $v = X(x)Y(y)$. Then,

$$X''Y + XY'' + \lambda XY = 0 \Rightarrow \frac{X''}{X} + \frac{Y''}{Y} = -\lambda.$$

This implies

$$\frac{X''}{X} = -\mu \Rightarrow X'' + \mu X = 0, \tag{7.1.8}$$

$$\frac{Y''}{Y} = -\nu \Rightarrow Y'' + \nu Y = 0, \tag{7.1.9}$$

where $\nu = \lambda - \mu$. The boundary conditions for X and Y are obtained from (7.1.2) and (7.1.3), respectively

$$X'(0) = 0, \quad X'(L) = 0. \tag{7.1.10}$$

$$Y(0) = 0, \quad Y(H) = 0. \tag{7.1.11}$$

We separated the space variables and time variable in (7.1.5). We can also separate X, Y, and T at once if we assume $u = X(x)Y(y)T(t)$.

Step 2 (Solving ODE's): For the initial boundary value problems we solve the eigenvalue problems for the space variables X and Y and the initial value problem for T. For X we solve (7.1.8) with the boundary conditions (7.1.10). The eigenvalues and eigenfunctions are

$$\mu_m = (\frac{m\pi}{L})^2, \quad X_m = \cos\frac{m\pi x}{L}, \quad m = 0, 1, \ldots. \tag{7.1.12}$$

Similarly, for Y we solve (7.1.9) with the boundary conditions (7.1.11). The eigenvalues and eigenfunctions are

$$\nu_n = (\frac{n\pi}{H})^2, \quad Y_n = \sin\frac{n\pi y}{H}, \quad n = 1, 2, \ldots. \tag{7.1.13}$$

So, the eigenvalues are

$$\lambda_{mn} = \mu_m + \nu_n = (\frac{m\pi}{L})^2 + (\frac{n\pi}{H})^2 \tag{7.1.14}$$

and the eigenfunctions are

$$v_{mn} = \cos\frac{m\pi x}{L}\sin\frac{n\pi y}{H}, \quad m = 0, 1, \ldots\; n = 1, 2, \ldots.$$

For T we solve the initial value problem for (7.1.7) with λ_{mn} given in (7.1.14). Then, we have

$$T_{mn} = (a_{mn}\cos c\sqrt{\lambda_{mn}}t + b_{mn}\sin c\sqrt{\lambda_{mn}}t).$$

Step 3 (Finding the Solution): Now we assemble the solution and find the coefficients. Since we assume that the form of solution is $u = X(x)Y(y)T(t)$, by the superposition principle the solution can be written as

$$u(x, y, t) = \sum_{n=1}^{\infty}(\frac{a_{0n}}{2}\cos c\sqrt{\lambda_{0n}}t + \frac{b_{0n}}{2}\sin c\sqrt{\lambda_{0n}}t)\sin\frac{n\pi y}{H}$$

$$+\sum_{m=1}^{\infty}\sum_{n=1}^{\infty}(a_{mn}\cos c\sqrt{\lambda_{mn}}t + b_{mn}\sin c\sqrt{\lambda_{mn}}t)\cos\frac{m\pi x}{L}\sin\frac{n\pi y}{H}. \tag{7.1.15}$$

Note that the first term on the right hand side corresponds to the case where $m = 0$. To find a_{mn} and b_{mn} we use the initial conditions. At $t = 0$, from (7.1.4) and (7.1.15) we see

$$u(x, y, 0) = \sum_{n=1}^{\infty}\frac{a_{0n}}{2}\sin\frac{n\pi y}{H} + \sum_{m=1}^{\infty}\sum_{n=1}^{\infty}a_{mn}\cos\frac{m\pi x}{L}\sin\frac{n\pi y}{H} = f(x, y). \tag{7.1.16}$$

As in the one dimensional problems we use the mutual orthogonality of eigenfunctions. Multiplying (7.1.16) by the eigenfunctions $v_{kj}(x,y) = \cos(k\pi x/L)\sin(j\pi y/W)$ and integrating over $0 < x < L$ and $0 < y < W$, we have

$$a_{kj} = \frac{4}{LH}\int_0^H \int_0^L f(x,y)\cos\frac{k\pi x}{L}\sin\frac{j\pi y}{H}dxdy, \quad k = 0,1,2,\ldots, \ j=1,2,\ldots.$$

To find b_{mn} we differentiate (7.1.15) in t and set $t = 0$. Then, by (7.1.4)

$$\begin{aligned} &u_t(x,y,0) \\ &= \sum_{n=1}^{\infty} c\sqrt{\lambda_{0n}}\frac{b_{0n}}{2}\sin\frac{n\pi y}{H} + \sum_{m=1}^{\infty}\sum_{n=1}^{\infty} c\sqrt{\lambda_{mn}}b_{mn}\cos\frac{m\pi x}{L}\sin\frac{n\pi y}{H} \\ &= g(x,y). \end{aligned}$$

Performing the same procedure, we obtain

$$\begin{aligned} b_{mn} &= \frac{1}{c\sqrt{\lambda_{mn}}}\frac{4}{LH}\int_0^H \int_0^L g(x,y)\cos\frac{m\pi x}{L}\sin\frac{n\pi y}{H}dxdy, \\ &m = 0,1,2,\ldots, \ n = 1,2,\ldots. \end{aligned}$$

If we define the inner product to be $(p,q) = \int_0^H \int_0^L pq dxdy$, the Fourier coefficients a_{mn} and b_{mn} have simpler representation and are given, respectively, by

$$a_{mn} = \frac{(f, v_{mn})}{(v_{mn}, v_{mn})}, \quad b_{mn} = \frac{(g, v_{mn})}{c\sqrt{\lambda_{mn}}(v_{mn}, v_{mn})}.$$

In the next example, we apply the separation of variables to the three or higher dimensional problems for the Laplace equation. The idea is similar to the higher dimensional heat or wave equation.

Example 7.1.2. Solve the Laplace equation

$$\Delta u = u_{xx} + u_{yy} + u_{zz} = 0,$$

in a rectangular region $0 < x < L$, $0 < y < W$, $0 < z < H$ with the boundary conditions

$$\begin{aligned} &u_x(0,y,z) = 0, \quad u(x,0,z) = 0, \quad u(x,y,0) = f(x,y), \\ &u_x(L,y,z) = 0, \quad u(x,W,z) = 0, \quad u(x,y,H) = 0. \end{aligned}$$

Solution: Step 1 (Separating Variables): In this step we assume $u = X(x)Y(y)Z(z)$ and derive the equations and the boundary conditions for X, Y, and Z. For the Laplace equation often it is better to start from the boundary conditions. The appropriate boundary conditions for X, Y, and Z are

$$X'(0) = 0, \quad X'(L) = 0, \tag{7.1.17}$$

$$Y(0) = 0, \quad Y(W) = 0, \tag{7.1.18}$$

$$Z(0) = ?, \quad Z(H) = 0, \tag{7.1.19}$$

where the question mark means that we do not know how to specify $Z(0)$ at this moment. Also, from the equation, we obtain

$$\frac{X''}{X} + \frac{Y''}{Y} + \frac{Z''}{Z} = 0. \tag{7.1.20}$$

We use the same idea to separate the three variables. For example for X, fix y and z, and change x. Then from (7.1.20), we see that X''/X is a constant. Similarly, Y''/Y and Z''/Z are also constants. The boundary conditions for X and Y are known and homogeneous. This suggests that we should set $X''/X = -\mu$ and $Y''/Y = -\nu$. Then,

$$X'' + \mu X = 0, \tag{7.1.21}$$

$$Y'' + \nu Y = 0, \tag{7.1.22}$$

$$Z'' - \lambda Z = 0, \tag{7.1.23}$$

where $\lambda = \mu + \nu$.

Step 2 (Solving ODE's): For the time independent problems we need to distinguish the variables for which we solve the eigenvalue problems and a variable for which we solve the boundary value problem. The boundary conditions suggest that we solve the eigenvalue problems for X and Y and the boundary value problem for Z. For X we solve (7.1.21) with the boundary conditions (7.1.17). The eigenvalues and eigenfunctions are

$$\mu_m = (\frac{m\pi}{L})^2, \quad X_m = \cos\frac{m\pi x}{L}, \quad m = 0, 1, \ldots. \tag{7.1.24}$$

For Y we solve (7.1.22) with the boundary conditions (7.1.18). The eigenvalues and eigenfunctions are

$$\nu_n = (\frac{n\pi}{W})^2, \quad Y_n = \sin\frac{n\pi y}{W}, \quad n = 1, 2, \ldots. \tag{7.1.25}$$

For Z, we have

$$\lambda_{mn} = (\frac{m\pi}{L})^2 + (\frac{n\pi}{W})^2, \quad Z'' - \lambda_{mn} Z = 0, \quad Z(H) = 0.$$

Since the boundary condition at $z = H$ is $Z(H) = 0$, it is more convenient to write the general solution for Z in the form of

$$Z(z) = a_{mn} \cosh \sqrt{\lambda_{mn}}(z - H) + b_{mn} \sinh \sqrt{\lambda_{mn}}(z - H).$$

Then from the boundary condition $Z(H) = 0$ we see that $a_{mn} = 0$ and

$$Z = b_{mn} \sinh \sqrt{\lambda_{mn}}(z - H).$$

Step 3 (Finding the Solution): The solution is given by

$$u(x, y, z) = \sum_{m=0}^{\infty} \sum_{n=1}^{\infty} b_{mn} \sinh \sqrt{\lambda_{mn}}(z - H) \sin \frac{m\pi y}{W} \cos \frac{n\pi x}{L}.$$

At $z = 0$,

$$u(x, y, 0) = -\sum_{m=0}^{\infty} \sum_{n=1}^{\infty} b_{mn} \sinh(\sqrt{\lambda_{mn}}H) \sin \frac{m\pi y}{W} \cos \frac{n\pi x}{L} = f(x, y).$$

Multiplying the above equation by the eigenfunctions $v_{kj}(x, y) = \cos(k\pi x/L) \sin(j\pi y/W)$ and integrating over $0 < x < L$ and $0 < y < W$, we have the following results. If $k = 0$,

$$-\frac{LW}{2} \sinh(\sqrt{\lambda_{0j}}H) b_{0j} = \int_0^W \int_0^L f(x, y) \sin \frac{j\pi y}{W} dxdy,$$

and for $k \geq 1$

$$-\frac{LW}{4} \sinh(\sqrt{\lambda_{kj}}H) b_{kj} = \int_0^W \int_0^L f(x, y) \cos \frac{k\pi x}{L} \sin \frac{j\pi y}{W} dxdy.$$

Exercises

1. Solve the heat equation

$$u_t = a^2(u_{xx} + u_{yy}), \quad 0 < x < L,\ 0 < y < H,\ 0 < t,$$

$$u(x, y, 0) = f(x, y)$$

with the following boundary conditions.

   (a) $u(0, y, t) = 0,\ u(L, y, t) = 0,\ u_y(x, 0, t) = 0,\ u_y(x, H, t) = 0$
   (b) $u(0, y, t) = 0,\ u(L, y, t) = 0,\ u(x, 0, t) = 0,\ u_y(x, H, t) = 0$
   (c) $u_x(0, y, t) = 0,\ u_x(L, y, t) = 0,\ u_y(x, 0, t) = 0,\ u_y(x, H, t) = 0$

2. Solve the Laplace equation

$$\Delta u = u_{xx} + u_{yy} + u_{zz} = 0,$$

in a rectangular region $0 < x < L$, $0 < y < W$, $0 < z < H$ with the following boundary conditions.

(a) $u_x(0,y,z)=0,\quad u(x,0,z)=0,\quad u(x,y,0)=0,$
$u_x(L,y,z)=f(y,z),\quad u(x,W,z)=0,\quad u(x,y,H)=0.$
(b) $u_x(0,y,z)=0,\quad u_y(x,0,z)=f(x,z),\quad u_z(x,y,0)=0,$
$u_x(L,y,z)=0,\quad u_y(x,W,z)=0,\quad u_z(x,y,H)=0.$
(c) For (b) find the solvability condition.

3. Consider the Laplace equation

$$\Delta u = u_{xx}+u_{yy}+u_{zz}=0,$$

in a rectangular region $0<x<L$, $0<y<W$, $0<z<H$ with the boundary condition

$$u_x(0,y,z)=0,\quad u(x,0,z)=0,\quad u(x,y,0)=f(x,y),$$
$$u_x(L,y,z)=0,\quad u_y(x,W,z)=g(x,z),\quad u(x,y,H)=0.$$

(a) Use the superposition principle to decompose the problem.
(b) Find the solution to the problem.

4. Solve the wave equation

$$u_{tt}=c^2\Delta u = c^2(u_{xx}+u_{yy}+u_{zz}),$$
$$0<x<L,\ 0<y<H,\ 0<z<H,\ t>0, \quad (7.1.26)$$

$$u_x(0,y,z,t)=0,\quad u_x(L,y,z,t)=0, \quad (7.1.27)$$
$$u(x,0,z,t)=0,\quad u(x,W,z,t)=0, \quad (7.1.28)$$
$$u(x,y,0,t)=0,\quad u_z(x,y,H,t)=0, \quad (7.1.29)$$

$$u(x,y,z,0)=f(x,y,z),\quad u_t(x,y,z,0)=g(x,y,z). \quad (7.1.30)$$

5. Solve the heat equation

$$u_t=c^2\Delta u = c^2(u_{xx}+u_{yy}+u_{zz}),$$
$$0<x<L,\ 0<y<W,\ 0<z<H,\ t>0, \quad (7.1.31)$$

$$u(0,y,z,t)=0,\quad u(L,y,z,t)=0, \quad (7.1.32)$$
$$u_y(x,0,z,t)=0,\quad u(x,W,z,t)=0, \quad (7.1.33)$$
$$u(x,y,0,t)=0,\quad u_z(x,y,H,t)=0, \quad (7.1.34)$$

$$u(x,y,0)=f(x,y). \quad (7.1.35)$$

7.2 Eigenvalue Problems

7.2.1 *Multidimensional Case*

The aim of the eigenvalue problems for the multidimensional case is essentially the same as the one-dimensional case. We find the values of λ for which there are nontrivial solutions for the boundary value problem

$$\mathcal{L}u = -\nabla \cdot (p\nabla u) + qu = \lambda hu, \quad \mathbf{x} \in \Omega,$$

$$\alpha u + \beta \frac{du}{dn} = 0, \quad \mathbf{x} \in \partial\Omega.$$

Some of the properties in the one dimensional case apply and some do not. One main difference is that Property 3 in Theorem 6.5.3 may not hold. There may be many eigenfunctions for one eigenvalue. For example, in Example 7.1.1, the eigenvalues and eigenfunctions are given by

$$\lambda_{mn} = (\frac{m\pi}{L})^2 + (\frac{n\pi}{H})^2,$$

$$v_{mn} = \cos\frac{m\pi x}{L}\sin\frac{n\pi y}{H}, \quad m = 0, 1,\ n = 1, 2, \ldots.$$

For simplicity consider the case where $L = H$ and $\lambda_{mn} = \lambda_{12} = \lambda_{21} = 5\pi^2/L^2$. Then, there are two eigenfunctions

$$v_{12} = \cos\frac{\pi x}{L}\sin\frac{2\pi y}{L}, \quad v_{21} = \cos\frac{2\pi x}{L}\sin\frac{\pi y}{L}.$$

Also, Property 4 in Theorem 6.5.3 should be adjusted after the definition of orthogonality in the multi-dimension is given.

Definition 7.2.1. We say that two functions ϕ_1 and ϕ_2 are orthogonal with respect to the weight $h > 0$ provided that

$$\int_\Omega h\phi_1\phi_2 d\mathbf{x} = 0,$$

where Ω is the region where the eigenvalue problem is defined.

If there are more than one eigenfunction for a given eigenvalue and they are not orthogonal, we may invoke the Gram-Schmidt orthogonalization procedure, explained in the next subsection, to find a set of orthogonal eigenfunctions for that eigenvalue. In the above example, v_{12} and v_{21} are already orthogonal.

7.2.2 *Gram-Schmidt Orthogonalization Procedure*

As explained in the previous subsection, in the multi-dimensional case there could be more than one eigenfunctions for an eigenvalue. They may or may not be orthogonal. It is more convenient if they are orthogonal. We explain briefly a procedure to make them orthogonal. The procedure is called the Gram-Schmidt orthogonalization. The procedure is similar to the case of vectors. Suppose $\psi_1, \psi_2, \ldots, \psi_n$ are the eigenfunctions for the same eigenvalue and assume that they are not orthogonal. We construct the orthogonal eigenfunctions $\phi_1, \phi_2, \ldots, \phi_n$ from $\psi_1, \psi_2, \ldots, \psi_n$. For ϕ_1 we choose $\phi_1 = \psi_1$. The choice of ψ_1 is arbitrary. For ϕ_2 we assume that ϕ_2 is given by $\phi_2 = \psi_2 - c_{21}\psi_1$, where c_{21} is a constant, and determine c_{21}. Since we want ϕ_2 to be orthogonal to ϕ_1, we choose c_1 so that

$$\int_\Omega h\phi_1\phi_2 d\mathbf{x} = \int_\Omega h\phi_1(\psi_2 - c_{21}\phi_1) d\mathbf{x} = 0.$$

From this we obtain

$$c_{21} = \frac{\int_\Omega h\phi_1\psi_2 d\mathbf{x}}{\int_\Omega h\phi_1^2 d\mathbf{x}}.$$

In general, we see that

$$\phi_k = \psi_k - \sum_{i=1}^{k-1} c_{ki}\phi_i, \quad k = 2, 3, \ldots, n,$$

where

$$c_{ki} = \frac{\int_\Omega h\phi_i\psi_k d\mathbf{x}}{\int_\Omega h\phi_i^2 d\mathbf{x}}, \quad i = 1, \ldots, k-1.$$

Example 7.2.2. Suppose $h = 1$, and $\psi_1 = 1$ and $\psi_2 = \sin(x/2)$ are given on $[0, \pi]$. Find the Gram-Schmidt orthogonalization.

Solution: We can set $\phi_1 = 1$. We need to find ϕ_2. Then,

$$\phi_2 = \psi_2 - c_{21}\phi_1. \tag{7.2.1}$$

To find c_{21}, multiply (7.2.1) by $\phi_1 = 1$ and integrate the resulting equation over $[0, \pi]$. Then,

$$\int_0^\pi \phi_1\phi_2 dx = 0 = \int_0^\pi 1 \sin\frac{x}{2} dx - c_{21}\int_0^\pi 1^2 dx,$$

$$c_{21} = \frac{\int_0^\pi 1 \sin\frac{x}{2} dx}{\int_0^\pi 1^2 dx} = \frac{2}{\pi}.$$

Therefore, $\phi_2 = \sin(x/2) - 2/\pi$.

7.2.3 *Rayleigh Quotient*

As an example of Rayleigh quotient, consider the eigenvalue problem for

$$\mathcal{L}u = -\nabla \cdot (p\nabla u) + qu = \lambda hu, \quad \mathbf{x} \in \Omega \tag{7.2.2}$$

with the Dirichlet boundary condition

$$u(\mathbf{x}) = 0, \quad \mathbf{x} \in \partial\Omega, \tag{7.2.3}$$

where the coefficients p, q, and h are functions of $\mathbf{x}$. We assume that $p(\mathbf{x}) > 0$ and $h(\mathbf{x}) > 0$. If $p = 1$, $q = 0$, $h = 1$, we have the eigenvalue problem for the Laplace equation.

To find the eigenvalues we multiply (7.2.2) by u and apply the Green's identity with the boundary condition (7.2.3). Then, we have

$$\lambda = \frac{\int_\Omega [p|\nabla u|^2 + qu^2] d\mathbf{x}}{\int_\Omega hu^2 d\mathbf{x}}.$$

This motivates to consider the minimization problem

$$m = \min \left\{ \frac{\int_\Omega [p|\nabla w|^2 + qw^2] d\mathbf{x}}{\int_\Omega hw^2 d\mathbf{x}} \mid w = 0 \text{ on } \partial\Omega,\ w \neq 0 \right\}. \tag{7.2.4}$$

The quotient inside (7.2.4) is called the Rayleigh quotient. We say that C^2-function u is a solution of (7.2.4) if u is not identically equal to zero, satisfies the boundary condition $w = 0$ on $\partial\Omega$, and satisfies

$$\frac{\int_\Omega [p|\nabla u|^2 + qu^2] d\mathbf{x}}{\int_\Omega hu^2 d\mathbf{x}} \leq \frac{\int_\Omega [p|\nabla w|^2 + qw^2] d\mathbf{x}}{\int_\Omega hw^2 d\mathbf{x}}$$

for all w with $w \neq 0$ and $w = 0$ on $\partial\Omega$.

Theorem 7.2.3. *Assume that $u(\mathbf{x})$ is a solution of (7.2.4). Then the value of the minimum is equal to the smallest (first) eigenvalue λ_1 to (7.2.2) and $u(\mathbf{x})$ is its eigenfunction.*

Proof. Let $u(\mathbf{x})$ be a solution of (7.2.4). By assumption

$$m = \frac{\int_\Omega [p|\nabla u|^2 + qu^2] d\mathbf{x}}{\int_\Omega hu^2 d\mathbf{x}} \leq \frac{\int_\Omega [p|\nabla w|^2 + qw^2] d\mathbf{x}}{\int_\Omega hw^2 d\mathbf{x}}$$

for all $w(\mathbf{x})$. Let $w = u + \varepsilon v$ where ε is any constant and v is any C^2 function satisfying the boundary condition. Then

$$f(\varepsilon) = \frac{\int_\Omega [p|\nabla (u + \varepsilon v)|^2 + q(u + \varepsilon v)^2] d\mathbf{x}}{\int_\Omega h(u + \varepsilon v)^2 d\mathbf{x}}$$

has a minimum at $\varepsilon = 0$. Therefore, $f'(0) = 0$. Since

$$f(\varepsilon) = \frac{\int_\Omega [p\{|\nabla u|^2 + 2\varepsilon \nabla u \cdot \nabla v + \varepsilon^2 |\nabla v|^2\} + q(u^2 + 2\varepsilon uv + \varepsilon^2 v^2)]d\mathbf{x}}{\int_\Omega h(u^2 + 2\varepsilon uv + \varepsilon^2 v^2)d\mathbf{x}}, \tag{7.2.5}$$

$$f'(0) = \frac{\int_\Omega [2p\nabla u \cdot \nabla v + 2quv]d\mathbf{x} \int_\Omega hu^2 d\mathbf{x}}{[\int_\Omega hu^2 d\mathbf{x}]^2} - \frac{\int_\Omega [p|\nabla u|^2 + qu^2]d\mathbf{x} \int_\Omega 2huvd\mathbf{x}}{[\int_\Omega hu^2 d\mathbf{x}]^2} = 0. \tag{7.2.6}$$

So,

$$\int_\Omega [p\nabla u \cdot \nabla v + quv]d\mathbf{x} = \frac{\int_\Omega [p|\nabla u|^2 + qu^2]d\mathbf{x} \int_\Omega huvd\mathbf{x}}{\int_\Omega hu^2 d\mathbf{x}} = m\int_\Omega huvd\mathbf{x}.$$

Using the Green's identity, we see that

$$\int_\Omega [-\nabla \cdot (p\nabla u) + qu - mhu]vd\mathbf{x} = 0.$$

Since this is valid for any function v, we conclude that

$$\mathcal{L}u = -\nabla \cdot (p\nabla u) + qu = mhu.$$

This show that the minimum value m of the Rayleigh quotient is an eigenvalue of $\mathcal{L}$ and $u(\mathbf{x})$ is its eigenfunction.

To show that m is the smallest eigenvalue of $\mathcal{L}$, let λ_i and v_i be the eigenvalue and the corresponding eigenfunction for $\mathcal{L}$. Then,

$$\mathcal{L}v_i = -\nabla \cdot (p\nabla v_i) + qv_i = \lambda_i h v_i.$$

By the definition of m as the minimum of the Rayleigh quotient

$$m \le \frac{\int_\Omega [p|\nabla v_i|^2 + qv_i^2]d\mathbf{x}}{\int_\Omega hv_i^2 d\mathbf{x}} = \frac{\int_\Omega [-\nabla \cdot (p\nabla v_i) + qv_i]v_i d\mathbf{x}}{\int_\Omega hv_i^2 d\mathbf{x}}$$
$$= \frac{\int_\Omega \lambda_i h v_i v_i d\mathbf{x}}{\int_\Omega hv_i^2 d\mathbf{x}} = \lambda_i.$$

Therefore, m is smaller than any other eigenvalue. □

Concerning the other eigenvalues we have

Theorem 7.2.4. *Suppose that the eigenvalues $\lambda_1, \ldots, \lambda_{n-1}$ and the corresponding eigenfunctions $v_1(\mathbf{x}), \ldots, v_{n-1}(\mathbf{x})$ are already determined. Then*

$$\lambda_n = \min\{\frac{\int_\Omega [p|\nabla w|^2 + qw^2]d\mathbf{x}}{\int_\Omega hw^2 d\mathbf{x}} \mid w = 0 \text{ on } \partial\Omega,\ w \neq 0,\ w \in C^2,$$
$$(w, v_1) = (w, v_2) = \cdots = (w, v_{n-1}) = 0\}, \tag{7.2.7}$$

provided that the minimum exists. Moreover, the minimizing function is the nth eigenfunction $v_n(\mathbf{x})$.

Proof. Suppose $u(\mathbf{x})$ is the minimizing function for (7.2.7). This exists by assumption. Let m denote the (minimum) value of the Rayleigh quotient for $w = u(\mathbf{x})$. We substitute $w = u + \varepsilon v$, where v satisfies the same constraints as w. Following the same procedure as in Theorem 7.2.3, we obtain

$$\int_\Omega [-\nabla \cdot (p\nabla u) + qu - mhu]v d\mathbf{x} = 0 \tag{7.2.8}$$

for all v satisfying the orthogonality relations

$$(v, v_1) = (v, v_2) = \cdots = (v, v_{n-1}) = 0. \tag{7.2.9}$$

The Green's identity and the fact that u is orthogonal to v_i, $i = 1, \ldots, (n-1)$ yield

$$\begin{aligned}
&\int_\Omega [-\nabla \cdot (p\nabla u) + qu - mhu]v_i d\mathbf{x} \\
&= \int_\Omega u[-\nabla \cdot (p\nabla v_i) + qv_i - mhv_i] d\mathbf{x} \\
&= (\lambda_i - m) \int_\Omega huv_i d\mathbf{x} = 0.
\end{aligned} \tag{7.2.10}$$

Now let $g(\mathbf{x})$ be any C^2 function satisfying $g \neq 0$ and $g = 0$ on $\partial\Omega$. Let

$$v(\mathbf{x}) = g(\mathbf{x}) - \sum_{k=1}^{n-1} c_k v_k(\mathbf{x}), \quad c_k = \frac{(g, v_k)}{(v_k, v_k)}.$$

Since this v satisfies (7.2.9), (7.2.8) is valid for this v. The linear combination of (7.2.8) and (7.2.10) lead to

$$\int_\Omega [-\nabla \cdot (p\nabla u) + qu - mhu] g d\mathbf{x} = 0.$$

This implies

$$-\nabla \cdot (p\nabla u) + qu = mhu,$$

which shows that m is an eigenvalue of $\mathcal{L}$. □

Often it is important to know the smallest eigenvalue. However, it is not easy to find its value. In such cases it is possible to estimate the smallest eigenvalue by Theorem 7.2.3. As an example consider

Example 7.2.5. Find an upper bound of the smallest eigenvalue.

$$u_{xx} + \lambda u = 0, \quad 0 < x < L,$$

$$u(0) = 0, \quad u(L) = 0.$$

Solution: We know that the smallest eigenvalue is $\lambda = (\pi/L)^2 = 9.87/L^2$. Since we are guessing the first eigenvalue, we choose u satisfying Property 3 in Theorem 6.5.3. If we choose

$$u = \begin{cases} x, & 0 < x < \frac{L}{2}, \\ L - x, & \frac{L}{2} < x < L, \end{cases}$$

Then, the Rayleigh quotient is

$$\lambda = \frac{\int_0^L (u')^2 dx}{\int_0^L u^2 dx} = \frac{\int_0^{L/2} dx + \int_{L/2}^L dx}{\int_0^{L/2} x^2 dx + \int_{L/2}^L (L-x)^2 dx} = \frac{12}{L^2}.$$

If we choose $u = x(L - x)$,

$$\lambda = \frac{\int_0^L (L-2x)^2 dx}{\int_0^L x^2(L-x)^2 dx} = \frac{L^3 - 2L^3 + \frac{4}{3}L^3}{\frac{1}{3}L^5 - \frac{1}{2}L^5 + \frac{1}{5}L^5} = \frac{10}{L^2}.$$

The second choice of u gives a good approximation for the smallest eigenvalue.

Exercises

1. Suppose $\psi_1 = 1$ and $\psi_2 = \cos(x/2)$ are given on $[0, \pi]$. Find the Gram-Schmidt orthogonalization if $h = 1$.
2. Suppose $\psi_1 = 1$, $\psi_2 = \cos(x/2)$, and $\psi_3 = \sin(x/2)$ are given on $[0, \pi]$. Find the Gram-Schmidt orthogonalization if $h = 1$.
3. Compute $f'(\varepsilon)$ from (7.2.5) and set $\varepsilon = 0$ to find $f'(0)$ in (7.2.6).
4. Use the Rayleigh quotient to obtain an upper bound of the lowest eigenvalue.

   (a) $\frac{d}{dx}[(1+x^2)\frac{du}{dx}] + (\lambda - x^2)u = 0, \quad u_x(0) = 0, \quad u_x(L) = 0.$
   (b) $\frac{d}{dx}[(1+x^2)\frac{du}{dx}] + (\lambda - x^2)u = 0, \quad u(0) = 0, \quad u_x(1) = 0.$

5. Consider

   $$\frac{d}{dx}[(1+x^2)\frac{du}{dx}] + (\lambda - x^2)u = 0, \quad u_x(0) = 0, \quad u_x(L) = 0.$$

   (a) Show that $\lambda \geq 0$.
   (b) Is $\lambda = 0$ an eigenvalue?

6. Consider the eigenvalue problem for

   $$\Delta u + \lambda u = 0, \quad 0 < x < L,\ 0 < y < H,$$

   $$u(0, y) = 0, \quad u(L, y) = 0,$$

   $$u_y(x, 0) = 0, \quad u_y(x, H) = 0.$$

(a) Find the eigenvalues λ and eigenfunctions.
(b) Show that eigenfunctions are orthogonal.
(c) If $L = H$, show that there are eigenvalues with more than one eigenfunctions.

7.3 Eigenfunction Expansions

In Sections 6.5 and 7.2 we learned that the eigenfunctions form a complete set and they are mutually orthogonal. With these results at hand, we extend the eigenfunction expansions (or the separation of variables) to non-homogeneous problems. There are roughly speaking two cases to discuss depending on whether the boundary conditions are homogeneous or not. In Subsection 7.3.1 we first discuss the non-homogeneous boundary conditions. Then, in Subsection 7.3.2 we study the homogeneous boundary conditions. In the third subsection we extend the idea of Section 2.4 and decompose the problem to two sub-problems.

7.3.1 *Non-homogeneous Boundary Conditions*

In this section we use the eigenfunction expansions to deal with the non-homogeneity in both the equation and the boundary conditions without decomposing into the two problems. We use the Green's identity or the integration by parts to handle both the equation and the boundary conditions. This is treated in Examples 7.3.1 and 7.3.3. For the eigenfunction expansions we use the eigenfunctions for the corresponding homogeneous boundary conditions. The drawback of this method is that the boundary conditions may not be satisfied. However, with this choice of eigenfunctions we do not create the unknown boundary conditions. Also, the Gibbs phenomenon might happen due to the non-homogeneous boundaries. To illustrate the method we start from the following examples.

Example 7.3.1. Find the solution to the

$$u_t = ku_{xx} + h(x,t), \quad 0 < x < L,\ 0 < t, \tag{7.3.1}$$

$$u_x(0,t) = b_l(t), \quad u_x(L,t) = b_r(t), \tag{7.3.2}$$

$$u(x,0) = f(x). \tag{7.3.3}$$

Solution: The eigenfunctions for the corresponding homogeneous problem are $\phi_m = \cos\frac{n\pi x}{L}$, $n = 0, 1, \ldots$. So, we assume

$$u(x,t) = \sum_{n=0}^{\infty} a_n(t) \cos\frac{n\pi x}{L}. \tag{7.3.4}$$

There are two observations for this form of solution. One is that u in (7.3.4) does not satisfy the boundary conditions. However, as we will see, this does not create unknown boundary terms. Another is the differentiability of (7.3.4). Since u_x is not equal at $x = 0$ and L, as we learned in Section 6.4, we expect that

$$u_{xx} \neq \sum_{n=1}^{\infty} a_n(t) \frac{d^2}{dx^2} \cos\frac{n\pi x}{L}. \tag{7.3.5}$$

Therefore, obviously we cannot use the right hand side of (7.3.5) for u_{xx} in (7.3.1). In the following procedure we circumvent such a problem by integration by parts (or the Green's identity). Since we expect that $a_n(t)$ is differentiable for $0 \leq t < \infty$, substituting (7.3.5) to the left hand side of (7.3.1), we obtain

$$\sum_{n=0}^{\infty} a_n'(t) \cos\frac{n\pi x}{L} = ku_{xx} + h(x,t).$$

The trick is to leave u_{xx} in the equation. Multiply the equation by the eigenfunctions $\cos\frac{m\pi x}{L}$, $m = 0, 1, \ldots$ and integrate in x over $[0, L]$. We have

$$\frac{L}{2} a_m'(t) = \int_0^L ku_{xx} \cos\frac{m\pi x}{L} dx + \int_0^L h(x,t) \cos\frac{m\pi x}{L} dx. \tag{7.3.6}$$

Performing the integrations by parts for the first term on the right hand side leads to

$$\begin{aligned}
\int_0^L ku_{xx} \cos\frac{m\pi x}{L} dx &= [ku_x \cos\frac{m\pi x}{L}]_0^L + \int_0^L ku_x \frac{m\pi}{L} \sin\frac{m\pi x}{L} dx \\
&= [kb_r(t)(-1)^m - kb_l(t)] + [ku\frac{m\pi}{L} \sin\frac{m\pi x}{L}]_0^L \\
&\quad - \int_0^L ku(\frac{m\pi}{L})^2 \cos\frac{m\pi x}{L} dx.
\end{aligned}$$

We now substitute (7.3.4) in the above u. Then, from (7.3.6) we obtain

$$a_m'(t) + k(\frac{m\pi}{L})^2 a_m(t) = \frac{2}{L}[kb_r(t)(-1)^m - kb_l(t)] + \frac{2}{L}\int_0^L h(x,t) \cos\frac{m\pi x}{L} dx. \tag{7.3.7}$$

For simplicity of presentation set

$$h_m(s) = \frac{2}{L}[kb_r(t)(-1)^m - kb_l(t)] + \frac{2}{L}\int_0^L h(x,t)\cos\frac{m\pi x}{L}dx.$$

Since (7.3.7) is a first order linear equation, Appendix A.3.1 implies that

$$a_m(t) = a_m(0)e^{-k(\frac{m\pi}{L})^2 t} + e^{-k(\frac{m\pi}{L})^2 t}\int_0^t h_m(s)e^{k(\frac{m\pi}{L})^2 s}ds.$$

To determine $a_m(0)$, we use the initial data. At $t = 0$,

$$\sum_{m=0}^{\infty} a_m(0)\cos\frac{m\pi x}{L} = f(x). \tag{7.3.8}$$

As usual we treat $m = 0$ and $m \geq 1$ separately. For $m = 0$ integration (7.3.8) over $[0.L]$ yields

$$a_0(0) = \frac{1}{L}\int_0^L f(x)dx,$$

and for $m \geq 1$ multiply (7.3.8) by $\cos(n\pi x/L)$ and integrate. Then, we obtain

$$a_n(0) = \frac{2}{L}\int_0^L f(x)\cos\frac{n\pi x}{L}dx, \quad n = 1, 2, \ldots.$$

Remark 7.3.2. Note that if we had differentiated (7.3.4) twice in x and substituted in (7.3.6), we would not have had the term $\frac{2}{L}[kb_r(t)(-1)^m - kb_l(t)]$ in (7.3.7). We would have been solving a different problem, the problem with the homogeneous Neumann boundary conditions.

In the next problem we consider the Poisson equation with non-homogeneous boundary conditions.

Example 7.3.3. Find the solution to

$$\nabla^2 u = h(x,y), \quad 0 < x < L,\ 0 < y < H,$$

$$u(0,y) = 0, \quad u(L,y) = 1,$$

$$u(x,0) = 0, \quad u(x,H) = 0.$$

Solution: We use the eigenfunction expansion. The question is what are the appropriate eigenfunctions to use. One choice is the eigenfunctions for $\nabla^2\phi = -\lambda\phi$ with the corresponding homogeneous boundary condition $\phi = 0$. For this choice the eigenvalue eigenfunction pairs are

$$\lambda_{mn} = (\frac{m\pi}{H})^2 + (\frac{n\pi}{L})^2, \quad \phi_{mn} = \sin\frac{m\pi y}{H}\sin\frac{n\pi x}{L}.$$

We assume that u is given by

$$u(x,y)=\sum_{m=1}^{\infty}\sum_{n=1}^{\infty}a_{mn}\phi_{mn}(x,y).$$

Then, by the Green's identity

$$\begin{aligned}
a_{mn} &= \frac{4}{LH}\int_0^L\int_0^H u\phi_{mn}dydx \\
&= -\frac{4}{LH\{(\frac{m\pi}{H})^2+(\frac{n\pi}{L})^2\}}\int_0^L\int_0^H u\nabla^2\phi_{mn}dydx \\
&= -\frac{4}{LH\{(\frac{m\pi}{H})^2+(\frac{n\pi}{L})^2\}}[\int_0^L\int_0^H \nabla^2 u\phi_{mn}dydx \\
&\quad+\int_{\partial\Omega}(u\frac{\partial\phi_{mn}}{\partial n}-\phi_{mn}\frac{\partial u}{\partial n})ds] \\
&= -\frac{4}{LH\{(\frac{m\pi}{H})^2+(\frac{n\pi}{L})^2\}}[\int_0^L\int_0^H h(x,t)\phi_{mn}dydx \\
&\quad+\int_0^H\begin{bmatrix}\frac{n\pi}{L}\sin\frac{m\pi y}{H}\cos n\pi \\ \frac{m\pi}{H}\cos\frac{m\pi y}{H}\sin n\pi\end{bmatrix}\cdot\begin{bmatrix}1\\0\end{bmatrix}dy] \\
&= -\frac{4}{LH\{(\frac{m\pi}{H})^2+(\frac{n\pi}{L})^2\}}[\int_0^L\int_0^H h(x,t)\phi_{mn}dydx \\
&\quad+(\frac{n\pi}{L})(\frac{H}{m\pi})((-1)^m-1)(-1)^n],
\end{aligned}$$

where the line integral along the boundary is evaluated as in Equation (4.2.4). See also Figure 4.1.

7.3.2 *Homogeneous Boundary Conditions*

We now consider the case where the boundary conditions are homogeneous. Remark 7.3.2 suggests that if the boundary conditions are homogeneous, we do not have a danger of losing the boundary conditions when we differentiate the eigenfunction expansions in the spacial variables. This simplifies the procedure significantly.

Example 7.3.4. Find the solution to the heat equation

$$u_t=\kappa^2u_{xx}+h(x,t),\quad 0<x<L,\ 0<t, \tag{7.3.9}$$

$$u(0,t)=0,\quad u(L,t)=0,$$

$$u(x,0) = f(x).$$

Solution: Since the eigenfunctions for the corresponding homogeneous problem are given by

$$\sin\frac{n\pi x}{L}, \quad n = 1, 2, \ldots,$$

it is reasonable to assume that u is given by

$$u(x,t) = \sum_{n=1}^{\infty} a_n(t) \sin\frac{n\pi x}{L}. \tag{7.3.10}$$

Substituting this in (7.3.9), we find that $a_n(t)$ satisfies

$$\sum_{n=1}^{\infty} a_n'(t) \sin\frac{n\pi x}{L} = -\kappa^2 \sum_{n=1}^{\infty} (\frac{n\pi}{L})^2 a_n(t) \sin\frac{n\pi x}{L} + h(x,t).$$

Therefore, multiplying by $\sin(m\pi x/L)$ and integrating over $[0, L]$, we have

$$a_m'(t) = -\kappa^2 (\frac{m\pi}{L})^2 a_m(t) + h_m(t), \quad m = 1, 2, \ldots, \tag{7.3.11}$$

where

$$h_m(t) = \frac{2}{L}\int_0^L h(x,t) \sin\frac{m\pi x}{L} dx.$$

The equations in (7.3.11) are the first order linear ODE's. By Appendix A.3.1, we see

$$a_m(t) = a_m(0) e^{-\kappa^2 (\frac{m\pi}{L})^2 t} + e^{-\kappa^2 (\frac{m\pi}{L})^2 t} \int_0^t h_m(s) e^{\kappa^2 (\frac{m\pi}{L})^2 s} ds.$$

From (7.3.10) we have

$$u(x,0) = \sum_{n=1}^{\infty} a_n(0) \sin\frac{n\pi x}{L} = f(x). \tag{7.3.12}$$

Therefore, multiplying (7.3.21) by $\sin(m\pi x/L)$ and integrating over $(0, L)$, we obtain

$$a_m(0) = \frac{2}{L}\int_0^L f(x) \sin\frac{m\pi x}{L} dx.$$

In the following example, the above procedure is worked out for specific functions with slight change in the boundary condition at $x = L$.

Example 7.3.5. Solve

$$u_t = \kappa^2 u_{xx} + xe^t, \quad 0 < x < L,\ 0 < t,$$
$$u(0,t) = 0, \quad u_x(L,t) = 0,$$
$$u(x,0) = 1.$$

Solution: Notice that the boundary condition at $x = L$ is different from the one in Example 7.3.4. However, the procedure is the same. Since the eigenfunctions for the corresponding homogeneous problem are given by

$$\sin\frac{(n-\frac{1}{2})\pi x}{L}, \quad n = 1, 2, \ldots,$$

we assume that u is given by

$$u(x,t) = \sum_{n=1}^{\infty} a_n(t) \sin\frac{(n-\frac{1}{2})\pi x}{L}.$$

Substituting this in (7.3.9) we obtain

$$\sum_{n=1}^{\infty} a_n'(t) \sin\frac{(n-\frac{1}{2})\pi x}{L}$$
$$= -\kappa^2 \sum_{n=1}^{\infty} (\frac{(n-\frac{1}{2})\pi}{L})^2 a_n(t) \sin\frac{(n-\frac{1}{2})\pi x}{L} + xe^t.$$

Therefore, multiplying by $\sin[(m-1/2)\pi x/L]$ and integrating over $[0, L]$, we have

$$a_m'(t) = -\kappa^2 (\frac{(m-\frac{1}{2})\pi}{L})^2 a_m(t) + h_m(t),\ m = 1, 2, \ldots, \tag{7.3.13}$$

where

$$h_m(t) = \frac{2}{L}\int_0^L xe^t \sin\frac{(m-\frac{1}{2})\pi x}{L} dx = \frac{2}{L}[-xe^t \frac{L}{(m-\frac{1}{2})\pi} \cos\frac{(m-\frac{1}{2})\pi x}{L}]_0^L$$
$$+\frac{2}{L}\int_0^L e^t \frac{L}{(m-\frac{1}{2})\pi} \cos\frac{(m-\frac{1}{2})\pi x}{L} dx$$
$$= e^t \frac{2}{L}(\frac{L}{(m-\frac{1}{2})\pi})^2 (-1)^{m+1}.$$

Since (7.3.13) is the first order linear ODE in $a_m(t)$, we see

$$a_m(t) = a_m(0)e^{-\kappa^2(\frac{(m-\frac{1}{2})\pi}{L})^2 t} + e^{-\kappa^2(\frac{(m-\frac{1}{2})\pi}{L})^2 t}\int_0^t h_m(s)e^{\kappa^2(\frac{(m-\frac{1}{2})\pi}{L})^2 s} ds$$
$$= a_m(0)e^{-\kappa^2(\frac{(m-\frac{1}{2})\pi}{L})^2 t}$$
$$+\frac{2}{L}\frac{(\frac{L}{(m-\frac{1}{2})\pi})^2(-1)^{m+1}}{\kappa^2(\frac{(m-\frac{1}{2})\pi}{L})^2 + 1} e^{-\kappa^2(\frac{(m-\frac{1}{2})\pi}{L})^2 t}[e^{\kappa^2(\frac{(m-\frac{1}{2})\pi}{L})^2 t + t} - 1]. \tag{7.3.14}$$

At $t = 0$,

$$u(x,0) = \sum_{n=1}^{\infty} a_n(0) \sin \frac{(n-\frac{1}{2})\pi x}{L} = 1 \tag{7.3.15}$$

holds. Therefore, to find $a_m(0)$ we multiply (7.3.15) by $\sin[(m-1/2)\pi x/L]$ and integrating on $(0, L)$. Then,

$$\begin{aligned} a_m(0) &= \frac{2}{L}\int_0^L \sin \frac{(m-\frac{1}{2})\pi x}{L} dx \\ &= \frac{2}{L}[-\frac{L}{(m-\frac{1}{2})\pi} \cos \frac{(m-\frac{1}{2})\pi x}{L}]_0^L = \frac{2}{(m-\frac{1}{2})\pi}. \end{aligned} \tag{7.3.16}$$

Therefore, the solution is

$$u(x,t) = \sum_{m=1}^{\infty} a_m(t) \sin \frac{(m-\frac{1}{2})\pi x}{L},$$

where $a_m(t)$ and $a_m(0)$ are given, respectively, by (7.3.14) and (7.3.16).

7.3.3 *Hybrid Method*

The method here is an extension of Section 2.4. We use the superposition principle to decompose the problem to two problems. We find a function satisfying the boundary conditions and use it to reduce the problem to the one with the homogeneous boundary conditions. This method is illustrated in Example 7.3.6. Also as an example of higher dimensional problems, we consider the Poisson equation in Example 7.3.7. These two problems are the same as Examples 7.3.1 and 7.3.3, respectively. The method discussed in Subsection 7.3.1 has less work to do since we solve the non-homogeneous boundary conditions directly. As indicated, the drawback is that the eigenfunctions do not satisfy the boundary conditions. Also, the Gibbs phenomenon might happen. On the other hand, with the method in this section we use the eigenfunction expansion to the homogeneous problems. Therefore, the Gibbs phenomenon is less likely to happen.

Example 7.3.6. Solve

$$u_t = ku_{xx} + h(x,t), \quad 0 < x < L,\ 0 < t,$$

$$u_x(0,t) = b_l(t), \quad u_x(L,t) = b_r(t),$$

$$u(x,0) = f(x).$$

Solution: Step 1: As described in the beginning of this subsection, we first find a function $r(x,t)$ satisfying the boundary conditions. For this purpose we assume $u = v(x,t) + r(x,t)$, where r is a known function satisfying the boundary conditions. Set $r_x(x,t) = c(t)x + b_l(t)$. Then, at $x = L$, $r_x(L,t) = Lc(t) + b_l(t) = b_r(t)$. $c(t) = (b_r(t) - b_l(t))/L$. Therefore,

$$r_x(x,t) = b_l(t) + \frac{x}{L}(b_r(t) - b_l(t)).$$

Integrating in x, we have

$$r(x,t) = b_l(t)x + (b_r(t) - b_l(t))\frac{x^2}{2L} + d(t),$$

where $d(t)$ is a constant of integration. Since for $r(x,t)$ we are seeking a known function satisfying the boundary conditions, the simplest choice is $d(t) = 0$.

Step 2: In this step we find v. Substitute u with this choice of $d(t)$ in the equation. Then,

$$\begin{aligned} v_t + b_l'(t)x + (b_r'(t) - b_l'(t))\frac{x^2}{2L} &= kv_{xx} + k(b_r(t) - b_l(t))\frac{1}{L} + h(x,t), \\ v_t &= kv_{xx} + \bar{h}(x,t), \end{aligned}$$

where $\bar{h}(x,t) = h(x,t) + k(b_r(t) - b_l(t))\frac{1}{L} - b_l'(t)x - (b_r'(t) - b_l'(t))\frac{x^2}{2L}$. The initial condition for v is

$$\begin{aligned} v(x,0) + r(x,0) &= f(x), \\ v(x,0) &= f(x) - b_l(0)x - (b_r(0) - b_l(0))\frac{x^2}{2L}. \end{aligned}$$

The boundary conditions for v is

$$v_x(0,t) = 0, \quad v_x(L,t) = 0.$$

We apply the eigenfunction expansion for v. Assume

$$v(x,t) = \sum_{n=0}^{\infty} a_n(t) \cos\frac{n\pi x}{L}.$$

Then, $a_n(t)$ satisfies

$$\sum_{n=0}^{\infty} a_n'(t) \cos\frac{n\pi x}{L} = -k\sum_{n=0}^{\infty} a_n(t)(\frac{n\pi}{L})^2 \cos\frac{n\pi x}{L} + \bar{h}(x,t). \qquad (7.3.17)$$

Multiply (7.3.17) by $\cos(m\pi x/L)$ and integrate

$$a_m'(t) = -ka_m(t)(\frac{m\pi}{L})^2 + h_m(t), \quad m = 0, 1, \ldots,$$

where

$$h_m(t) = \frac{2}{L}\int_0^L \bar{h}(x,t)\cos\frac{m\pi x}{L}dx.$$

The above ODE is a linear first order equation. By Appendix A.3.1 we multiply the equation by the integrating factor $\exp[k(m\pi/L)^2 t]$, where $\exp(a) = e^a$, and integrate in t. Then, the solution is

$$a_m(t) = a_m(0)e^{-k(\frac{m\pi}{L})^2 t} + e^{-k(\frac{m\pi}{L})^2 t}\int_0^t h_m(s)e^{k(\frac{m\pi}{L})^2 s}ds.$$

To determine $a_m(0)$, we use the initial data. At $t = 0$,

$$\sum_{m=0}^{\infty} a_m(0)\cos\frac{m\pi x}{L} = f(x) - b_l(0)x - (b_r(0) - b_l(0))\frac{x^2}{2L}. \quad (7.3.18)$$

As usual we treat $m = 0$ and $m \geq 1$ separately. For $m = 0$ integration (7.3.18) over $[0.L]$ yields

$$a_0(0) = \frac{1}{L}\int_0^L \{f(x) - b_l(0)x - (b_r(0) - b_l(0))\frac{x^2}{2L}\}dx,$$

and for $m \geq 1$ multiply (7.3.18) by $\cos(n\pi x/L)$ and integrate. Then, we obtain

$$a_n(0) = \frac{2}{L}\int_0^L \{f(x) - b_l(0)x - (b_r(0) - b_l(0))\frac{x^2}{2L}\}\cos\frac{n\pi x}{L}dx, \quad n = 1, 2, \ldots.$$

Example 7.3.7. Consider

$$\nabla^2 u = h(x,y), \quad 0 < x < L,\ 0 < y < H,$$

$$u(0,y) = 0,\ u(L,y) = 1,\ u(x,0) = 0,\ u(x,H) = 0.$$

Solution: We decompose the problem into two sub-problems $u = v + w$, where v and w satisfy

$$\nabla^2 v = h, \quad v(0,y) = 0,\ v(L,y) = 0,\ v(x,0) = 0,\ v(x,H) = 0, \quad (7.3.19)$$

$$\nabla^2 w = 0, \quad w(0,y) = 0,\ w(L,y) = 1,\ w(x,0) = 0,\ w(x,H) = 0. \quad (7.3.20)$$

Step 1: We find the solution v for (7.3.19). We have the homogeneous Dirichlet boundary conditions and this suggests that we assume the eigenfunction expansions of the form

$$v(x,y) = \sum_{m=1}^{\infty}\sum_{n=1}^{\infty} a_{mn}\sin\frac{m\pi y}{H}\sin\frac{n\pi x}{L} \quad (7.3.21)$$

as in Example 7.3.3. Since the boundary conditions are homogeneous, we proceed as in Subsection 7.3.2. Substituting (7.3.21) in (7.3.19), we have

$$-\sum_{m=1}^{\infty}\sum_{n=1}^{\infty} a_{mn}\{(\frac{m\pi}{H})^2+(\frac{n\pi}{L})^2\}\sin\frac{m\pi y}{H}\sin\frac{n\pi x}{L}=h(x,y). \tag{7.3.22}$$

Multiply (7.3.22) by $\sin\frac{m\pi y}{H}\sin\frac{n\pi x}{L}$ and integrate the resulting equation. Then,

$$a_{mn}=-\frac{4}{LH\{(\frac{m\pi}{H})^2+(\frac{n\pi}{L})^2\}}\int_0^L\int_0^H h\sin\frac{m\pi y}{H}\sin\frac{n\pi x}{L}dydx.$$

Step 2: We find the solution w for (7.3.20). As in Section 4.2 we assume $w(x,y)=X(x)Y(y)$. The boundary conditions suggest that we solve the eigenvalue problem (7.3.24) for Y and then solve (7.3.23) for each eigenvalue.

$$X''-\lambda X=0,\quad X(0)=0, \tag{7.3.23}$$

$$Y''+\lambda Y=0,\quad Y(0)=0,\ Y(H)=0. \tag{7.3.24}$$

Then, the eigenvalues and eigenfunctions for Y are

$$\lambda_n=(\frac{n\pi}{H})^2,\quad Y_n(y)=\sin\frac{n\pi y}{H}$$

and the corresponding solution for X is

$$X_n(x)=b_n\sinh\frac{n\pi x}{H}.$$

Therefore,

$$w(x,y)=\sum_{n=1}^{\infty}b_n\sinh\frac{n\pi x}{H}\sin\frac{n\pi y}{H}.$$

The boundary condition $w(L,y)=1$ leads to

$$w(L,y)=\sum_{n=1}^{\infty}b_n\sinh\frac{n\pi L}{H}\sin\frac{n\pi y}{H}=1.$$

Multiplying this by $\sin\frac{m\pi y}{H}$ and integrating over $[0,H]$, we obtain

$$b_m=\frac{2}{H\sinh\frac{m\pi L}{H}}\int_0^H\sin\frac{m\pi y}{H}dy=\frac{2(1-(-1)^m)}{m\pi\sinh\frac{m\pi L}{H}},\ m=1,2,\dots.$$

Exercises

1. Find the solution to the initial boundary value problem in two different ways.

$$u_t = a^2 u_{xx} + h(x,t), \quad 0 < x < L,\ 0 < t,$$

$$u(0,t) = b_l(t), \quad u_x(L,t) = b_r(t),$$

$$u(x,0) = f(x).$$

   (a) Follow Example 7.3.6.
   (b) Follow Example 7.3.1.

2. Find the solution to the Poisson equation on a rectangle in two different ways.

$$\Delta u = h(x,y), \quad 0 < x < L,\ 0 < y < H,$$

$$u(0,y) = 0,\ u(L,y) = 0,\ u(x,0) = 0,\ u_y(x,H) = 1.$$

   (a) Decompose the problem to two sub-problems.
   (b) Handle both non-homogeneous terms at the same time.

3. Solve the initial boundary value problem for the wave equation if $\Omega = \{\mathbf{x} = (x,y) \mid 0 < x < L,\ 0 < y < H\}$.

$$u_{tt} = c^2 \Delta u + h(\mathbf{x},t), \quad \mathbf{x} \in \Omega,\ 0 < t,$$

$$u(\mathbf{x}) = 0, \quad \mathbf{x} \in \partial\Omega,$$

$$u(\mathbf{x},0) = f(\mathbf{x}), \quad u_t(\mathbf{x},0) = g(\mathbf{x}).$$

Chapter 8

More Separation of Variables

In this chapter we continue the separation of variables in higher dimensions and study the effects of geometry. In this chapter we treat the cases where the domains are circular, cylindrical, and spherical. These are the domains more complex than the rectangular ones, yet the separation of variables is applicable. ODE's we obtain are not so simple as in the rectangular case. For the circular, cylindrical, and spherical cases we need to introduce the Bessel's equation to handle the radial components in the separation of variables. Also, in the spherical case we introduce the associated Legendre equation to handle one of the angular components.

8.1 Circular Domains

We already did the Laplace equation in a circular disk in Chapter 4. In this section we use the heat equation as an example to study how the separation of variables is carried out in the polar coordinates for the time dependent problems.

8.1.1 *Initial Boundary Value Problems*

We discuss the heat equation in a circular disk with radius a. So, the domain is $\Omega = \{(r, \theta) \mid 0 \leq r < a, \ -\pi < \theta < \pi\}$. We use the Laplacian in the polar coordinates, whose derivation is discussed in Subsection 4.2.2. The initial boundary value problem is formulated as follows.

$$u_t = \alpha^2 \Delta u = \alpha^2 \{\frac{1}{r}\frac{\partial}{\partial r}(r\frac{\partial u}{\partial r}) + \frac{1}{r^2}\frac{\partial^2 u}{\partial \theta^2}\}, \quad (r, \theta) \in \Omega, \ 0 < t \tag{8.1.1}$$

with the boundary conditions

$$u(a,\theta,t)=0, \tag{8.1.2}$$

$$u(r,-\pi,t)=u(r,\pi,t), \quad u_\theta(r,-\pi,t)=u_\theta(r,\pi,t), \tag{8.1.3}$$

and the initial condition

$$u(r,\theta,0)=f(r,\theta). \tag{8.1.4}$$

The boundary condition for r at $r=a$ is given in (8.1.2). The boundary condition for r at $r=0$ is not given explicitly. However, it is physically reasonable to require that the solution is finite at $r=0$ and this provides the boundary condition at $r=0$. The boundary conditions for θ are given in (8.1.3). They are the same periodic boundary conditions as in Subsection 4.2.2, which guarantee that the solution is smoothly connected across $\theta = \pm\pi$.

Separating Variables

We use the separation of variables to solve the initial boundary value problem. So, we assume that the solution is given by

$$u=v(r,\theta)T(t). \tag{8.1.5}$$

As in the rectangular case we substitute (8.1.5) in (8.1.1) and divide the resulting equation by $\alpha^2 v(r,\theta)T(t)$. Then, we have

$$\frac{\frac{dT}{dt}}{\alpha^2 T}=\frac{\{\frac{1}{r}\frac{\partial}{\partial r}(r\frac{\partial v}{\partial r})+\frac{1}{r^2}\frac{\partial^2 v}{\partial \theta^2}\}}{v}.$$

We separate the variables to have

$$\frac{dT}{dt}=-\lambda\alpha^2 T, \tag{8.1.6}$$

$$\frac{1}{r}\frac{\partial}{\partial r}(r\frac{\partial v}{\partial r})+\frac{1}{r^2}\frac{\partial^2 v}{\partial \theta^2}=-\lambda v, \tag{8.1.7}$$

where λ is a separation constant to be determined. To separate r and θ, we assume that $v=R(r)\Theta(\theta)$ and divide the resulting equation by $R(r)\Theta(\theta)/r^2$. Then,

$$\frac{r\frac{d}{dr}(r\frac{dR}{dr})}{R}+\frac{\frac{d^2\Theta}{d\theta^2}}{\Theta}=-\lambda r^2. \tag{8.1.8}$$

Solving ODE's

We solve the ODE's for Θ, R, and T in (8.1.8) and (8.1.6). We first separate Θ and R in (8.1.8) and solve two eigenvalue problems. For Θ we have the eigenvalue problem

$$\frac{d^2\Theta}{d\theta^2} = -\mu\Theta \tag{8.1.9}$$

with the periodic boundary conditions

$$\Theta(-\pi) = \Theta(\pi), \quad \frac{d\Theta}{d\theta}(-\pi) = \frac{d\Theta}{d\theta}(\pi). \tag{8.1.10}$$

Both ODE and the boundary conditions are exactly the same as in Subsection 4.2.2. The eigenvalues μ are given by

$$\mu_n = n^2, \quad n = 0, 1, 2, \ldots \tag{8.1.11}$$

and the corresponding eigenfunctions Θ are

$$\Theta_0(\theta) = a_0, \quad \Theta_n(\theta) = a_n \cos n\theta + b_n \sin n\theta, \quad n = 1, 2, \ldots. \tag{8.1.12}$$

The ODE associated with the eigenvalue problem R is called the Bessel's equation and from (8.1.8) and (8.1.9) with $\mu = n^2$, it is given by

$$r^2\frac{d^2R}{dr^2} + r\frac{dR}{dr} + (\lambda r^2 - n^2)R = 0. \tag{8.1.13}$$

The boundary conditions are

$$R(0) \text{ is finite}, \quad R(a) = 0.$$

If we prefer a form consistent with (6.5.1) in the Sturm-Liouville theory, we have

$$-\frac{d}{dr}(r\frac{dR}{dr}) + \frac{n^2}{r}R = \lambda r R. \tag{8.1.14}$$

Unfortunately (8.1.13) is not exactly the same as (4.2.26) in Subsection 4.2.2. Because of $\lambda r^2 R$ term we do not have $R = cr^k$ type solutions. To find the solutions we set $s = \sqrt{\lambda}r$ to normalize the equation. Then, $dR/dr = \sqrt{\lambda}dR/ds$, $d^2R/dr^2 = \frac{d^2R}{dr^2} = \lambda d^2R/ds^2$, and

$$s^2\frac{d^2R}{ds^2} + s\frac{dR}{ds} + (s^2 - n^2)R = 0, \quad n = 0, 1, \ldots. \tag{8.1.15}$$

There are two linearly independent solutions J_n and Y_n to (8.1.15) for each n. They are called the Bessel functions of the first and second kind of order n, respectively, and given by

$$J_n(s) = \frac{s^n}{2^n n!}[1 - \frac{s^2}{2^2(n+1)} + \frac{s^4}{2!2^2(n+1)(n+2)} - \cdots]$$
$$= \sum_{k=0}^{\infty}(-1)^k \frac{1}{k!(n+k)!}(\frac{s}{2})^{n+2k}, \tag{8.1.16}$$

and

$$Y_n(s) = \frac{2}{\pi}(\ln\frac{s}{2} + \eta)J_n(s) - \frac{1}{\pi}\sum_{k=0}^{n-1}\frac{(n-k-1)!}{k!}(\frac{s}{2})^{2k-n}$$
$$+\frac{1}{\pi}\sum_{k=0}^{\infty}(-1)^{k+1}\frac{\psi(k)+\psi(k+n)}{k!(k+n)!}(\frac{s}{2})^{2k+n}, \tag{8.1.17}$$

where

$$\psi(k) = 1 + \frac{1}{2} + \cdots + \frac{1}{k},\ \psi(0) = 0,$$

$$\eta = \lim_{k\to\infty}[\psi(k) - \ln k] = 0.5772157\ldots,$$

$$\text{if } n = 0,\ \sum_{k=0}^{n-1} = 0.$$

Their derivation is given in Subsection 8.1.2. With J_n and Y_n, the solution to (8.1.15) for each n is expressed as

$$R_n(s) = c_1 J_n(s) + c_2 Y_n(s).$$

Since Y_n has a singularity at $s = 0$, we set $c_2 = 0$ so that $R(0)$ is finite. Denote the mth zero of $J_n(s)$ as s_{nm}. As seen in Figure 8.1 J_n's are oscillatory and have infinite number of zeros. The values of λ are chosen so that $J_n(\sqrt{\lambda}a) = 0$. Therefore, the eigenvalues for λ are

$$\lambda_{nm} = (\frac{s_{nm}}{a})^2,\quad m = 1, 2, \ldots,\ n = 0, 1, \ldots,$$

and by the Sturm-Liouville theory, for the same n we have

$$\int_0^a J_n(\sqrt{\lambda_{nm}}r)J_n(\sqrt{\lambda_{nk}}r)rdr = 0,\quad m \neq k.$$

For T we solve the initial value problem (8.1.6) for each λ_{nm}. Then,

$$T_{nm}(t) = T_0 e^{-\alpha^2\lambda_{nm}t}.$$

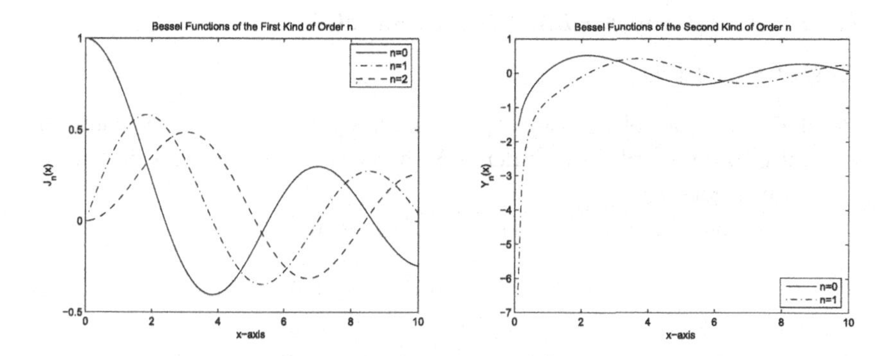

Fig. 8.1 Bessel functions of the first and second kind of order n.

Solution to the Initial Boundary Value Problem

To find the solution to the initial boundary value problem (8.1.1) to (8.1.4), we combine what we have done. From $v(r,\theta) = R(r)\Theta(\theta)$ we have

$$v_{nm}(r,\theta) = J_n(\sqrt{\lambda_{nm}}r)(a_{nm}\cos n\theta + b_{nm}\sin n\theta).$$

We denote the eigenfunctions by

$$v^c_{nm} = J_n(\sqrt{\lambda_{nm}}r)\cos n\theta, \quad v^s_{nm} = J_n(\sqrt{\lambda_{nm}}r)\sin n\theta.$$

Since we assume $u = v(r,\theta)T(t)$, the solution is

$$\begin{aligned} u(r,\theta,t) &= \sum_{n=0}^{\infty}\sum_{m=1}^{\infty} e^{-\alpha^2\lambda_{nm}t} J_n(\sqrt{\lambda_{nm}}r)(a_{nm}\cos n\theta + b_{nm}\sin n\theta) \\ &= \sum_{n=0}^{\infty}\sum_{m=1}^{\infty} e^{-\alpha^2\lambda_{nm}t}(a_{nm}v^c_{nm} + b_{nm}v^s_{nm}). \end{aligned}$$

At $t = 0$

$$u(r,\theta,0) = f(r,\theta) = \sum_{n=0}^{\infty}\sum_{m=1}^{\infty}(a_{nm}v^c_{nm} + b_{nm}v^s_{nm}). \tag{8.1.18}$$

Therefore, using the orthogonality relations, we obtain

$$a_{nm} = \frac{\int_{-\pi}^{\pi}\int_0^a f(r,\theta)v^c_{nm}rdrd\theta}{\int_{-\pi}^{\pi}\int_0^a (v^c_{nm})^2 rdrd\theta},$$

$$b_{nm} = \frac{\int_{-\pi}^{\pi}\int_0^a f(r,\theta)v^s_{nm}rdrd\theta}{\int_{-\pi}^{\pi}\int_0^a (v^s_{nm})^2 rdrd\theta}.$$

Note that for example the denominator b_{nm} is the same as

$$\int_{-\pi}^{\pi}\int_0^a (J_n(\sqrt{\lambda_{nm}}r)\sin n\theta)^2 rdrd\theta = \int_0^a J_n^2(\sqrt{\lambda_{nm}}r)rdr\int_{-\pi}^{\pi}\sin^2 n\theta d\theta.$$

The evaluation of the normalizing integral $\int_0^a J_n^2(\sqrt{\lambda_{nm}}r)rdr$ is not trivial and given in Subsection 8.1.2.

8.1.2 *Bessel and Modified Bessel Functions*

Bessel Functions

We derive the Bessel function J_n as an example of power series solution. Since it is more complicated to derive Y_n as we see in (8.1.17), the derivation is not given for Y_n.

We assume a power series solution of the form

$$R(s) = s^p \sum_{k=0}^{\infty} a_k s^k, \quad a_0 \neq 0, \tag{8.1.19}$$

and substitute (8.1.19) to (8.1.15) to determine p and a_k. Since

$$R'(s) = \sum_{k=0}^{\infty}(k+p)a_k s^{k+p-1}, \; R''(s) = \sum_{k=0}^{\infty}(k+p)(k+p-1)a_k s^{k+p-2},$$

(8.1.15) yields

$$\sum_{k=0}^{\infty}(k+p)(k+p-1)a_k s^{k+p} + \sum_{k=0}^{\infty}(k+p)a_k s^{k+p} + \sum_{k=0}^{\infty} a_k s^{k+p+2} - \sum_{k=0}^{\infty} n^2 a_k s^{k+p} = 0.$$

Changing the index for the third term, we have

$$\sum_{k=0}^{\infty}(k+p)(k+p-1)a_k s^{k+p} + \sum_{k=0}^{\infty}(k+p)a_k s^{k+p} + \sum_{k=2}^{\infty} a_{k-2} s^{k+p} - \sum_{k=0}^{\infty} n^2 a_k s^{k+p} = 0.$$

We collect the like powers of k. Then, for $k = 0$,

$$p(p-1)a_0 + pa_0 - n^2 a_0 = 0 \Rightarrow p^2 - n^2 = 0,$$

from which we obtain $p = \pm n$.

If we choose $p = n$, we obtain J_n, the Bessel function of the first kind of order n. For $k = 1$,

$$(1+n)na_1 + (1+n)a_1 - n^2 a_1 = 0 \Rightarrow a_1 = 0.$$

For $k = 2$,

$$(2+n)(1+n)a_2 + (2+n)a_2 + a_0 - n^2 a_2 = 0.$$

So, if k is odd, $a_k = 0$ and if k is even,

$$a_k = -\frac{a_{k-2}}{(k+n)^2 - n^2}. \tag{8.1.20}$$

This is called a recursive relation and every a_k ($k \geq 2$ and even) is expressed in terms of a_0. Substituting (8.1.20) in (8.1.19), we obtain (8.1.16).

If we choose $p = -n$, we obtain the second linearly independent solution Y_n given in (8.1.17), the Bessel function of the second kind of order n. As mentioned earlier this has a singularity at $x = 0$.

Modified Bessel Functions

The modified Bessel functions are the solutions to

$$s^2 \frac{d^2R}{ds^2} + s\frac{dR}{ds} + (-s^2 - n^2)R = 0. \tag{8.1.21}$$

The difference is the sign of the coefficient for R term. There are two linearly independent solutions denoted as $I_n(s)$ and $K_n(s)$. They are well tabulated and referred as the modified Bessel function of order n of the first and second kind, respectively. Their graphs for a few values of n are shown in Figure 8.2. The first kind is well defined at $s = 0$ while the second kind is singular at $s = 0$. Both the first and second kind are non-oscillatory and they are not zero for $s > 0$. Therefore, they would not be used as the eigenfunctions. The modified Bessel functions will appear in the next section when we study the Laplace equation in the cylindrical coordinates.

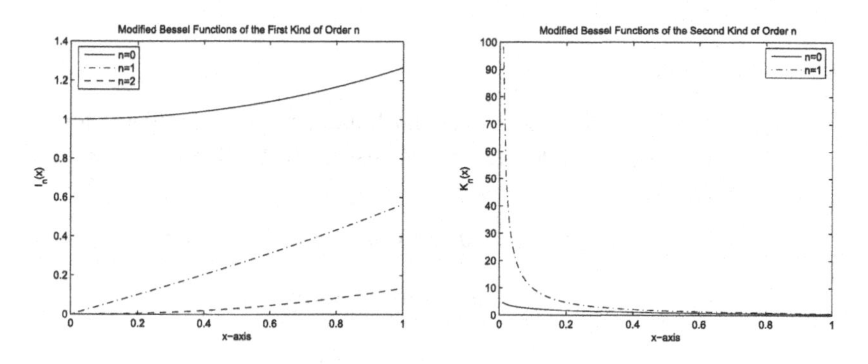

Fig. 8.2 Modified Bessel functions of the first and second kind of order n.

Evaluation of Normalizing Integrals

It is not trivial to the evaluate the normalizing integral $\int_0^a J_n^2(\sqrt{\lambda_{nm}}r)rdr$. To find the integral, we observe that $J_n(s)$ satisfies

$$s^2 \frac{d^2R}{ds^2} + s\frac{dR}{ds} + (s^2 - n^2)R = (sR')' + s^{-1}(s^2 - n^2)R = 0.$$

Multiplying the equation by $2sR'$, we obtain the identity

$$[(sR')^2 + (s^2 - n^2)R^2]' = 2sR^2.$$

Integrating this over $0 < s < \sqrt{\lambda_{nm}}a$, we have

$$\begin{aligned} 2\int_0^{\sqrt{\lambda_{mn}}a} sR^2 ds &= [(sR')^2 + (s^2 - n^2)R^2]_0^{\sqrt{\lambda_{nm}}a} \\ &= \lambda_{nm}a^2 R'(\sqrt{\lambda_{nm}}a)^2 + (\lambda_{nm}a^2 - n^2)R(\sqrt{\lambda_{nm}}a)^2. \end{aligned}$$

After changing the variable to r with $s = \sqrt{\lambda_{nm}}r$, we obtain

$$2\lambda_{nm}\int_0^a R^2(\sqrt{\lambda_{nm}}r)rdr = \lambda_{nm}a^2 R'(\sqrt{\lambda_{nm}}a)^2 + (\lambda_{nm}a^2 - n^2)R(\sqrt{\lambda_{nm}}a)^2.$$

In particular if $\sqrt{\lambda_{nm}}a$ is a zero of $J_n(s)$,

$$2\int_0^a J_n^2(\sqrt{\lambda_{nm}}r)rdr = a^2 J_n'(\sqrt{\lambda_{nm}}a)^2 = a^2 J_{n+1}(\sqrt{\lambda_{nm}}a)^2. \quad (8.1.22)$$

The last equality is derived in Exercise 5 at the end of this section. We use this relation to evaluate the normalizing integrals. More details about the Bessel functions including the evaluation of normalizing integrals are available in [Abramowitz and Stegun (1974); Churchill (1972); Watson (1995)].

Exercises

1. Find the solutions to the initial boundary value problems for the heat equation inside the circular disk with the Neumann boundary condition.

$$u_t = \alpha^2 \Delta u, \quad 0 < r < a, \ -\pi < \theta < \pi, \ 0 < t,$$

$$u(r, \theta, 0) = f(r, \theta),$$

$$u_r(a, \theta, t) = 0.$$

2. Find the solutions to the initial boundary value problems for the wave equation inside the circular disk

$$u_{tt} = \alpha^2 \Delta u, \quad 0 < r < a, \ -\pi < \theta < \pi, \ 0 < t,$$

$$u(r, \theta, 0) = f(r, \theta), \quad u_t(r, \theta, 0) = g(r, \theta),$$

with the following boundary conditions.

(a) $u(a, \theta, t) = 0$

(b) $u_r(a, \theta, t) = 0$

3. Find the solutions to the initial boundary value problems for the heat equation inside the semi-circular disk

$$u_t = \alpha^2 \Delta u, \quad 0 < r < a, \ 0 < \theta < \pi, \ 0 < t,$$

$$u(r, \theta, 0) = f(r, \theta),$$

with the following boundary conditions.

(a) $u(a, \theta, t) = 0, \quad u(r, 0, t) = 0, \quad u(r, \pi, t) = 0$
(b) $u_r(a, \theta, t) = 0, \quad u(r, 0, t) = 0, \quad u(r, \pi, t) = 0$

4. Find the solutions to the initial boundary value problems for the wave equation inside the semi-circular disk

$$u_{tt} = c^2 \Delta u, \quad 0 < r < a, \ 0 < \theta < \pi, \ 0 < t,$$

$$u(r, \theta, 0) = f(r, \theta), \quad u_t(r, \theta, 0) = g(r, \theta),$$

with the following boundary conditions.

(a) $u(a, \theta, t) = 0, \quad u(r, 0, t) = 0, \quad u(r, \pi, t) = 0$
(b) $u_r(a, \theta, t) = 0, \quad u(r, 0, t) = 0, \quad u(r, \pi, t) = 0$

5. In this exercise we obtain (8.1.22).

(a) Substitute (8.1.16) in $\frac{d}{dx}[x^{-n} J_n(x)]$ to obtain

$$\frac{d}{dx}[x^{-n} J_n(x)] = \frac{d}{dx}[\sum_{k=0}^{\infty} (-1)^k \frac{1}{k!(n+k)!} (\frac{s}{2})^{2k}]. \tag{8.1.23}$$

(b) Differentiate both sides of (8.1.23) to show that

$$J_n'(x) - \frac{n}{x} J_n(x) = -J_{n+1}(x) \quad \text{for } n \geq 0.$$

8.2 Cylindrical Domains

In this section we discuss two examples of the separation of variables in the cylindrical coordinates. Specifically, we study the heat conduction and the equilibrium heat distribution in the circular cylinder placed on $z \geq 0$ as in Figure 8.3. The domain is $\Omega = \{(r, \theta, z) \mid 0 \leq r < a, \ -\pi < \theta < \pi, \ 0 < z < H\}$. The modified Bessel functions discussed in the previous section appear in this section.

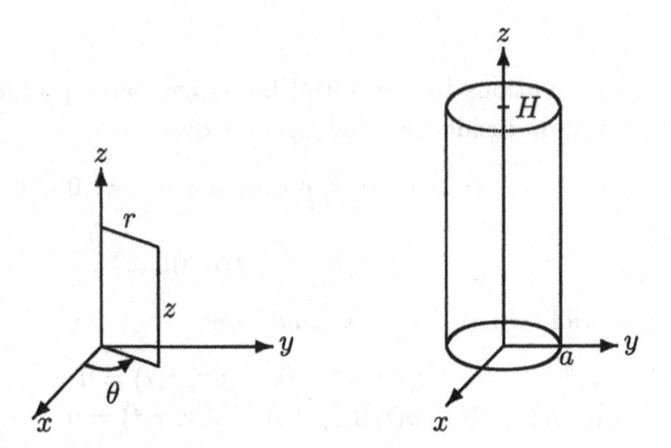

Fig. 8.3 (Left) Cylindrical coordinates (r, θ, z). (Right) Circular cylinder.

8.2.1 *Initial Boundary Value Problems*

We consider the heat equation in a circular cylinder with radius a, height H, and with the Dirichlet boundary conditions. The cylindrical coordinates and the circular cylinder are drawn in Figure 8.3. The initial boundary value problem for the temperature $u(r, \theta, z, t)$ is given by

$$u_t = \alpha^2 \Delta u = \alpha^2 \{\frac{1}{r}\frac{\partial}{\partial r}(r\frac{\partial u}{\partial r}) + \frac{1}{r^2}\frac{\partial^2 u}{\partial \theta^2} + \frac{\partial^2 u}{\partial z^2}\}, \quad (r, \theta, z) \in \Omega,\ 0 < t,$$
$$u(r, \theta, 0, t) = 0, \quad u(r, \theta, H, t) = 0,$$
$$u(a, \theta, z, t) = 0,$$
$$u(r, \theta, z, 0) = f(r, \theta, z).$$

Separating Variables

We assume that the solution is given by

$$u = T(t)R(r)\Theta(\theta)Z(z).$$

Then, we separate the time and space variables and have

$$\frac{\frac{dT}{dt}}{\alpha^2 T} = \frac{\frac{1}{r}\frac{d}{dr}\left(r\frac{dR}{dr}\right)}{R} + \frac{\frac{1}{r^2}\frac{d^2\Theta}{d\theta^2}}{\Theta} + \frac{\frac{d^2 Z}{dz^2}}{Z} = -\lambda. \tag{8.2.1}$$

This suggests that we set

$$\frac{\frac{d^2 Z}{dz^2}}{Z} = -\gamma, \qquad \frac{\frac{d^2\Theta}{d\theta^2}}{\Theta} = -\mu,$$

where γ and μ will be eigenvalues. The boundary conditions are

$$u(r, \theta, 0, t) = 0 \Rightarrow T(t)R(r)\Theta(\theta)Z(0) = 0 \Rightarrow Z(0) = 0,$$
$$u(r, \theta, H, t) = 0 \Rightarrow T(t)R(r)\Theta(\theta)Z(H) = 0 \Rightarrow Z(H) = 0,$$
$$u(a, \theta, z, t) = 0 \Rightarrow T(t)R(a)\Theta(\theta)Z(H) = 0 \Rightarrow R(a) = 0.$$

Solving ODE's

In this step we solve the ODE's in (8.2.1). We solve the eigenvalue problems for the space variables Z, Θ, and R, and solve the initial value problem for T. For Z we solve

$$\frac{d^2Z}{dz^2} + \gamma Z = 0, \quad Z(0) = 0, \ Z(H) = 0.$$

The eigenvalues and eigenfunctions are

$$\gamma_m = (\frac{m\pi}{H})^2, \quad Z_m(z) = \sin\frac{m\pi z}{H}, \ m = 1, 2, \ldots.$$

For Θ we solve

$$\frac{d^2\Theta}{d\theta^2} + \mu\Theta = 0 \tag{8.2.2}$$

with the periodic boundary conditions. Then,

$$\mu_n = n^2, \quad \Theta_n = a_n \cos n\theta + b_n \sin n\theta, \quad n = 0, 1, \ldots \tag{8.2.3}$$

with the restriction that $b_0 = 0$. The problem has been treated in Section 4.2.2. For $R(r)$ we solve

$$r^2\frac{d^2R}{dr^2} + r\frac{dR}{dr} + [\{\lambda - (\frac{m\pi}{H})^2\}r^2 - n^2]R = 0,$$

with the boundary conditions

$$R(0) \text{ is finite.} \quad R(a) = 0.$$

Denote the lth zero of $J_n(s)$ as s_{nl}. The values of λ are chosen so that $J_n(\sqrt{\lambda}a) = 0$. Therefore, the eigenvalues for λ are

$$\lambda_{nml} = (\frac{s_{nl}}{a})^2 + (\frac{m\pi}{H})^2, \quad l = 1, 2, \ldots, \ m = 1, 2, \ldots, \ n = 0, 1, \ldots.$$

It is important that $\lambda - (m\pi/H)^2 \geq 0$ is satisfied. The above λ_{nml} obviously satisfies the condition. Note that if $\lambda < (m\pi/H)^2$ we have a modified Bessel function for which there are no zeros for $r > 0$. As explained in the previous section we cannot use it as eigenfunctions since they do not satisfy the boundary condition at $r = a$. Finally solving the ODE for the time variable with $\lambda = \lambda_{nml}$

$$\frac{dT}{dt} + \lambda_{nml}\alpha^2 T = 0,$$

we have

$$T(t) = e^{-\lambda_{nml}t}.$$

Solution to the Initial Boundary Value Problem

We now assemble the solution to the initial boundary value problem. We denote the eigenfunctions $v = R(r)\Theta(\theta)Z(z)$ as

$$v^c_{nml} = J_n(\sqrt{\lambda_{nml}}r)\cos n\theta \sin\frac{m\pi z}{H},$$
$$v^s_{nml} = J_n(\sqrt{\lambda_{nml}}r)\sin n\theta \sin\frac{m\pi z}{H}.$$

Then, since we assume $u = T(t)R(r)\Theta(\theta)Z(z)$, by the superposition principle we have

$$\begin{aligned} &u(r,\theta,z,t) \\ &= \sum_{n=0}^{\infty}\sum_{m=1}^{\infty}\sum_{l=1}^{\infty} e^{-\lambda_{nml}\alpha^2 t} J_n(\sqrt{\lambda_{nml}}r)(a_{nml}\cos n\theta + b_{nml}\sin n\theta)\sin\frac{m\pi z}{H} \\ &= \sum_{n=0}^{\infty}\sum_{m=1}^{\infty}\sum_{l=1}^{\infty} e^{-\lambda_{nml}\alpha^2 t}(a_{nml}v^c_{nml} + b_{nml}v^s_{nml}), \end{aligned}$$

where $b_{0ml} = 0$. At $t = 0$

$$u(r,\theta,z,0) = f(r,\theta,z) = \sum_{n=0}^{\infty}\sum_{m=1}^{\infty}\sum_{l=1}^{\infty}(a_{nml}v^c_{nml} + b_{nml}v^s_{nml}).$$

Using the orthogonality relations, we have

$$a_{nml} = \frac{\int_0^H\int_{-\pi}^{\pi}\int_0^a f(r,\theta)v^c_{nml}rdrd\theta dz}{\int_0^H\int_{-\pi}^{\pi}\int_0^a (v^c_{nml})^2 rdrd\theta dz}, \quad n = 0,1,\ldots,$$

$$b_{nml} = \frac{\int_0^H\int_{-\pi}^{\pi}\int_0^a f(r,\theta)v^s_{nml}rdrd\theta dz}{\int_0^H\int_{-\pi}^{\pi}\int_0^a (v^s_{nml})^2 rdrd\theta dz}, \quad n = 1,2,\ldots,$$

where we use 1 for $\cos n\theta$ if $n = 0$. Note that, for example, the denominator of b_{nml} is the same as

$$\begin{aligned} &\int_0^H\int_{-\pi}^{\pi}\int_0^a (v^s_{nml})^2 rdrd\theta dz \\ &= \int_0^a J_n^2(\sqrt{\lambda_{nml}}r)rdr\int_{-\pi}^{\pi}\sin^2 n\theta d\theta\int_0^H \sin^2\frac{m\pi z}{H}dz. \end{aligned}$$

8.2.2 *Laplace Equation*

As in Subsection 4.2.1, if necessary we use the superposition principle to decompose the boundary value problems. In the following we treat two typical boundary value problems we solve after such decomposition. Whether we use the Bessel's or the modified Bessel's equation for the radial variable depends on the boundary conditions. If the boundary condition for the radial variable is non-homogeneous, we use the modified Bessel's equation since we solve the boundary value problem for the radial variable. On the other the hand if the top or bottom boundary condition is non-homogeneous, we solve the eigenvalue problem for the radial variable and consequently we use the Bessel's equation.

8.2.2.1 *Non-homogeneous Radial Boundary Conditions*

As an example consider

$$\triangle u = \frac{1}{r}\frac{\partial}{\partial r}(r\frac{\partial u}{\partial r}) + \frac{1}{r^2}\frac{\partial^2 u}{\partial \theta^2} + \frac{\partial^2 u}{\partial z^2} = 0,$$
$$0 \le r < a, \ -\pi < \theta < \pi, \ 0 < z < H,$$
$$u(r,\theta,0) = 0, \quad u(r,\theta,H) = 0,$$
$$u(a,\theta,z) = f(\theta,z).$$

Separating Variables

We assume that the solution is given by

$$u = R(r)\Theta(\theta)Z(z). \tag{8.2.4}$$

$$\frac{\frac{1}{r}\frac{d}{dr}(r\frac{dR}{dr})}{R} + \frac{\frac{1}{r^2}\frac{d^2\Theta}{d\theta^2}}{\Theta} + \frac{\frac{d^2 Z}{dz^2}}{Z} = 0. \tag{8.2.5}$$

This is the same as the case where $\lambda = 0$ in (8.2.1). Check the boundary conditions first. The boundary conditions are

$$u(r,\theta,0) = 0 \Rightarrow R(r)\Theta(\theta)Z(0) = 0 \Rightarrow Z(0) = 0,$$
$$u(r,\theta,H) = 0 \Rightarrow R(r)\Theta(\theta)Z(H) = 0 \Rightarrow Z(H) = 0,$$
$$u(a,\theta,z) = f(\theta,z) \Rightarrow R(a)\Theta(\theta)Z(H) = f(\theta,z) \Rightarrow R(a) = ?.$$

This suggests that we solve the eigenvalue problems for Z and Θ and use them to solve the ODE for R.

Solving ODE's

We solve the ODE's resulting from (8.2.5). For Z and Θ we solve the eigenvalue problems. We set

$$\frac{\frac{d^2Z}{dz^2}}{Z} = -\gamma, \frac{\frac{d^2\Theta}{d\theta^2}}{\Theta} = -\mu,$$

where γ and μ will be eigenvalues. For Z the eigenvalue problem is

$$\frac{d^2Z}{dz^2} + \gamma Z = 0, \quad Z(0) = 0, \ Z(H) = 0.$$

Therefore, the eigenvalues and eigenfunctions are

$$\gamma_m = (\frac{m\pi}{H})^2, \quad Z_m(z) = \sin\frac{m\pi z}{H}, \ m = 1, 2, \ldots.$$

For Θ we solve the eigenvalue problem

$$\frac{d^2\Theta}{d\theta^2} + \mu\Theta = 0 \tag{8.2.6}$$

with the periodic boundary conditions

$$\Theta(-\pi) = \Theta(\pi), \quad \Theta'(-\pi) = \Theta'(\pi). \tag{8.2.7}$$

Then,

$$\mu_n = n^2, \quad \Phi_n = a_n \cos n\theta + b_n \sin n\theta, \quad n = 0, 1, \ldots \tag{8.2.8}$$

with the restriction that $b_0 = 0$.

For R we solve the boundary value problem

$$r^2\frac{d^2R}{dr^2} + r\frac{dR}{dr} - [(\frac{m\pi}{H})^2 r^2 + n^2]R = 0 \tag{8.2.9}$$

with the restriction that R is bounded at $r = 0$. This is a modified Bessel's equation but it is not the same form as (8.1.21). We make a change of variable $s = m\pi r/H$ similar to the one used in deriving (8.1.15). Then, we have

$$s^2\frac{d^2R}{ds^2} + s\frac{dR}{ds} - [s^2 + n^2]R = 0.$$

As discussed in Subsection 8.1.2 there are two linearly independent solutions denoted as $I_n(s)$ and $K_n(s)$. They are referred as the modified Bessel function of order n of the first and second kind, respectively. The first kind is well defined at $s = 0$ while the second kind is singular at $s = 0$. We should make note of the fact that both first and second kind are not oscillatory and are not zero for $s > 0$. Therefore, for each m and n we use

$$R_{nm}(s) = I_n(\frac{m\pi}{H}r)$$

as the solution to (8.2.9).

Solution to the Boundary Value Problem

We now assemble the solution. The eigenfunctions for this problem are

$$v_{nm}^c = \cos n\theta \sin \frac{m\pi z}{H}, \quad v_{nm}^s = \sin n\theta \sin \frac{m\pi z}{H}.$$

Since we assume that the solution is given by (8.2.4), by the superposition principle the solution has the form

$$\begin{aligned} u(r,\theta,z) &= \sum_{n=0}^{\infty}\sum_{m=1}^{\infty} I_n(\frac{m\pi}{H}r)(a_{nm}\cos n\theta + b_{nm}\sin n\theta)\sin\frac{m\pi z}{H} \\ &= \sum_{n=0}^{\infty}\sum_{m=1}^{\infty} I_n(\frac{m\pi}{H}r)(a_{nm}v_{nm}^c + b_{nm}v_{nm}^s) \end{aligned}$$

where $b_{0m} = 0$. At $r = a$ from the boundary conditions

$$u(a,\theta,z) = f(\theta,z) = \sum_{n=0}^{\infty}\sum_{m=1}^{\infty} I_n(\frac{m\pi}{H}a)(a_{nm}v_{nm}^c + b_{nm}v_{nm}^s). \tag{8.2.10}$$

Multiplying (8.2.10) by the eigenfunctions and using the orthogonality conditions, we have $b_{0m} = 0$ and

$$a_{nm} = \frac{\int_{-\pi}^{\pi}\int_0^H f(\theta,z)v_{nm}^c dz d\theta}{I_n(\frac{m\pi}{H}a)\int_{-\pi}^{\pi}\int_0^H (v_{nm}^c)^2 dz d\theta},$$

$$b_{nm} = \frac{\int_{-\pi}^{\pi}\int_0^H f(\theta,z)v_{nm}^s dz d\theta}{I_n(\frac{m\pi}{H}a)\int_{-\pi}^{\pi}\int_0^H (v_{nm}^s)^2 dz d\theta}.$$

8.2.2.2 *Homogeneous Radial Boundary Conditions*

As an example we consider the case where the boundary on the top is non-homogeneous.

$$\Delta u = \frac{1}{r}\frac{\partial}{\partial r}(r\frac{\partial u}{\partial r}) + \frac{1}{r^2}\frac{\partial^2 u}{\partial \theta^2} + \frac{\partial^2 u}{\partial z^2} = 0, \tag{8.2.11}$$
$$0 \le r < a, \ -\pi < \theta < \pi, \ 0 < z < H,$$

$$u(r,\theta,0) = 0, \quad u(r,\theta,H) = g(r,\theta), \tag{8.2.12}$$

$$u(a,\theta,z) = 0. \tag{8.2.13}$$

As before we assume that u is given by $u = v(r,\theta)Z(z)$ where $v(r,\theta) = R(r)\Theta(\theta)$. This problem is very similar to the one discussed in Section 8.1

except that we solve the boundary value problem for Z instead of the initial value problem. First separating (8.2.11) to v and Z, we have

$$\frac{1}{r}\frac{\partial}{\partial r}(r\frac{\partial v}{\partial r}) + \frac{1}{r^2}\frac{\partial^2 v}{\partial \theta^2} = -\lambda v, \tag{8.2.14}$$

$$\frac{d^2 Z}{dz^2} = \lambda Z. \tag{8.2.15}$$

The equation (8.2.14) is the same as (8.1.7). All the result obtained in Section 8.1 applies. If we denote the eigenfunctions to be

$$v_{nm}^c = J_n(\sqrt{\lambda_{nm}}r)\cos n\theta, \quad v_{nm}^s = J_n(\sqrt{\lambda_{nm}}r)\sin n\theta$$

$$\begin{aligned} v_{nm}(r,\theta) &= J_n(\sqrt{\lambda_{nm}}r)[a_{nm}\cos n\theta + b_{nm}\sin n\theta] \\ &= a_{nm}v_{nm}^c + b_{nm}v_{nm}^s. \end{aligned} \tag{8.2.16}$$

For Z we solve (8.2.15) with $\lambda = \lambda_{nm}$. Since λ_{nm} is positive, the general solution is

$$Z_{nm}(z) = c_1 \cosh\sqrt{\lambda_{nm}}z + c_2 \sinh\sqrt{\lambda_{nm}}z.$$

From the boundary condition at $z = 0$ in (8.2.12) we have $Z(0) = 0$. Therefore, $c_2 = 0$ and

$$Z_{nm}(z) = \sinh\sqrt{\lambda_{nm}}z. \tag{8.2.17}$$

Combining (8.2.16) and (8.2.17), we obtain

$$u(r,\theta,z) = \sum_{n=0}^{\infty}\sum_{m=1}^{\infty}\sinh\sqrt{\lambda_{nm}}z[a_{nm}v_{nm}^c + b_{nm}v_{nm}^s].$$

The values of a_{nm} and b_{nm} ($b_{0m} = 0$) are determined from the boundary condition at $z = H$ in (8.2.12) and

$$a_{nm} = \frac{\int_{-\pi}^{\pi}\int_0^a g(r,\theta)v_{nm}^c r dr d\theta}{\sinh\sqrt{\lambda_{nm}}H \int_{-\pi}^{\pi}\int_0^a (v_{nm}^c)^2 r dr d\theta},$$

$$b_{nm} = \frac{\int_{-\pi}^{\pi}\int_0^a g(r,\theta)v_{nm}^s r dr d\theta}{\sinh\sqrt{\lambda_{nm}}H \int_{-\pi}^{\pi}\int_0^a (v_{nm}^s)^2 r dr d\theta}.$$

Exercises

1. The following is the most general Dirichlet boundary condition for the Laplace equation in a circular cylinder.
$$u(r,\theta,0)=f(r,\theta),\quad u(r,\theta,H)=g(r,\theta),\quad u(a,\theta,z)=h(\theta,z).$$
Suggest how we decompose the problem.
2. Solve the Laplace equation in a circular cylinder with the following boundary conditions.
(a) $u_z(r,\theta,0)=0,\quad u_z(r,\theta,H)=0,\quad u(a,\theta,z)=h(\theta,z)$
(b) $u(r,\theta,0)=0,\quad u_z(r,\theta,H)=g((r,\theta),\quad u(a,\theta,z)=0$
(c) $u(r,\theta,0)=f(r,\theta),\quad u(r,\theta,H)=0,\quad u_r(a,\theta,z)=0$
(d) $u_z(r,\theta,0)=0,\quad u_z(r,\theta,H)=0,\quad u_r(a,\theta,z)=h(\theta,z)$
(e) For (d) find the solvability condition on $h(\theta,z)$.
3. Solve the heat equation in a circular cylinder
$$u_t=\alpha^2\Delta u$$
$$u(r,\theta,z,0)=f(r,\theta,z)$$
with the following boundary conditions.
(a) $u_z(r,\theta,0,t)=0,\quad u_z(r,\theta,H,t)=0,\quad u(a,\theta,z,t)=0$
(b) $u(r,\theta,0,t)=0,\quad u_z(r,\theta,H,t)=0,\quad u(a,\theta,z,t)=0$
(c) $u(r,\theta,0,t)=0,\quad u(r,\theta,H,t)=0,\quad u_r(a,\theta,z,t)=0$
(d) $u_z(r,\theta,0,t)=0,\quad u_z(r,\theta,H,t)=0,\quad u_r(a,\theta,z,t)=0$
4. Solve the wave equation in a circular cylinder
$$u_{tt}=c^2\Delta u$$
$$u(r,\theta,z,0)=f(r,\theta,z),\quad u_t(r,\theta,z,0)=g(r,\theta,z)$$
with the same boundary conditions as the previous problem.

8.3 Spherical Domains

When the domain is a ball $B(\mathbf{0},a)$, we often use the spherical coordinates $\mathbf{x}=(r,\phi,\theta)$. In Figure 8.4 the spherical coordinates are drawn. We first use the wave equation with the Dirichlet boundary conditions to illustrate how the separation of variables are performed for the initial boundary value problems in the spherical coordinates. ODE's we obtain are no longer simple as the ones for the rectangular domain. For ϕ-direction the ODE we solve is called the associated Legendre equation and for r-direction we obtain the Bessel's equation after some change of variable. The separation of variables for the Laplace equation is simpler and explained in Subsection 8.3.3.

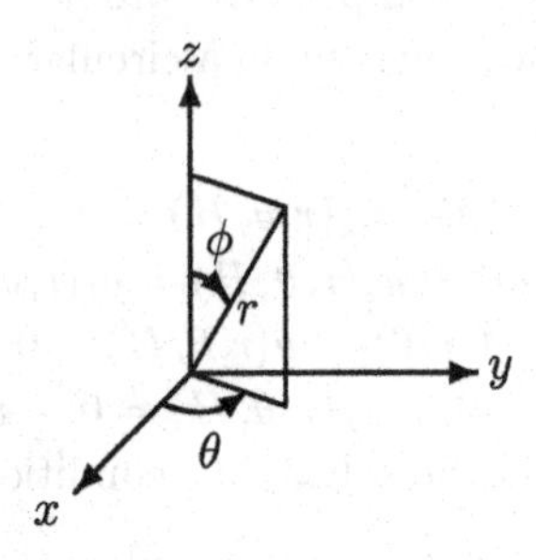

Fig. 8.4 The spherical coordinates (r, ϕ, θ).

8.3.1 *Initial Boundary Value Problems*

As an example we consider the initial boundary value problem for the wave equation on the sphere with radius a and center at the origin. For simplicity we treat the Dirichlet boundary conditions. The initial boundary value problem is given by

$$u_{tt} = c^2 \Delta u, \quad 0 \le r < a,\ 0 \le \phi \le \pi,\ -\pi \le \theta \le \pi,$$

with the boundary condition

$$u(a, \phi, \theta, t) = 0 \quad 0 \le \phi \le \pi,\ -\pi \le \theta \le \pi,$$

and the initial conditions

$$u(r, \phi, \theta, 0) = f(r, \phi, \theta), \quad u_t(r, \phi, \theta, 0) = g(r, \phi, \theta).$$

Note that the boundary condition is given for the r variable only.

Separating Variables

We use the separation of variables $u(\mathbf{x}, t) = T(t)v(\mathbf{x})$ and separate the time and space variables. Substituting this in the wave equation $u_{tt} = c^2 \Delta u$, we obtain

$$\frac{T''}{c^2 T} = \frac{\Delta v}{v} = -\lambda, \tag{8.3.1}$$

The Laplacian in the spherical coordinates (r, ϕ, θ) is given by

$$\begin{aligned}&\Delta v + \lambda v \\ &= \frac{1}{r^2}(r^2 v_r)_r + \frac{1}{r^2 \sin\phi}(\sin\phi v_\phi)_\phi + \frac{1}{r^2 \sin^2\phi} v_{\theta\theta} + \lambda v = 0. \quad (8.3.2)\end{aligned}$$

The derivation is very tedious and carried out in Exercises. Substituting $v = R(r)\Phi(\phi)\Theta(\theta)$ to (8.3.2), we obtain

$$\frac{(r^2 R')'}{R} + \lambda r^2 + \frac{1}{\sin\phi}\frac{(\sin\phi\Phi_\phi)_\phi}{\Phi} + \frac{1}{\sin^2\phi}\frac{\Theta_{\theta\theta}}{\Theta} = 0. \quad (8.3.3)$$

So, the equation (8.3.3) suggests the following separation of variables for Θ, R and Φ.

$$\Theta_{\theta\theta} + \mu\Theta = 0, \quad (8.3.4)$$

$$-\frac{1}{\sin^2\phi}\mu + \frac{1}{\sin\phi}\frac{(\sin\phi\Phi_\phi)_\phi}{\Phi} + \gamma = 0, \quad (8.3.5)$$

$$(r^2 R')' + (\lambda r^2 - \gamma)R = 0. \quad (8.3.6)$$

The boundary condition is

$$u(a, \phi, \theta) = R(a)\Phi(\phi)\Theta(\theta) = 0 \Rightarrow R(a) = 0. \quad (8.3.7)$$

Solving ODE's

We solve the eigenvalue problems for Θ, Φ, and R and solve the initial value problem for T. There is only one boundary condition available. We impose reasonable or appropriate boundary conditions and solve the eigenvalue problems for each variable. For Θ the appropriate boundary conditions are the periodic boundary conditions given by

$$\Theta(-\pi) = \Theta(\pi), \quad \Theta'(-\pi) = \Theta'(\pi).$$

The eigenvalues are $\mu = n^2$, $(n = 0, 1, 2, \ldots)$ and the eigenfunctions are $\cos n\theta$ and $\sin n\theta$.

For the Φ, we solve

$$(\sin\phi\Phi_\phi)_\phi + (\gamma\sin\phi - \frac{n^2}{\sin\phi})\Phi = 0, \quad n = 0, 1, \ldots. \quad (8.3.8)$$

with the boundary conditions

$$\Phi(0) \text{ and } \Phi(\pi) \text{ are bounded.}$$

To solve (8.3.8) we change the independent variable by $x = \cos\phi$ and denote Φ as $\Phi(\phi) = P(x)$. Then, since

$$\frac{d}{d\phi} = \frac{dx}{d\phi}\frac{d}{dx} = -\sin\phi\frac{d}{dx},$$

(8.3.8) is

$$\frac{d}{dx}[(1-x^2)\frac{dP}{dx}] + (\gamma - \frac{n^2}{1-x^2})P = 0, \quad n = 0, 1, \dots, \tag{8.3.9}$$

$$P(x) \text{ is bounded at } x = \pm 1. \tag{8.3.10}$$

The equation is called the Legendre equation if $n = 0$ and the associated Legendre equation if $n \geq 1$. The solutions can be found by the power series expansion and is discussed in Subsection 8.3.2. It is difficult to show, but it is known that the solutions bounded at $x = \pm 1$ exist only when

$$\gamma = l(l+1), \quad l \geq n, \tag{8.3.11}$$

where l are integers and the such power series solutions are given by

$$P(x) = P_l^n(x) = \frac{(-1)^n}{2^l l!}(1-x^2)^{n/2}\frac{d^{l+n}}{dx^{l+n}}[(x^2-1)^l]. \tag{8.3.12}$$

Therefore, γ satisfying (8.3.11) is an eigenvalue and (8.3.12) is the corresponding eigenfunction for (8.3.9) and (8.3.10). By the Sturm-Liouville Theorem, for the same n we have the orthogonality relations

$$\int_0^\pi P_l^n(\cos\phi)P_{l'}^n(\cos\phi)\sin\phi d\phi = 0, \quad l \neq l'.$$

For R we rewrite (8.3.6)

$$R'' + \frac{2}{r}R' + (\lambda - \frac{\gamma}{r^2})R = 0 \tag{8.3.13}$$

and the boundary conditions are

$$R(r) = 0, \quad R(0) \text{ is bounded.}$$

This equation is similar to (8.1.15). We change the dependent variable as $z(r) = \sqrt{r}R(r)$. Then, z satisfies

$$z'' + \frac{1}{r}z' + (\lambda - \frac{\gamma + \frac{1}{4}}{r^2})z = 0. \tag{8.3.14}$$

As we did in Section 8.1, we normalize the equation (8.3.14). Set $s = \sqrt{\lambda}r$. Then,

$$z'' + \frac{1}{s}z' + (1 - \frac{\gamma + \frac{1}{4}}{s^2})z = 0.$$

This is the same as the Bessel function we studied in Subsection 8.1.2. Since $\gamma = l(l+1)$, $\sqrt{\gamma + 1/4} = l + 1/2$, the solution is

$$z(r) = J_{l+\frac{1}{2}}(\sqrt{\lambda}r).$$

Going back to R, we obtain

$$R(r) = \frac{1}{\sqrt{r}} J_{l+\frac{1}{2}}(\sqrt{\lambda}r).$$

We choose the values of λ_{lk} so that $J_{l+1/2}(\sqrt{\lambda_{lk}}a) = 0$, $(k = 1, 2, \ldots)$. The existence of such λ_{lk} is well-known.

For each λ_{lk} we solve $T'' + c^2\lambda_{lk}T = 0$ in (8.3.1) for T. Then,

$$T = c_k \cos c\sqrt{\lambda_{lk}}t + d_k \sin c\sqrt{\lambda_{lk}}t.$$

Solution to the Initial Boundary Value Problem

We put the eigenfunctions together.

$$\begin{aligned} v &= R(r)P_l^n(x)\Theta(\theta) \\ &= \frac{J_{l+\frac{1}{2}}(\sqrt{\lambda_{lk}}r)}{\sqrt{r}} P_l^n(\cos\phi)(a_n \cos n\theta + b_n \sin n\theta). \end{aligned}$$

Set

$$\begin{aligned} v_{nlk}^c &= \frac{J_{l+\frac{1}{2}}(\sqrt{\lambda_{lk}}r)}{\sqrt{r}} P_l^n(\cos\phi)\cos n\theta, \\ v_{nlk}^s &= \frac{J_{l+\frac{1}{2}}(\sqrt{\lambda_{lk}}r)}{\sqrt{r}} P_l^n(\cos\phi)\sin n\theta. \end{aligned} \tag{8.3.15}$$

Then the general form of the solution is

$$\begin{aligned} u = &\sum_{n=0}^{\infty}\sum_{l=n}^{\infty}\sum_{k=0}^{\infty}[a_{nlk}(\cos c\sqrt{\lambda_{lk}}t)v_{nlk}^c + b_{nlk}(\cos c\sqrt{\lambda_{lk}}t)v_{nlk}^s] \\ &+\sum_{n=0}^{\infty}\sum_{l=n}^{\infty}\sum_{k=0}^{\infty}[c_{nlk}(\sin c\sqrt{\lambda_{lk}}t)v_{nlk}^c + d_{nlk}(\sin c\sqrt{\lambda_{lk}}t)v_{nlk}^s]. \end{aligned}$$

The coefficients a_{nlk} are given by

$$a_{nlk} = \frac{\int_{-\pi}^{\pi}\int_0^{\pi}\int_0^a f v_{nlk}^c r^2 \sin\phi dr d\phi d\theta}{\int_{-\pi}^{\pi}\int_0^{\pi}\int_0^a (v_{nlk}^c)^2 r^2 \sin\phi dr d\phi d\theta}$$

and the other coefficients are given similarly. Note that the denominator is the same as

$$\begin{aligned} &\int_{-\pi}^{\pi}\int_0^{\pi}\int_0^a (v_{nlk}^c)^2 r^2 \sin\phi dr d\phi d\theta \\ &= \int_0^a J_{l+\frac{1}{2}}^2(\sqrt{\lambda_{lk}}r) r dr \int_0^{\pi}[P_l^n(\cos\phi)]^2 \sin\phi d\phi \int_{-\pi}^{\pi}\cos^2 n\theta d\theta, \end{aligned}$$

where the normalizing integral for ϕ is given by

$$\int_0^{\pi} [P_l^n(\cos\phi)]^2 \sin\phi d\phi = \int_{-1}^{1} [P_l^n(s)]^2 ds = \frac{2(l+n)!}{(2l+1)(l-n)!}.$$

The details of the evaluation of the integrals are available in [Berg and McGregor (1966)].

8.3.2 *Legendre Equation*

The Legendre equation is

$$[(1-x^2)u']' + \gamma u = 0, \tag{8.3.16}$$

where $\gamma = l(l+1)$ for integers $l \geq 0$. The solutions to (8.3.16) are called the Legendre function and given by

$$P_l(x) = \frac{1}{2^l l!}\frac{d^l}{dx^l}(x^2-1)^l = \frac{1}{2^l}\sum_{i=0}^{k}\frac{(-1)^i}{i!}\frac{(2l-2i)!}{(l-2i)!(l-i)!}x^{l-2i}, \tag{8.3.17}$$

where $k = l/2$ if l is even and $k = (l-1)/2$ if l is odd. This formula is known as Rodrigues' formula. To find $P_l(x)$ we assume a power series solution

$$u(x) = \sum_{i=0}^{\infty} a_i x^i$$

and substitute this in (8.3.16) to obtain the relation

$$\sum_{i=2}^{\infty} i(i-1)a_i x^{i-2} - \sum_{i=0}^{\infty}(i^2+i-\gamma)a_i x^i = 0.$$

Changing the index i to $i+2$ in the first term and collecting the like terms, we have the recursive relation

$$a_{i+2} = \frac{i^2+i-\gamma}{(i+2)(i+1)}a_i. \tag{8.3.18}$$

As clearly seen, the series terminates when $\gamma = i(i+1)$ for a nonnegative integer i. When $i = l$, it is possible to show that we obtain (8.3.17) from (8.3.18).

The associated Legendre equation is

$$[(1-x^2)u']' + (\gamma - \frac{n^2}{1-x^2})u = 0, \tag{8.3.19}$$

where $\gamma = l(l+1)$ and $l \geq n$ with l and n integers. The solutions are called the associated Legendre functions and given by

$$P_l^n(x) = (1-x^2)^{n/2}\frac{d^n}{dx^n}P_l(x). \tag{8.3.20}$$

The derivation of the solution is given in Exercises.

8.3.3 *Laplace Equation*

The problem reduces to the case where $\lambda = 0$ in (8.3.2) and this simplifies the equation for the radial variable. As an example we consider the Dirichlet problem for the Laplace equation.

$$\Delta u(\mathbf{x}) = 0, \quad \mathbf{x} = (r, \phi, \theta) \in B(\mathbf{0}, a), \tag{8.3.21}$$

$$u(a, \phi, \theta) = f(\phi, \theta), \quad \mathbf{x} \in \partial B(\mathbf{0}, a). \tag{8.3.22}$$

Substituting $u = R(r)\Phi(\phi)\Theta(\theta)$ to (8.3.21), for Θ and Φ we obtain the same equations as (8.3.4) and (8.3.5), respectively. For R we get (8.3.6) with $\lambda = 0$, *i.e.*,

$$r^2 R'' + 2rR' - \gamma R = 0. \tag{8.3.23}$$

This is the Euler equation. We have the solution of the form $R(r) = r^p$. Substituting this in the above equation and using $\gamma = l(l+1)$, we obtain

$$p(p-1) + 2p - l(l+1) = (p-l)(p+l+1) = 0.$$

Since the solution should be bounded at $r = 0$, we use $R(r) = r^l$ as the solution of (8.3.23). The eigenfunctions for the problem are

$$v_{nl}^c = P_l^n(\cos\phi)\cos n\theta, \ v_{nl}^s = P_l^n(\cos\phi)\sin n\theta.$$

Then, the solution is

$$u = \sum_{n=0}^{\infty}\sum_{l=n}^{\infty} r^l (a_{nl}v_{nl}^c + b_{nl}v_{nl}^s),$$

where

$$a_{nl} = \frac{\int_{-\pi}^{\pi}\int_0^{\pi} f(\phi,\theta)v_{nl}^c \sin\phi d\phi d\theta}{a^l \int_{-\pi}^{\pi}\int_0^{\pi} (v_{nl}^c)^2 \sin\phi d\phi d\theta},$$

$$b_{nl} = \frac{\int_{-\pi}^{\pi}\int_0^{\pi} f(\phi,\theta)v_{nl}^s \sin\phi d\phi d\theta}{a^l \int_{-\pi}^{\pi}\int_0^{\pi} (v_{nl}^s)^2 \sin\phi d\phi d\theta}.$$

Exercises

1. Derive the Laplacian (8.3.2) in the spherical coordinates.
2. Derive (8.3.14) from (8.3.13).
3. We show that (8.3.20) satisfies (8.3.19) in the following way.
   (a) Show that $w = \frac{d^n}{dx^n}P_l(x)$ satisfies the n times differentiated Legendre equation

$$(1-x^2)w'' - 2(n+1)xw' + [\gamma - n(n+1)]w = 0. \tag{8.3.24}$$

(b) Substitute $w = (1 - x^2)^{-n/2}u(x)$ to (8.3.24) to show that u satisfies (8.3.19).

4. Solve the initial boundary value problems for the wave equation $u_{tt} = c^2 \Delta u$ in a ball $B(\mathbf{0}, a)$ with the boundary condition $u(a, \phi, \theta, t) = 0$ and with the following initial conditions.

   (a) $u(r, \phi, \theta, 0) = f(r, \phi, \theta)$, $u_t(r, \phi, \theta, 0) = 0$
   (b) $u(r, \phi, \theta, 0) = 0$, $u_t(r, \phi, \theta, 0) = g(r, \phi, \theta)$

5. Solve the initial boundary value problems for the heat equation $u_t = a^2 \Delta u$ in a ball $B(\mathbf{0}, a)$ with the boundary condition $u(a, \phi, \theta, t) = 0$ and with the following initial conditions.

   (a) $u(r, \phi, \theta, 0) = f(r, \phi, \theta)$ (b) $u(r, \phi, \theta, 0) = 1$

6. Solve the Laplace equation inside a spherical annulus Ω with the following boundary conditions. The region is given by $\Omega = \{(r, \phi, \theta) \mid a \le r \le b,\ 0 \le \phi \le \pi,\ 0 \le \theta \le 2\pi\}$, where a and b are positive constants. If there is a solvability condition, state it.

   (a) $u_r(a, \phi, \theta) = f(\phi, \theta)$, $u_r(b, \phi, \theta) = g(\phi, \theta)$
   (b) $u(a, \phi, \theta) = f(\phi, \theta)$, $u_r(b, \phi, \theta) = g(\phi, \theta)$

7. Suppose that $\Omega = \{\mathbf{x} = (r, \phi, \theta) \mid 0 \le r \le a,\ 0 \le \phi \le \frac{\pi}{2},\ -\pi \le \theta \le \pi\}$. Solve

$$\Delta u(\mathbf{x}) = 0, \quad \mathbf{x} \in \Omega,$$

$$u(a, \phi, \theta) = f(\phi, \theta), \quad u(r, \frac{\pi}{2}, \theta) = 0, \quad \mathbf{x} \in \partial\Omega.$$

Chapter 9

Fourier Transform

The Fourier transform is a useful tool in linear PDE's. In this chapter we first introduce the delta function in two different ways. In Section 9.1.1 we use a classical way and define it as a limit of a sequence of functions whose peak approaches infinity and whose area is one. In Section 9.1.2 we use a modern way and define it as a special functional called the distribution. In Section 9.2 we define the Fourier transform and its inverse. The properties of Fourier transform are discussed in Section 9.3. The applications of the Fourier transform is studied in Section 9.4. In the first three examples we see how the Fourier transform is applied to the heat, wave and Laplace equations. In the fourth example, the more advanced application is carried out for the Black-Scholes-Merton equation in finance.

9.1 Delta Functions

9.1.1 *Classical Introduction*

We define the delta function δ_a to be a mapping which assigns for a function ϕ

$$\int_{-\infty}^{\infty} \phi(x)\delta_a(x)dx = \phi(a),$$

i.e., the value of ϕ at $x = a$. The delta function δ_0 is denoted as δ. A classical argument justifying the delta function is given by the limit as $\varepsilon \to 0$ of ψ_ε, where

$$\psi_\varepsilon(x) = \frac{1}{2\varepsilon}\chi_{\{x-\varepsilon,x+\varepsilon\}}(s) = \frac{1}{2\varepsilon}\begin{cases} 1, & x-\varepsilon < s < x+\varepsilon, \\ 0, & \text{otherwise,} \end{cases}$$

where $\chi_{\{x-\varepsilon,x+\varepsilon\}}(s)$ is called the characteristic function. We take the limit after the integration, *i.e.*,

$$\int_{-\infty}^{\infty} \phi(x)\delta_a(x)dx = \lim_{\varepsilon\to 0}\int_{-\infty}^{\infty} \phi(x)\psi_\varepsilon(a)dx = \lim_{\varepsilon\to 0}\frac{1}{2\varepsilon}\int_{a-\varepsilon}^{a+\varepsilon} \phi(x)dx = \phi(a).$$

For this we cannot change the order of integration and limit because if we do so,

$$\lim_{\varepsilon\to 0}\psi_\varepsilon(x) = \begin{cases} \infty, & x, \\ 0, & \text{otherwise} \end{cases}$$

and

$$\int_{-\infty}^{\infty} \phi(x)\lim_{\varepsilon\to 0}\psi_\varepsilon(a)dx = \int_{-\infty}^{a} 0dx + \int_{a}^{\infty} 0dx = 0.$$

There are other functions whose limits give the delta function. The Dirichlet kernel studied in Section 6.3 is an example of the delta function as (6.3.9) and (6.3.10) imply

$$\lim_{N\to\infty}\frac{1}{2\pi}\int_{-\pi}^{\pi} D_N(\theta)\phi(x+\frac{L\theta}{\pi})d\theta = \phi(x). \tag{9.1.1}$$

Therefore, $D_N(\theta)/(2\pi)$ is a δ_0. The fact that the integral is over $(-\pi,\pi)$ is not a problem. Also, the Gaussian function (the heat kernel)

$$\psi_\varepsilon(x) = \frac{1}{\varepsilon\sqrt{\pi}}e^{-x^2/\varepsilon^2} \text{ as } \varepsilon\to 0 \tag{9.1.2}$$

defines a delta function. Common characteristics for $\psi_\varepsilon(x)$ are

1. $\int_{-\infty}^{\infty}\psi_\varepsilon(x)dx = 1$.
2. $\psi_\varepsilon(x)$ is an even function.

The derivatives of the delta functions are defined by the integration by parts. If we use (9.1.2) as the delta function

$$\begin{aligned}\int_{-\infty}^{\infty} \phi(x)\delta_a'(x)dx &= \lim_{\varepsilon\to 0}\int_{-\infty}^{\infty} \phi(x)\psi_\varepsilon'(x-a)dx.\\ &= \lim_{\varepsilon\to 0}\left\{[\phi(x)\psi_\varepsilon(x-a)]_{-\infty}^{\infty} - \int_{-\infty}^{\infty}\phi'(x)\psi_\varepsilon(x-a)dx\right\}\\ &= -\phi'(a).\end{aligned} \tag{9.1.3}$$

Therefore, the nth derivative of the delta function is

$$\int_{-\infty}^{\infty} \phi(x)\delta_a^{(n)}(x)dx = (-1)^n\phi^{(n)}(a). \tag{9.1.4}$$

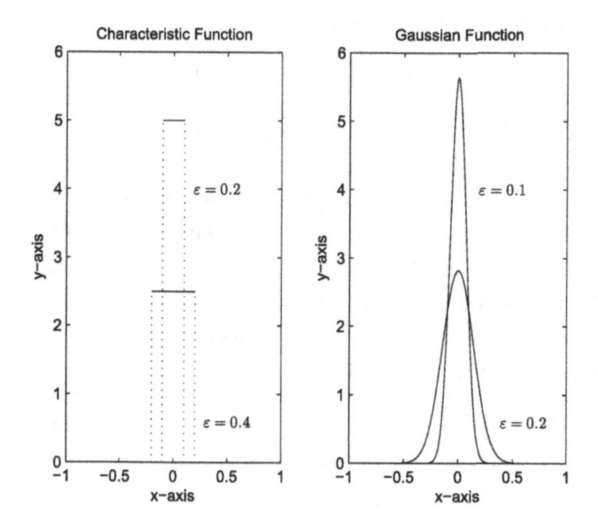

Fig. 9.1 Graphs of the characteristic and Gaussian functions.

Example 9.1.1. Show that $\delta(x-a) = \delta_a(x)$.

Solution: Since

$$\int_{-\infty}^{\infty} \phi(x)\delta(x-a)dx = \int_{-\infty}^{\infty} \phi(s+a)\delta(s)ds = \phi(a) = \int_{-\infty}^{\infty} \phi(x)\delta_a(x)dx,$$

we conclude that $\delta(x-a) = \delta_a(x)$.

In practice $\delta(x-a)$ may be easier to use. If we use $\delta(x-a)$ instead $\delta_a(x)$, (9.1.4) is written as

$$\int_{-\infty}^{\infty} \phi(x)\delta^{(n)}(x-a)dx = (-1)^n \phi^{(n)}(a).$$

9.1.2 *Modern Introduction*

Distributions

We define the delta function as a special case of distribution. To do so we need to introduce new terminologies and definitions including distribution. We start from the definition of functional. A functional is a rule that assigns numbers to functions or a mapping from functions to numbers. A function assigns numbers to numbers but a functional assigns a number to a function. So, we use a slightly different terminology and call it a functional.

An example of functional is an integral

$$F(\phi) = \int_a^b f(x)\phi(x)dx, \tag{9.1.5}$$

where f is a given function and $a \leq b$ are constants.

A distribution is a special linear functional where ϕ is a C^∞ function which vanishes outside a closed bounded subset of a domain Ω of ϕ. Here, C^∞ means we can differentiate ϕ as many times we like. The collection of such C^∞ functions is called the test functions and we use the notation $\mathcal{D}$. Now consider the set of functions f that give a finite number in the integral (9.1.5) when ϕ is a test function. Such collection of functions is called the distribution and use the notation $\mathcal{D}'$.

Definition 9.1.2. A distribution f is a continuous linear functional $\mathcal{D} \mapsto R$ defined on test functions. We denote it by (f, ϕ).

We can think that this is a functional on ϕ because this gives a rule that assigns a number for a function ϕ. We could also say that the distribution is a linear mapping with domain $\mathcal{D}$ and range R, the real number. A functional (f, ϕ) is not necessarily equal to $\int f(x)\phi(x)dx$, but it is convenient to use this notation. This is a linear functional since

$$(f, \phi_1+\phi_2) = \int f(x)(\phi_1(x)+\phi_2(x))dx = \int f(x)\phi_1(x)dx + \int f(x)\phi_2(x)dx.$$

Before defining continuity, we need to define the uniform convergence. A sequence of functions $\{\phi_n\}$ converges uniformly on $[a, b]$ to ϕ provided that

$$\max_{a \leq x \leq b} |\phi_n(x) - \phi(x)| \to 0 \quad \text{as } n \to \infty.$$

In other words, for each n we compute the maximum of the difference $|\phi_n(x) - \phi(x)|$ on $a \leq x \leq b$ and take the limit of the maximum as n goes to infinity. Now we define continuity in the following way. If $\{\phi_n\}$ is a sequence of test functions that vanish outside a common finite interval and, including their derivatives, converge uniformly to a test function ϕ, then

$$(f, \phi_n) \to (f, \phi) \quad \text{as } n \to \infty.$$

Rigorous Definition of the Delta Function

Now we can discuss the delta function.

Definition 9.1.3. The delta function is a distribution $\delta(x)$ that assigns a number $\phi(0)$ to a function $\phi(x)$, *i.e.*,

$$(\delta, \phi) = \phi(0). \tag{9.1.6}$$

Example 9.1.4. Show that for any function $f(x)$ for which $\int_{-\infty}^{\infty} |f(x)|dx$ is finite

$$\phi \mapsto \int_{-\infty}^{\infty} f(x)\phi(x)dx$$

is a distribution.

Solution: We need to show the linearity and continuity. The linearity is simple since

$$\int_{-\infty}^{\infty} f(x)(\phi_1(x) + \phi_2(x))dx = \int_{-\infty}^{\infty} f(x)\phi_1(x)dx + \int_{-\infty}^{\infty} f(x)\phi_2(x)dx.$$

To show the continuity, we consider the sequence $\{\phi_n\}$ of test functions converging uniformly to a test function ϕ.

$$\begin{aligned}
&\left|\int_{-\infty}^{\infty} f(x)\phi_n(x)dx - \int_{-\infty}^{\infty} f(x)\phi(x)dx\right| \\
&\leq \left|\int_{-\infty}^{\infty} f(x)(\phi_n(x) - \phi(x))dx\right| \\
&\leq \int_{-\infty}^{\infty} |f(x)(\phi_n(x) - \phi(x))|dx \\
&\leq \max_{a\leq x\leq b} |(\phi_n(x) - \phi(x))| \int_{-\infty}^{\infty} |f(x)|dx.
\end{aligned}$$

Since $\{\phi_n\}$ converges uniformly to ϕ, $\max_{a\leq x\leq b} |(\phi_n(x) - \phi(x))| \to 0$ as $n \to \infty$. Therefore,

$$\left|\int_{-\infty}^{\infty} f(x)\phi_n(x)dx - \int_{-\infty}^{\infty} f(x)\phi(x)dx\right| \to 0 \quad \text{as } n \to \infty.$$

Example 9.1.5. Although there is a rigorous definition of the delta functions as given in (9.1.6), it is a common practice to write the delta function as

$$\int_{-\infty}^{\infty} \delta(x)\phi(x)dx = \phi(0)$$

or as

$$\phi(x) = \int_{-\infty}^{\infty} \delta(x-y)\phi(y)dy.$$

In the second case setting $u = x - y$, using the definition we can show that

$$\int_{-\infty}^{\infty} \delta(x-y)\phi(y)dy = \int_{-\infty}^{\infty} \delta(u)\phi(x-u)du = \phi(x).$$

Derivatives of Distributions

The derivative of distribution can be justified as follows.

$$\int_{-\infty}^{\infty} f'(x)\phi(x)dx = -\int_{-\infty}^{\infty} f(x)\phi'(x)dx.$$

We pass the derivatives to the test function by integration by parts. Therefore, the derivatives of distributions always exist and they are distributions. Using the definition of the distribution, we write the above as

$$(f', \phi) = -(f, \phi'), \quad \phi \in \mathcal{D}$$

and this defines the derivative of distribution. We may call the above derivatives the distributional derivative. For example, in the case of delta function the first and second derivatives are

$$(\delta', \phi) = -(\delta, \phi') = -\phi'(0), \tag{9.1.7}$$

$$(\delta'', \phi) = -(\delta, \phi') = (\delta, \phi'') = \phi''(0).$$

So, the nth derivative of the delta function is

$$(\delta^{(n)}, \phi) = (-1)^n(\delta, \phi^{(n)}) = (-1)^n\phi^{(n)}(0).$$

The first derivative of the delta function given in (9.1.7) agrees with the result in (9.1.3).

Example 9.1.6. Find the derivative of the Heaviside function $H(x)$ defined by

$$H(x) = \begin{cases} 0 & x < 0 \\ 1 & 0 \le x. \end{cases}$$

Solution: We pass the differentiation of the Heaviside function to the test function. Then,

$$(H', \phi) = -(H, \phi') = -\int_{-\infty}^{\infty} H(x)\phi'(x)dx = -\int_{0}^{\infty} \phi'(x)dx = \phi(0) = (\delta, \phi).$$

As we see, the derivative of a Heaviside function is the delta function.

Exercises

1. Show that δ is even, *i.e.*, show that $\delta(-x) = \delta(x)$. More precisely, show that
$$\int_{-\infty}^{\infty} f(x)\delta(-x)dx = \int_{-\infty}^{\infty} f(x)\delta(x)dx$$
for every f continuous at $x = 0$.
2. Show that $\psi_\varepsilon(x) = \mathrm{e}^{-x^2/\varepsilon^2}/(\varepsilon\sqrt{\pi})$ satisfies the properties listed in the common characteristics for $\psi_\varepsilon(x)$.

9.2 Fourier Transform

9.2.1 *Complex Form of the Fourier Series*

In order to motivate the Fourier transform, we derive the complex form of the Fourier series. As we learned in Section 6.2, the Fourier series of $f(x)$ specified on $-L \leq x \leq L$ is given by

$$f(x) = \frac{a_0}{2} + \sum_{n=1}^{\infty} [a_n \cos \frac{n\pi x}{L} + b_n \sin \frac{n\pi x}{L}], \tag{9.2.1}$$

where the Fourier coefficients are given by

$$a_n = \frac{1}{L} \int_{-L}^{L} f(x) \cos \frac{n\pi x}{L} dx, \; n = 0, 1, \ldots, \tag{9.2.2}$$

$$b_n = \frac{1}{L} \int_{-L}^{L} f(x) \sin \frac{n\pi x}{L} dx, \; n = 1, 2, \ldots. \tag{9.2.3}$$

We now obtain the complex form of the Fourier series. First, substituting

$$\cos \frac{n\pi x}{L} = \frac{1}{2}(e^{in\pi x/L} + e^{-in\pi x/L}),$$

$$\sin \frac{n\pi x}{L} = \frac{1}{2i}(e^{in\pi x/L} - e^{-in\pi x/L})$$

to (9.2.1), we have

$$\begin{aligned} f(x) &= \frac{a_0}{2} + \sum_{n=1}^{\infty} [a_n \frac{1}{2}(e^{in\pi x/L} + e^{-in\pi x/L}) + b_n \frac{1}{2i}(e^{in\pi x/L} - e^{-in\pi x/L})] \\ &= \sum_{n=1}^{\infty} \frac{1}{2}(a_n + ib_n)e^{-in\pi x/L} + \frac{a_0}{2} + \sum_{n=1}^{\infty} \frac{1}{2}(a_n - ib_n)e^{in\pi x/L}. \end{aligned}$$

We also express $\frac{1}{2}(a_n \pm ib_n)$ in the complex form. Then, since

$$a_n = \frac{1}{L} \int_{-L}^{L} f(x) \cos \frac{n\pi x}{L} dx = \frac{1}{2L} \int_{-L}^{L} f(x)(e^{in\pi x/L} + e^{-in\pi x/L}) dx,$$

$$b_n = \frac{1}{2iL} \int_{-L}^{L} f(x)(e^{in\pi x/L} - e^{-in\pi x/L}) dx,$$

we define

$$c_{(-n)} = \frac{1}{2}(a_n + ib_n) = \frac{1}{2L} \int_{-L}^{L} f(x) e^{in\pi x/L} dx, \; n = 1, 2,$$

$$c_n = \frac{1}{2}(a_n - ib_n) = \frac{1}{2L}\int_{-L}^{L} f(x)e^{-in\pi x/L}dx, \ n = 0, 1, 2.$$

Then, the Fourier series of $f(x)$ in its complex form is given by

$$f(x) = \sum_{n=-\infty}^{\infty} c_n e^{in\pi x/L}, \tag{9.2.4}$$

where

$$c_n = \frac{1}{2L}\int_{-L}^{L} f(y)e^{-in\pi y/L}dy. \tag{9.2.5}$$

9.2.2 *Fourier Transform and Inverse*

We now obtain the Fourier transform and inverse. First, substituting (9.2.5) in (9.2.4), we have

$$f(x) = \sum_{n=-\infty}^{\infty} \frac{1}{2L}[\int_{-L}^{L} f(y)e^{-in\pi y/L}dy]e^{in\pi x/L}.$$

We already took the limit as $n \to \infty$ in Theorem 6.3.13, and in (9.1.1) we have observed that the Dirichlet kernel converges to the delta function. Here, we also take the limit as $L \to \infty$. Set $\omega = n\pi/L$ and $\Delta\omega = \pi/L$. Then, we expect that

$$\begin{aligned} f(x) &= \lim_{L\to\infty} \frac{1}{2\pi} \sum_{n=-\infty}^{\infty} \frac{\pi}{L}\int_{-L}^{L} f(y)e^{-i\omega y}dye^{i\omega x} \\ &= \frac{1}{2\pi}\int_{-\infty}^{\infty}\int_{-\infty}^{\infty} f(y)e^{-i\omega y}e^{i\omega x}dyd\omega. \end{aligned} \tag{9.2.6}$$

Based on this result, we define the Fourier transform of $f(x)$ and the inverse transform as follows.

Definition 9.2.1. The Fourier transform $\mathcal{F}(f)$ of $f(x)$ is defined by

$$\mathcal{F}(f) = F(\omega) = \int_{-\infty}^{\infty} f(y)e^{-i\omega y}dy$$

and the inverse Fourier transform by

$$\mathcal{F}^{-1}(F) = f(x) = \frac{1}{2\pi}\int_{-\infty}^{\infty} F(\omega)e^{i\omega x}d\omega.$$

In the multi-dimensional case, for a given $f(\mathbf{x})$, $\mathbf{x} \in R^n$, the Fourier transform of f is defined as

$$\mathcal{F}(f) = F(\boldsymbol{\omega}) = \int_{R^n} f(x)e^{-i\boldsymbol{\omega}\cdot\mathbf{x}}d\mathbf{x}.$$

The inverse Fourier transform is given by

$$\mathcal{F}^{-1}(F) = f(\mathbf{x}) = \frac{1}{(2\pi)^n}\int_{R^n} F(\omega)e^{i\omega\cdot\mathbf{x}}d\omega.$$

Basically, in (9.2.6) after the change of the order of integrations

$$\begin{aligned} f(\mathbf{x}) &= \int_{R^n}\Big(\frac{1}{(2\pi)^n}\int_{R^n} f(\mathbf{y})e^{-i\omega\cdot\mathbf{y}}dy\Big)e^{i\omega\cdot\mathbf{x}}d\omega \\ &= \int_{R^n} f(\mathbf{y})\Big(\frac{1}{(2\pi)^n}\int_{R^n} e^{i\omega\cdot(\mathbf{x}-\mathbf{y})}d\omega\Big)d\mathbf{y} \end{aligned}$$

holds. Therefore, making the observation that

$$\frac{1}{(2\pi)^n}\int_{R^n} e^{i\omega\cdot(\mathbf{x}-\mathbf{y})}d\omega = \delta(\mathbf{x}-\mathbf{y}),$$

where $\delta(\mathbf{x}-\mathbf{y}) = \delta(x_1-y_1)\cdots\delta(x_n-y_n)$, we see that

$$f(\mathbf{x}) = \int_{R^n} f(\mathbf{y})\delta(\mathbf{x}-\mathbf{y})d\mathbf{y},$$

which agrees with what we learned for the δ-function.

Table 9.1 Fourier transform.

	$f(x)$	$F(\omega)$
1.	$\delta(x-x_0)$	$e^{-i\omega x_0}$
2.	$H(x)^*$	$\pi\delta(\omega)+\frac{1}{i\omega}$
3.	$H(a-\lvert x\rvert),\ a>0$	$\frac{2}{\omega}\sin a\omega = \frac{e^{ia\omega}-e^{-ia\omega}}{i\omega}$
4.	e^{-ax^2}	$\sqrt{\frac{\pi}{a}}e^{-\omega^2/(4a)}$
5.	$\frac{\partial f}{\partial x}$	$i\omega F(\omega)$
6.	$xf(x)$	$i\frac{dF}{d\omega}$
7.	$f(x-a)$	$e^{-ia\omega}F(\omega)$
8.	$\int_{-\infty}^{\infty} f(y)g(x-y)dy$	$F(\omega)G(\omega)$

*The Fourier transform of the Heaviside function $H(x)$ is complicated. See [Howell (2001)] for the details.

Example 9.2.2. Find the Fourier transform

(a) δ and δ_a

(b) $e^{-a|x|}$, $a>0$.

Solution: (a) We compute

$$\mathcal{F}(\delta_a) = \int_{-\infty}^{\infty}\delta_a(y)e^{-i\omega y}dy = e^{-i\omega a}.$$

Then, $\mathcal{F}(\delta) = 1$.

(b) We compute

$$\int_{-\infty}^{\infty} e^{-a|y|} e^{-i\omega y} dy = \int_{-\infty}^{0} e^{(a-i\omega)y} dy + \int_{0}^{\infty} e^{-(a+i\omega)y} dy$$
$$= [\frac{e^{(a-i\omega)y}}{(a-i\omega)}]_{-\infty}^{0} + [\frac{e^{-(a+i\omega)y}}{-(a+i\omega)}]_{0}^{\infty}$$
$$= \frac{1}{(a-i\omega)} + \frac{1}{(a+i\omega)} = \frac{2a}{a^2+\omega^2}.$$

Exercises

1. Find the Fourier transform

   (a) $f(x) = \begin{cases} 0 & |x| > a \\ 1 & |x| < a \end{cases}$

   (b) $f(x) = e^{-ax^2}$, $a > 0$

2. Find the inverse of $(\sin a\omega)/\omega$.
3. Show that if f is odd (or even), F is odd (or even), i.e., show the following.

   (a) If $f(-x) = -f(x)$, $F(-\omega) = -F(\omega)$.

   (b) If $f(-x) = f(x)$, $F(-\omega) = F(\omega)$.

9.3 Properties of Fourier Transform

In this section we summarize the properties of Fourier transform and show several examples of Fourier transform.

9.3.1 *Fourier Transform of Derivatives*

Since we use the Fourier transform to solve PDE's, it is useful to examine the Fourier transform of the derivatives of functions. For simplicity we first consider the one dimensional case. Then,

$$\mathcal{F}(\frac{\partial f}{\partial x}) = \int_{-\infty}^{\infty} \frac{\partial f}{\partial x} e^{-i\omega x} dx$$
$$= [fe^{-i\omega x}]_{-\infty}^{\infty} + i\omega \int_{R^n} fe^{-i\omega x} dx = i\omega \mathcal{F}(f).$$

Therefore, if $f \to 0$ as $x \to \pm\infty$, there is no contribution from $x = \pm\infty$. Hence,

$$\mathcal{F}(\frac{\partial f}{\partial x}) = i\omega \mathcal{F}(f).$$

To find the Fourier transform of the second derivatives, we repeat the procedure.

$$\mathcal{F}(\frac{\partial^2 f}{\partial x^2}) = \int_{-\infty}^{\infty} \frac{\partial^2 f}{\partial x^2} e^{-i\omega x} dx$$
$$= [\frac{\partial f}{\partial x} e^{-i\omega x}]_{-\infty}^{\infty} + i\omega \int_{R^n} \frac{\partial f}{\partial x} e^{-i\omega x} dx = i\omega \mathcal{F}(\frac{\partial f}{\partial x}).$$

Using the previous result, we have

$$\mathcal{F}(\frac{\partial^2 f}{\partial x^2}) = (i\omega)^2 \mathcal{F}(f).$$

The multi-dimensional case is similar. For example,

$$\mathcal{F}(\frac{\partial f}{\partial x_k}) = \int_{R^n} \frac{\partial f}{\partial x_k} e^{-i\omega \cdot \mathbf{x}} d\mathbf{x}$$
$$= i\omega_k \int_{R^n} f e^{-i\omega \cdot \mathbf{x}} d\mathbf{x} = i\omega_k \mathcal{F}(f).$$

9.3.2 *Convolution*

The convolution of two functions f and g is defined as

$$(f * g)(x) = \int_{-\infty}^{\infty} f(x-y) g(y) dy.$$

Suppose that the Fourier transform of $f(x)$ and $g(x)$ are $F(\omega)$ and $G(\omega)$, respectively, then the Fourier transform of the convolution is given as follows.

$$\int (f * g)(x) e^{-i\omega x} dx = \int_{-\infty}^{\infty} \int_{-\infty}^{\infty} f(x-y) g(y) dy e^{-i\omega x} dx.$$

If we set $z = x - y$ and change one of the integration variables from x to z, then

$$\int_{-\infty}^{\infty} \int_{-\infty}^{\infty} f(x-y) g(y) dy e^{-i\omega x} dx = \int_{-\infty}^{\infty} \int_{-\infty}^{\infty} f(z) g(y) dy e^{-i\omega(y+z)} dz$$
$$= \int_{-\infty}^{\infty} f(z) e^{-i\omega z} dz \int_{-\infty}^{\infty} g(y) e^{-i\omega y} dy$$
$$= F(\omega) G(\omega).$$

9.3.3 Plancherel Formula

We first prove the following lemma.

Lemma 9.3.1. *Function f, g and their Fourier transforms F, G satisfy*

$$\int_{-\infty}^{\infty} f(x)\overline{g(x)}dx = \frac{1}{2\pi}\int_{-\infty}^{\infty} F(\omega)\overline{G(\omega)}d\omega.$$

Proof. The right hand side is rewritten as

$$\begin{aligned}
&\frac{1}{2\pi}\int_{-\infty}^{\infty} F(\omega)\overline{G(\omega)}d\omega \\
&= \frac{1}{2\pi}\int_{-\infty}^{\infty}\int_{-\infty}^{\infty} f(x)e^{-i\omega x}dx \int_{-\infty}^{\infty} \overline{g(y)e^{-i\omega y}}dyd\omega \\
&= \frac{1}{2\pi}\int_{-\infty}^{\infty} f(x)\int_{-\infty}^{\infty} \overline{g(y)}\int_{-\infty}^{\infty} e^{-i\omega(x-y)}d\omega dydx \\
&= \int_{-\infty}^{\infty} f(x)\int_{-\infty}^{\infty} \overline{g(y)}\delta(x-y)dydx \\
&= \int_{-\infty}^{\infty} f(x)\overline{g(x)}dx.
\end{aligned}$$

□

The following Plancherel formula is a special case of Lemma 9.3.1 where g and G are replaced with f and its Fourier transform F.

$$\int_{-\infty}^{\infty} |f(x)|^2 dx = \frac{1}{2\pi}\int_{-\infty}^{\infty} |F(\omega)|^2 d\omega.$$

Exercises

1. Show that $\mathcal{F}(xf(x)) = i\frac{dF}{d\omega}$, where $F(\omega) = \mathcal{F}(f(x))$.
2. Verify the Plancherel formula for $f(x) = e^{-ax^2}$, $a > 0$.
3. Find the inverse.

   (a) $F(\omega) = e^{-i\omega x_0}$
   (b) $F(\omega) = G(\omega)e^{-a|\omega|}$, where a is a positive constant and $G(\omega) = \int_{-\infty}^{\infty} g(y)e^{-i\omega y}dy$.

9.4 Applications of Fourier Transform

In this section we use the Fourier transform to solve the initial value problems for the heat, wave, and Laplace equations. We also study the Black-Scholes-Merton equation as an example of the convection diffusion equation where both the convection and diffusion terms are present.

The basic idea is as follows. First apply the Fourier transform to a partial differential equations and convert to an ODE in time where ω appears passively as a parameter. Integrate the ODE in t to find the Fourier transform of solution at t. Then, use the inverse Fourier transform to go back to the original variables and obtain the solution in the original variables.

9.4.1 *Heat Equation*

In Subsection 2.6.2 we studied the solutions to the initial value problems for the one-dimensional heat equation using the Fourier transform. In this subsection we extend to the higher dimensional case and derive the solution to the initial value problem of the heat equation in R^n.

$$u_t - \Delta u = 0, \quad \mathbf{x} \in R^n,\ t > 0, \tag{9.4.1}$$

$$u(\mathbf{x}, 0) = f(\mathbf{x}). \tag{9.4.2}$$

Step 1: As explained in the beginning of the section, first we apply the Fourier transform, *i.e.*, we multiply the equation (9.4.1) and the initial condition (9.4.2) by $e^{-i\boldsymbol{\omega}\cdot\mathbf{x}}$ and integrate over R^n. Denote the Fourier transform of u by

$$U(\boldsymbol{\omega}, t) = \int_{R^n} u(y, t) e^{-i\boldsymbol{\omega}\cdot\mathbf{y}} d\mathbf{y}.$$

Then, after performing the integration by parts, we obtain

$$U_t(\boldsymbol{\omega}, t) + (\sum \omega_i^2) U(\boldsymbol{\omega}, t) = 0, \tag{9.4.3}$$

$$U(\boldsymbol{\omega}, 0) = F(\boldsymbol{\omega}) = \int_{R^n} f(\mathbf{y}) e^{-i\boldsymbol{\omega}\cdot\mathbf{y}} d\mathbf{y}.$$

This is an ODE in t. Integrating (9.4.3) in t, we obtain

$$U(\boldsymbol{\omega}, t) = F(\boldsymbol{\omega}) e^{-t(\sum \omega_i^2)}.$$

Step 2: Now we apply the inverse Fourier transform. Then,

$$\begin{aligned} u(\mathbf{x},t) &= \frac{1}{(2\pi)^n}\int_{R^n} F(\omega)e^{-t(\sum \omega_i^2)}e^{i\omega\cdot\mathbf{x}}d\omega \\ &= \frac{1}{(2\pi)^n}\int_{R^n}(\int_{R^n} f(\mathbf{y})e^{-i\omega\cdot\mathbf{y}}d\mathbf{y})e^{-t(\sum \omega_i^2)}e^{i\omega\cdot\mathbf{x}}d\omega \\ &= \frac{1}{(2\pi)^n}\int_{R^n} f(\mathbf{y})\int_{R^n} e^{-t(\sum \omega_i^2)}e^{i\omega\cdot(\mathbf{x}-\mathbf{y})}d\omega d\mathbf{y}. \end{aligned} \tag{9.4.4}$$

In (9.4.4) ω variables in the inside of the exponents is combined as

$$-t(\sum_{i=1}^n \omega_i^2) + i\omega\cdot(\mathbf{x}-\mathbf{y}) = -t\sum_{i=1}^n[\omega_i - \frac{i(x_i-y_i)}{2t}]^2 - \frac{1}{4t}\sum_{i=1}^n (x_i-y_i)^2.$$

Therefore, (9.4.4) is rewritten as

$$u(\mathbf{x},t) = \frac{1}{(2\pi)^n}[\int_{R^n} f(\mathbf{y})e^{-\frac{1}{4t}\sum_{i=1}^n (x_i-y_i)^2} \int_{R^n} e^{-t\sum_{i=1}^n[\omega_i - \frac{i(x_i-y_i)}{2t}]^2}d\omega d\mathbf{y}]. \tag{9.4.5}$$

In (9.4.5) $z_i = \omega_i - i(x_i-y_i)/(2t)$ is a horizontal line in the complex plane. We use the Cauchy theorem in the complex analysis to effectively "shift" the integration path to the real line (See Remark 2.6.5). Using also the result of Exercise 2 in Section 2.6

$$\int_{-\infty}^{\infty} e^{-t[\omega_i - \frac{i(x_i-y_i)}{2t}]^2}d\omega_i = \int_{-\infty}^{\infty} e^{-tz_i^2}dz_i = \sqrt{\frac{\pi}{t}} \tag{9.4.6}$$

n times, we have

$$\int_{R^n} e^{-t\sum_{i=1}^n[\omega_i - \frac{i(x_i-y_i)}{2t}]^2}d\omega = \int_{R^n} e^{-t\sum_{i=1}^n z_i^2}d\mathbf{z} = (\frac{\pi}{t})^{\frac{n}{2}}.$$

Therefore, (9.4.5) is computed as

$$u(\mathbf{x},t) = \frac{1}{(4\pi t)^{\frac{n}{2}}}\int_{R^n} f(\mathbf{y})e^{-\frac{1}{4t}\sum_{i=1}^n (x_i-y_i)^2}d\mathbf{y}.$$

Note that

$$\frac{1}{(4\pi t)^{\frac{1}{2}}}\int_{-\infty}^{\infty} e^{-\frac{1}{4t}(x_i-y_i)^2}dy_i = 1. \tag{9.4.7}$$

We denote

$$G(\mathbf{x},t) = \frac{1}{(4\pi t)^{\frac{n}{2}}}e^{-\frac{1}{4t}\sum_{i=1}^n x_i^2}. \tag{9.4.8}$$

This is the heat kernel for the multidimensional case. The relation (9.4.7) implies

$$\int_{R^n} G(\mathbf{x},t)d\mathbf{x} = 1. \tag{9.4.9}$$

If $n=1$, it reduces to $G(x,t)$ in (2.6.7) of Section 2.6.

9.4.2 *Wave Equation*

In Chapter 3 we discussed the solutions to the wave equation using the characteristics. In this section we use the Fourier transform to find the solutions to the initial value problems for the wave equation

$$u_{tt} = c^2 u_{xx}, \quad -\infty < x < \infty, \quad 0 < t, \tag{9.4.10}$$

$$u(x, 0) = f(x), \quad u_t(x, 0) = g(x).$$

We denote the Fourier transform of $u(x, t)$ by

$$U(\omega, t) = \int_{-\infty}^{\infty} u(y, t) e^{-i\omega y} dy.$$

Step 1: First, apply the Fourier transform to (9.4.10) and obtain

$$U_{tt}(\omega, t) = -c^2 \omega^2 U, \tag{9.4.11}$$

where the initial conditions for U are

$$U(\omega, 0) = \int_{-\infty}^{\infty} u(y, 0) e^{-i\omega y} dy = \int_{-\infty}^{\infty} f(y) e^{-i\omega y} dy,$$

$$U_t(\omega, 0) = \int_{-\infty}^{\infty} u_t(y, 0) e^{-i\omega y} dy = \int_{-\infty}^{\infty} g(y) e^{-i\omega y} dy.$$

Solving (9.4.11), we see

$$U(\omega, t) = A(\omega) \cos c\omega t + B(\omega) \sin c\omega t.$$

From the initial conditions

$$A(\omega) = \int_{-\infty}^{\infty} f(y) e^{-i\omega y} dy,$$

$$B(\omega) = \frac{1}{c\omega} \int_{-\infty}^{\infty} g(y) e^{-i\omega y} dy.$$

Therefore,

$$U(\omega, t) = [\int_{-\infty}^{\infty} f(y) e^{-i\omega y} dy] \cos c\omega t + [\frac{1}{c\omega} \int_{-\infty}^{\infty} g(y) e^{-i\omega y} dy] \sin c\omega t. \tag{9.4.12}$$

Step 2: Taking the inverse transform, we should recover the solution formula (3.2.9) we studied in Chapter 3. Substituting (9.4.12) in the inverse and changing the order of integration, we have

$$\begin{aligned} u(x,t) &= \int_{-\infty}^{\infty} U(\omega,t)e^{i\omega x}d\omega \\ &= \frac{1}{2\pi}\int_{-\infty}^{\infty}[\int_{-\infty}^{\infty} f(y)e^{-i\omega y}dy]\cos c\omega t e^{i\omega x}d\omega \\ &+\frac{1}{2\pi}\int_{-\infty}^{\infty}[\frac{1}{c\omega}\int_{-\infty}^{\infty} g(y)e^{-i\omega y}dy]\sin c\omega t e^{i\omega x}d\omega \\ &= \frac{1}{2\pi}\int_{-\infty}^{\infty} f(y)\int_{-\infty}^{\infty}\frac{1}{2}(e^{i\omega(x-y+ct)}+e^{i\omega(x-y-ct)})d\omega dy \\ &+\frac{1}{2\pi}\int_{-\infty}^{\infty} g(y)\int_{-\infty}^{\infty}\frac{1}{c\omega}\frac{1}{2i}(e^{i\omega(x-y+ct)}-e^{i\omega(x-y-ct)})d\omega dy. \end{aligned}$$

The integrals for ω have the familiar forms of the delta and Heaviside functions. By Table 9.1 we obtain

$$\begin{aligned} u(x,t) &= \int_{-\infty}^{\infty} f(y)\frac{1}{2}[\delta(x-y+ct)+\delta(x-y-ct)]dy \\ &+\int_{-\infty}^{\infty} g(y)\frac{1}{2c}[H(x-y+ct)-H(x-y-ct)]dy \\ &= \frac{1}{2}[f(x+ct)+f(x-ct)]+\frac{1}{2c}\int_{x-ct}^{x+ct} g(y)dy. \end{aligned}$$

This is precisely the solution formula (3.2.9) we studied in Chapter 3.

9.4.3 *Laplace Equation in a Half Space*

We solve the Laplace equation in a half space by the Fourier transform. We look for a bonded solution in x and y.

$$u_{xx}+u_{yy}=0, \quad -\infty<x<\infty,\ 0<y, \tag{9.4.13}$$

$$u(x,0)=g(x), \quad -\infty<x<\infty.$$

Step 1: We need to find out to which variable(s) we apply the Fourier transform. Since y is restricted and x is not, it is reasonable to apply the Fourier transform to x and get an ODE for y. Therefore, we define

$$U(\omega,y)=\int_{-\infty}^{\infty} u(x,y)e^{-i\omega x}dx.$$

We apply the Fourier transform in x to (9.4.13). Then,

$$\int_{-\infty}^{\infty} (u_{xx} + u_{yy}) e^{-i\omega x} dx = -\omega^2 U + U_{yy} = 0, \tag{9.4.14}$$

$$U(\omega, 0) = \int_{-\infty}^{\infty} g(x) e^{-i\omega x} dx = G(\omega). \tag{9.4.15}$$

Equation (9.4.14) is a second order linear ODE. The general solution is

$$U = A(\omega) e^{-y\omega} + B(\omega) e^{y\omega}.$$

Since we are looking for a bounded solution in y, we have $U = B(\omega) e^{y\omega}$ if $\omega < 0$ and $U = A(\omega) e^{-y\omega}$ if $\omega > 0$. Also if $y = 0$, $U = G(\omega)$. So, we can write

$$U(\omega, y) = G(\omega) e^{-y|\omega|}.$$

Step 2: We now apply the inverse transform. As before, substituting the Fourier transform of the boundary condition (9.4.15) and changing the order of integrations, we have

$$\begin{aligned} u(x,y) &= \frac{1}{2\pi} \int_{-\infty}^{\infty} U(\omega, y) e^{i\omega x} d\omega \\ &= \frac{1}{2\pi} \int_{-\infty}^{\infty} [\int_{-\infty}^{\infty} g(t) e^{-i\omega t} dt] e^{-y|\omega|} e^{i\omega x} d\omega \\ &= \frac{1}{2\pi} \int_{-\infty}^{\infty} g(t) [\int_{-\infty}^{\infty} e^{-i\omega t} e^{-y|\omega|} e^{i\omega x} d\omega)] dt \\ &= \frac{1}{2\pi} \int_{-\infty}^{\infty} g(t) [\int_{-\infty}^{0} e^{(y - it + ix)\omega} d\omega)] dt \\ &+ \frac{1}{2\pi} \int_{-\infty}^{\infty} g(t) [\int_{0}^{\infty} e^{-(y + it - ix)\omega} d\omega)] dt. \end{aligned}$$

Performing the integration in ω, we obtain

$$\begin{aligned} u(x,y) &= \frac{1}{2\pi} \int_{-\infty}^{\infty} g(t) [\frac{1}{y + i(x-t)} + \frac{1}{y - i(x-t)}] dt \\ &= \frac{y}{\pi} \int_{-\infty}^{\infty} \frac{g(t)}{(x-t)^2 + y^2} dt. \end{aligned}$$

The solution is the same as (4.4.7).

9.4.4 *Black-Scholes-Merton Equation*

Most applications of PDE are physics or engineering. However they are also very useful in Economics or Finance. The Black-Scholes-Merton equation is perhaps the most celebrated equation in finance. This equation gives a way to price options and derivative securities.

An option gives a right but not an obligation to buy or sell stock at some specified price K called strike price. Using the simplest form of option called a European call option, we explain some basic terminologies. This is a contract giving a right to buy a specific stock at a specified time and at a specified price. The specific stock is called the underlying asset. The future specified time at which we buy stock is called the expiration date. The specified price is called the strike price.

The most well-known option is a European option where the expiration date T is fixed. A call option with the strike price K pays $(S(T)-K)^+$ and a put option with the strike price K pays $(K - S(T))^+$ at the expiration date. Here, for example,

$$(S(T)-K)^+ = \begin{cases} S(T)-K, & S(T) \geq K \\ 0, & S(T) < K. \end{cases}$$

A call option gives a right but not an obligation to buy stock at some specified price K called strike price. So, if the stock price at the expiration date is higher than K, the holder of the call option can exercises the right to buy the stock at price K and if he (or she) sells it immediately after the purchase, she gets the profit of $S(T) - K$ dollars per share. A put option gives a right to sell stock at a strike price K. For example, suppose you purchase a call option for IBM stock with strike price \$180 which expires in three months. If IBM stock price is \$190 at the expiration date, you exercise the option to get IBM stock at the strike price \$180 and immediately sell it at \$190 on the market. Then, you get a profit of \$10 per share minus the cost of the purchase of option.

So, a question is how to price such a contract. The derivation of option pricing equation is beyond the scope of this book. It is given in (9.4.16) with the terminal condition (9.4.17), where x is for the stock price and t is for time. As we see it has both convection and diffusion terms. So, it is a convection diffusion equation.

$$c_t + rxc_x + \frac{1}{2}\sigma^2 x^2 c_{xx} - rc = 0. \tag{9.4.16}$$

$$c(x,T) = (x-K)^+. \tag{9.4.17}$$

We need the boundary conditions. At $x = 0$,

$$c_t(0,t) = rc(0,t) \Rightarrow c(0,t) = e^{rt}c(0,0).$$

Since $c(T,0) = (0-K)^+ = 0$, this implies that

$$c(0,t) = 0. \tag{9.4.18}$$

Also, as $x \to \infty$ we specify the rate of growth

$$\lim_{x\to\infty} [c(x,t) - (x - e^{-r(T-t)}K)] = 0. \tag{9.4.19}$$

What this means is that the price of the call option is close to the stock price minus the discounted value of the strike price.

To solve the equation with the Fourier transform, we first set $\tau = T - t$ and $x = e^y$. Then,

$$c_t = -c_\tau, \quad c_x = \frac{1}{x}c_y, \quad c_{xx} = -\frac{1}{x^2}c_y + \frac{1}{x^2}c_{yy}.$$

Therefore, (9.4.16), (9.4.17), (9.4.18), and (9.4.19) are reduced to

$$c_\tau + (\frac{1}{2}\sigma^2 - r)c_y - \frac{1}{2}\sigma^2 c_{yy} + rc = 0, \tag{9.4.20}$$

$$c(y,0) = (e^y - K)^+, \tag{9.4.21}$$

$$\lim_{y\to-\infty} c(y,t) = 0, \tag{9.4.22}$$

$$\lim_{y\to\infty} [c(y,t) - (y - e^{-r(T-t)}K)] = 0. \tag{9.4.23}$$

Now we have an initial boundary value problem where the equation is constant coefficient and the boundary conditions (9.4.22) and (9.4.23) are specified at $y = \pm\infty$. We have a setting where the Fourier transform is applicable. We introduce the Fourier transform and the inverse transform.

$$C(\omega) = \int_{-\infty}^{\infty} c(y)e^{-i\omega y}dy,$$

$$c(x) = \frac{1}{2\pi}\int_{-\infty}^{\infty} C(\omega)e^{i\omega x}d\omega.$$

At first we neglect the boundary conditions and apply the Fourier transform to the equation (9.4.20) and the initial condition (9.4.21). Then,

$$C_\tau + i\omega(\frac{1}{2}\sigma^2 - r)C + \frac{1}{2}\sigma^2\omega^2 C + rC = 0, \tag{9.4.24}$$

$$C(\omega,0)=\int_{-\infty}^{\infty}(e^y-K)^+e^{-i\omega y}dy=\int_{\ln K}^{\infty}(e^y-K)e^{-i\omega y}dy. \quad (9.4.25)$$

The equation (9.4.20) is a linear first order or a separable ODE.

$$C(\omega,\tau)=C(\omega,0)e^{-\frac{1}{2}\sigma^2\omega^2\tau-i\omega(\frac{1}{2}\sigma^2-r)\tau-r\tau}.$$

Now we apply the inverse transform. Then, after substituting the initial data (9.4.25) and changing the order of integrations, we obtain

$$\begin{aligned} & c(z,\tau) \\ &= \frac{1}{2\pi}\int_{-\infty}^{\infty}C(\omega,0)e^{-\frac{1}{2}\sigma^2\omega^2\tau-i\omega(\frac{1}{2}\sigma^2-r)\tau-r\tau}e^{i\omega z}d\omega \\ &= \frac{1}{2\pi}\int_{-\infty}^{\infty}\int_{-\infty}^{\infty}(e^y-K)^+e^{-i\omega y}dye^{-\frac{1}{2}\sigma^2\omega^2\tau-i\omega(\frac{1}{2}\sigma^2-r)\tau-r\tau}e^{i\omega z}d\omega \\ &= \frac{1}{2\pi}\int_{-\infty}^{\infty}\int_{-\infty}^{\infty}(e^y-K)^+e^{-r\tau}e^{g(\omega)}d\omega dy, \qquad (9.4.26) \end{aligned}$$

where

$$\begin{aligned} g(\omega) &= -\frac{1}{2}\sigma^2\omega^2\tau+i\omega((r-\frac{1}{2}\sigma^2)\tau+z-y) \\ &= -\frac{1}{2}\sigma^2\tau(\omega-\frac{i((r-\frac{1}{2}\sigma^2)\tau+z-y)}{2\frac{1}{2}\sigma^2\tau})^2-\frac{((r-\frac{1}{2}\sigma^2)\tau+z-y)^2}{2\sigma^2\tau}. \end{aligned}$$

We complete the square for ω in $g(\omega)$ as above and integrate in ω first. Then,

$$\begin{aligned} c(z,\tau) &= \frac{1}{2\pi}\int_{-\infty}^{\infty}\int_{-\infty}^{\infty}[(e^y-K)^+e^{-r\tau-\frac{((r-\frac{1}{2}\sigma^2)\tau+z-y)^2}{2\sigma^2\tau}} \\ &\quad e^{-\frac{1}{2}\sigma^2\tau(\omega-\frac{i((r-\frac{1}{2}\sigma^2)\tau+z-y)}{2\frac{1}{2}\sigma^2\tau})^2}]d\omega dy \qquad (9.4.27) \\ &= \frac{1}{\sqrt{2\pi\sigma^2\tau}}\int_{\ln K}^{\infty}(e^y-K)e^{-r\tau-\frac{((r-\frac{1}{2}\sigma^2)\tau+z-y)^2}{2\sigma^2\tau}}dy. \qquad (9.4.28) \end{aligned}$$

See Remark 2.6.5 for the complex integral in (9.4.27).

The rest is to rewrite (9.4.28) so that the resulting form will match the standard form available in finance literature. Splitting the integral into two and completing the squares in the integrand yield

$$\begin{aligned} c(z,\tau) &= \frac{1}{\sqrt{2\pi\sigma^2\tau}}\int_{\ln K}^{\infty}e^{y-r\tau-\frac{((r-\frac{1}{2}\sigma^2)\tau+z-y)^2}{2\sigma^2\tau}}dy \\ &\quad -\frac{1}{\sqrt{2\pi\sigma^2\tau}}\int_{\ln K}^{\infty}Ke^{-r\tau-\frac{((r-\frac{1}{2}\sigma^2)\tau+z-y)^2}{2\sigma^2\tau}}dy \\ &= \frac{1}{\sqrt{2\pi\sigma^2\tau}}\int_{\ln K}^{\infty}e^{z-\frac{(y-((r+\frac{1}{2}\sigma^2)\tau+z))^2}{2\sigma^2\tau}}dy \\ &\quad -\frac{1}{\sqrt{2\pi\sigma^2\tau}}\int_{\ln K}^{\infty}Ke^{-r\tau-\frac{((r-\frac{1}{2}\sigma^2)\tau+z-y)^2}{2\sigma^2\tau}}dy. \end{aligned}$$

We go back to the original variable x and use the standard normal distribution. To do so we use $e^z = x$ or $z = \ln x$ and set

$$u = \frac{y - z - (r \pm \frac{1}{2}\sigma^2)\tau}{\sqrt{\sigma^2\tau}} = \frac{y - \ln x - (r \pm \frac{1}{2}\sigma^2)\tau}{\sigma\sqrt{\tau}}.$$

Then,

$$\begin{aligned} c(x,\tau) &= x\frac{1}{\sqrt{2\pi\sigma^2\tau}}\int_{\ln K}^{\infty} e^{-\frac{(y-z-(r+\frac{1}{2}\sigma^2)\tau)^2}{2\sigma^2\tau}}\,dy \\ &\quad - Ke^{-r\tau}\frac{1}{\sqrt{2\pi\sigma^2\tau}}\int_{\ln K}^{\infty} e^{-\frac{(y-z-(r-\frac{1}{2}\sigma^2)\tau)^2}{2\sigma^2\tau}}\,dy \qquad (9.4.29) \\ &= x\frac{1}{\sqrt{2\pi}}\int_{-\frac{1}{\sigma\sqrt{\tau}}[\ln\frac{x}{K}+(r+\frac{1}{2}\sigma^2)\tau]}^{\infty} e^{-\frac{u^2}{2}}\,du \\ &\quad - Ke^{-r\tau}\frac{1}{\sqrt{2\pi}}\int_{-\frac{1}{\sigma\sqrt{\tau}}[\ln\frac{x}{K}+(r-\frac{1}{2}\sigma^2)\tau]}^{\infty} e^{-\frac{u^2}{2}}\,du. \end{aligned}$$

In terms of t, we obtain

$$c(x,t) = xN(d_+(T-t,x)) - Ke^{-\tau(T-t)}N(d_-(T-t,x)), \qquad (9.4.30)$$

where N is the cumulative standard normal distribution (similar to the error function) given by

$$N(y) = \frac{1}{\sqrt{2\pi}}\int_{-y}^{\infty} e^{-\frac{u^2}{2}}\,du = \frac{1}{\sqrt{2\pi}}\int_{-\infty}^{y} e^{-\frac{u^2}{2}}\,du$$

and

$$d_\pm(\tau,x) = \frac{1}{\sigma\sqrt{\tau}}[\ln\frac{x}{K} + (r \pm \frac{1}{2}\sigma^2)\tau].$$

It is left as an exercise to verify that (9.4.30) satisfies the boundary conditions (9.4.22) and (9.4.23). The solution (9.4.30) is a very standard form in finance.

Exercises

1. Use the Fourier transform to solve

$$u_t - \Delta u = \delta(x - x_0)\delta(t - t_0), \quad -\infty < x < \infty,\ 0 < t,$$
$$u(x,0) = f(x).$$

2. Solve

$$u_t = a^2 u_{xx} - \gamma u, \quad -\infty < x < \infty, \quad 0 < t,$$
$$u(x,0) = f(x).$$

3. Consider

$$u_t = ku_{xx} + h(x,t), \quad -\infty < x < \infty, \quad 0 < t,$$

$$u(x,0) = f(x).$$

(a) Find the Fourier transform $\bar{U}$ of u.
(b) Find the solution $u(x,t)$.

4. Solve

$$u_t + \mathbf{v}_0 \cdot \nabla u = k\Delta u,$$

$$u(x,y,0) = f(x,y).$$

5. Solve

$$u_t - \Delta u = 0, \quad -\infty < x < \infty,\ 0 < y < H,\ 0 < t,$$

$$u(x,0,t) = 0, \quad u(x,H,t) = 0,$$

$$u(x,y,0) = f(x,y).$$

6. Solve the heat equation in a strip.

$$u_t = k(u_{xx} + u_{yy}), \quad -\infty < x < \infty, \quad 0 < y < H$$

$$u(x,0,t) = 0, \quad u(0,y,t) = 0,$$

$$u(x,y,0) = f(x,y).$$

7. Use the Fourier transform to solve

$$u_{tt} = c^2 u_{xx} + \delta(x - x_0)\delta(t - t_0), \quad -\infty < x < \infty,\ 0 < t,$$

$$u(x,0) = f(x), \quad u_t(x,0) = g(x).$$

8. Verify the result in (9.4.28).
9. Perform the change of variable and derive (9.4.30) from (9.4.29).
10. Verify that (9.4.30) satisfies the boundary conditions (9.4.22) and (9.4.23).
11. Price an European put using the Fourier transform.

Chapter 10

Laplace Transform

The Laplace transform is another transformation that is useful for solving DE's. Some students learn the Laplace transform in an elementary ODE course. In this chapter we apply the Laplace transform to solve PDE's. We define the Laplace and the inverse transforms in Section 10.1. In Section 10.2 we discuss the properties of the Laplace transform including the relation with the Fourier transform. The applications to DE's are discussed in Section 10.3.

10.1 Laplace Transform and the Inverse

10.1.1 *Laplace Transform*

The Laplace transform is defined as

$$F(s) = \mathcal{L}[f(t)] = \int_0^\infty f(t)e^{-st}dt, \tag{10.1.1}$$

where s is a parameter whose values are restricted so that the integral is finite.

We apply the transform to the time rather than the space variables. The inverse transform is given by

$$f(t) = \mathcal{L}^{-1}[F(s)] = \frac{1}{2\pi i}\int_{a-i\infty}^{a+i\infty} F(s)e^{st}ds, \tag{10.1.2}$$

where the integral is taken along a vertical line $s = a + ib$ in the complex plane, where a is a positive constant and $-\infty < b < \infty$. The definition of the inverse looks strange. The justification is given in Subsection 10.2.3 where we study the relation between the Laplace and Fourier transforms. From the definitions it is easy to see that both the Laplace and inverse transforms are linear, *i.e.*,

$$\mathcal{L}[c_1 f(t) + c_2 g(t)] = c_1\mathcal{L}[f(t)] + c_2\mathcal{L}[g(t)],$$

$$\mathcal{L}^{-1}[c_1F(s) + c_2G(s)] = c_1\mathcal{L}^{-1}[F(s)] + c_2\mathcal{L}^{-1}[G(s)],$$

where c_1 and c_2 are constants. The Laplace and the inverse transforms of typical functions are listed in Table 10.1.

Table 10.1 The Laplace and the inverse transforms.

	$f(x)$	$F(s) = \mathcal{L}[f(t)]$
1.	1	$\frac{1}{s}$
2.	$t^n \ (n > -1)$	$\frac{n!}{s^{(n+1)}}$
3.	$t^p \ (n > -1)$	$\frac{\Gamma(p+1)}{s^{(p+1)}}$
4.	e^{at}	$\frac{1}{s-a}$
5.	$t^n e^{at}$	$\frac{n!}{(s-a)^{(n+1)}}$
6.	$\sin bt$	$\frac{b}{s^2+b^2}$
7.	$\cos bt$	$\frac{s}{s^2+b^2}$
8.	$\sinh bt$	$\frac{b}{s^2-b^2}$
9.	$\cosh bt$	$\frac{s}{s^2-b^2}$
10.	$e^{at} \sin bt$	$\frac{b}{(s-a)^2+b^2}$
11.	$e^{at} \cos bt$	$\frac{s-a}{(s-a)^2+b^2}$
12.	$\delta(t-b)$	$e^{-bs} \ (b \geq 0)$
13.	$H(t-t_0)$	$\frac{e^{-t_0 s}}{s}$
14.	$(\pi t)^{-1/2} e^{-a^2/4t}$	$\frac{1}{\sqrt{s}} e^{-a\sqrt{s}}$
15.	$(4\pi)^{-1/2} t^{-3/2} e^{-a^2/4t}$	$\frac{1}{a} e^{-a\sqrt{s}}$
16.	$1 - \mathrm{erf}(\frac{a}{\sqrt{4t}})$	$\frac{1}{s} e^{-a\sqrt{s}}$
17.	$\int_0^t f(t-\tau)g(\tau)d\tau$	$F(s)G(s)$
18.	$e^{at} f(t)$	$F(s-a)$

Example 10.1.1. Find the Laplace transform.

(a) $f(x) = \cos \omega t$, (b) $f(x) = \delta(t-a)$, (c) $f(x) = \sinh \omega t$.

Solution: (a) We use the integration by parts twice.

$$F(s) = \int_0^\infty \cos \omega t e^{-st} dt = [-\frac{1}{s} \cos \omega t e^{-st}]_0^\infty - \int_0^\infty \frac{\omega}{s} \sin \omega t e^{-st} dt$$

$$= \frac{1}{s} + [\frac{\omega}{s^2} \sin \omega t e^{-st}]_0^\infty - \int_0^\infty \frac{\omega^2}{s^2} \cos \omega t e^{-st} dt = \frac{1}{s} - \frac{\omega^2}{s^2} F(s).$$

Therefore,

$$F(s) = \frac{s}{s^2 + \omega^2}.$$

(b) $F(s) = \int_0^\infty \delta(t-a)e^{-st}dt = e^{-as}$.

(c) Using $\sinh \omega t = (e^{\omega t} - e^{-\omega t})/2$, we compute

$$F(s) = \int_0^\infty \frac{1}{2}(e^{\omega t} - e^{-\omega t})e^{-st}dt = \int_0^\infty \frac{1}{2}(e^{(\omega-s)t} - e^{-(\omega+s)t})dt$$
$$= \frac{1}{2}[\frac{e^{(\omega-s)t}}{\omega - s} + \frac{e^{-(\omega+s)t}}{\omega+s}]_0^\infty = \frac{\omega}{s^2-\omega^2}.$$

10.1.2 *Inverse Transform*

It is rather difficult to use (10.1.2) to find the inverse as we need the knowledge of complex integrations. Usually the inverse is performed through Table 10.1. A basic idea is that we decompose $F(s)$ so that we can use the table. There are several useful tricks. They are adjusting constants, partial fractions, and completing the square. These tricks are explained in the following example.

Example 10.1.2. Find the inverse transforms using Table 10.1.

(a) $F(s) = \frac{2}{s(s+1)}$, (b) $F(s) = \frac{s}{s^2-2s+3}$.

Solution: To find the inverse transform, it is often good idea to check the denominator to see if which formula in Table 10.1 apply.

(a) First we look at the denominator and see which inverse transform could be applied. In Table 10.1 we see

$$\mathcal{L}^{-1}[\frac{1}{s}] = 1, \quad \mathcal{L}^{-1}[\frac{1}{s-a}] = e^{at}.$$

This suggests that we decompose $F(s)$ using the partial fractions and apply the above inverse transformations with $a = -1$. The partial fractions are

$$\frac{2}{s(s+1)} = \frac{A}{s} + \frac{B}{s+1} = \frac{A(s+1)+Bs}{s(s+1)},$$

where A and B are constants to be determined. We must have

$$2 = A(s+1) + Bs$$

for all s. There are two ways to determine A and B. One way is to assign convenient numbers to s. In this example $s = 0$ and -1 are convenient numbers. If we substitute $s = 0$, B is eliminated and we have $A = 2$. Setting $s = -1$, we can eliminate A and obtain $B = -2$. Another way is to arrange the right hand side in the like powers of s so that we have

$$2 = (A+B)s + A.$$

This equation must hold for all s. This tells us that the constant term A must be 2 and the s term is $A + B = 0$, which implies that $B = -2$. Therefore, using the linearity of the inverse, we obtain

$$\mathcal{L}^{-1}[\frac{2}{s(s+1)}] = \mathcal{L}^{-1}[\frac{2}{s} - \frac{2}{s+1}] = 2\mathcal{L}^{-1}[\frac{1}{s}] - 2\mathcal{L}^{-1}[\frac{1}{s+1}] = 2 - 2e^{-t}.$$

(b) This example deals with completing the square. The form of the denominator implies that the formulas 10 and 11 might apply. To use the formula we complete the square of the denominator, use the linearity of the inverse, and adjust the coefficients. Then, we obtain

$$\begin{aligned}\mathcal{L}^{-1}[\frac{s}{s^2-2s+3}] &= \mathcal{L}^{-1}[\frac{s-1+1}{(s-1)^2+2}] \\ &= \mathcal{L}^{-1}[\frac{s-1}{(s-1)^2+2}] + \frac{1}{\sqrt{2}}\mathcal{L}^{-1}[\frac{\sqrt{2}}{(s-1)^2+2}] \\ &= e^t \cos\sqrt{2}t + \frac{1}{\sqrt{2}}e^t \sin\sqrt{2}t.\end{aligned}$$

Exercises

1. Use the definition to find the Laplace transform of $f(t)$.

   (a) $f(x) = 1$
   (b) $f(x) = \sin at$
   (c) $f(x) = \cosh at$
   (d) $f(x) = H(t - t_0), \quad t_0 > 0$

2. Suppose $F(s) = \int_0^\infty f(t)e^{-st}dt$. Derive the following relations.

   (a) $\mathcal{L}[-tf(t)] = F'(s)$
   (b) $\mathcal{L}[H(t-b)f(t-b)] = e^{-bs}F(s)$

3. Find the inverse Laplace transform.

   (a) $F(s) = \frac{s}{(s^2+1)(s^2+2)}$
   (b) $F(s) = \frac{1}{(s^2+1)(s^2+2)}$
   (c) $F(s) = \frac{s}{(s+1)(s^2+2s+4)}$

4. We carry out the derivation of the formula 14 and 15 in Table 10.1. For this purpose define

$$I(a) = \mathcal{L}[\frac{1}{t^{1/2}}e^{-a^2/(4t)}], \quad J(a) = \mathcal{L}[\frac{1}{t^{3/2}}e^{-a^2/(4t)}].$$

   (a) Use the substitution $u = \sqrt{s}\sqrt{t} - \frac{a}{2}t^{-1/2}$ in $\int_{-\infty}^{\infty} e^{-u^2}du = \sqrt{\pi}$ to find the relation between I and J.

(b) Differentiate I in a and use (a) to obtain the ODE for I.

(c) Find $\lim_{a\to\infty} I(a)$ and use it to find the solution $I(a)$ to the ODE in (b).

(d) Use $I(a)$ and (a) to find $J(a)$.

5. The formula 16 in Table 10.1 can be obtained in the following way. Define

$$K(a) = \mathcal{L}[1 - \text{erf}(\frac{a}{\sqrt{4t}})].$$

(a) Differentiate $K(a)$ in a to show that

$$\frac{dK}{da} = -\frac{1}{\sqrt{s}}e^{-a\sqrt{s}}. \tag{10.1.3}$$

(b) Integrate (10.1.3) in a and use $\lim_{a\to\infty} K(a) = 0$ to show the formula 16 in Table 10.1.

10.2 Properties of the Laplace Transform

10.2.1 *Laplace Transform of Derivatives*

To solve the differential equations with the Laplace transform, we need to know how the derivatives are transformed. We begin with the first derivatives.

$$\mathcal{L}[f'(t)] = \int_0^\infty f'(t)e^{-st}dt = [f(t)e^{-st}]_0^\infty + s\int_0^\infty f(t)e^{-st}dt = sF(s) - f(0). \tag{10.2.1}$$

For the second derivatives we repeat the integration by parts twice.

$$\begin{aligned}\mathcal{L}[f''(t)] &= \int_0^\infty f''(t)e^{-st}dt = [f'(t)e^{-st}]_0^\infty + s\int_0^\infty f'(t)e^{-st}dt \\ &= -f'(0) + s[f(t)e^{-st}]_0^\infty + s^2\int_0^\infty f(t)e^{-st}dt \\ &= s^2F(s) - sf(0) - f'(0).\end{aligned} \tag{10.2.2}$$

Example 10.2.1. Find the Laplace transform.

(a) $g(t) = (te^{-2t})'$ (b) $g(t) = (te^{-2t})''$

Solution: (a) We use (10.2.1) with $f(t) = te^{-2t}$.

$$\mathcal{L}[f'(t)] = s\int_0^\infty te^{-2t}e^{-st}dt$$
$$= s[-\frac{1}{s+2}te^{-(s+2)t}]_0^\infty + s\int_0^\infty \frac{1}{s+2}e^{-(s+2)t}dt$$
$$= \frac{s}{(s+2)^2}.$$

(b) We use (10.2.2) with $f(t) = te^{-2t}$.

$$\mathcal{L}[f''(t)] = s^2\int_0^\infty te^{-2t}e^{-st}dt - 1$$
$$= s^2[-\frac{1}{s+2}te^{-(s+2)t}]_0^\infty + s^2\int_0^\infty \frac{1}{s+2}e^{-(s+2)t}dt - 1$$
$$= \frac{s^2}{(s+2)^2} - 1.$$

10.2.2 *Convolution Theorem*

The convolution of functions $f(t)$ and $g(t)$ is $\int_0^t f(\tau)g(t-\tau)d\tau$. The Laplace transform of the convolution is given by

$$\mathcal{L}[\int_0^t f(\tau)g(t-\tau)d\tau] = \int_0^\infty [\int_0^t f(\tau)g(t-\tau)d\tau]e^{-st}dt.$$

The key to evaluate this double integral is to change the order of integrations. For this purpose, examine the limits of integrations. In the above integral we integrate in τ and then in t. The limits of integrations for τ and t are given by

$$0 \le \tau \le t, \quad 0 \le t < \infty. \tag{10.2.3}$$

What (10.2.3) says is that for each t we integrate in τ over $0 \le \tau \le t$ and then integrate in t over $0 \le t < \infty$. This is illustrated as the vertical arrows in Figure 10.1 where each vertical arrow represents the integration in τ. The two inequalities in (10.2.3) gives the region of integration as depicted in Figure 10.1. If we change the order of integration, we integrate in t first. For each τ we integrate in t over $\tau \le t < \infty$. Then for τ we integrate over $0 \le \tau < \infty$. This is illustrated as the horizontal arrows in Figure 10.1 where each horizontal arrow represents the integration in t. Therefore,

$$\mathcal{L}[\int_0^t f(\tau)g(t-\tau)d\tau] = \int_0^\infty \int_\tau^\infty f(\tau)g(t-\tau)e^{-st}dtd\tau.$$

Set $p = t - \tau$. Then, p changes from 0 to ∞ as t changes from τ to ∞. Thus,

$$\begin{aligned}\int_0^\infty \int_\tau^\infty f(\tau)g(t-\tau)e^{-st}dtd\tau &= \int_0^\infty \int_0^\infty f(\tau)g(p)e^{-s(p+\tau)}dpd\tau \\ &= \int_0^\infty g(p)e^{-sp}dp \int_0^\infty f(\tau)e^{-s\tau}d\tau \\ &= F(s)G(s).\end{aligned}$$

Therefore, the Laplace transform of convolution $\int_0^t f(t)g(t-\tau)d\tau$ is given by

$$\mathcal{L}[\int_0^t f(t)g(t-\tau)d\tau] = F(s)G(s) \tag{10.2.4}$$

and the inverse is given by

$$\mathcal{L}^{-1}[F(s)G(s)] = \int_0^t f(\tau)g(t-\tau)d\tau. \tag{10.2.5}$$

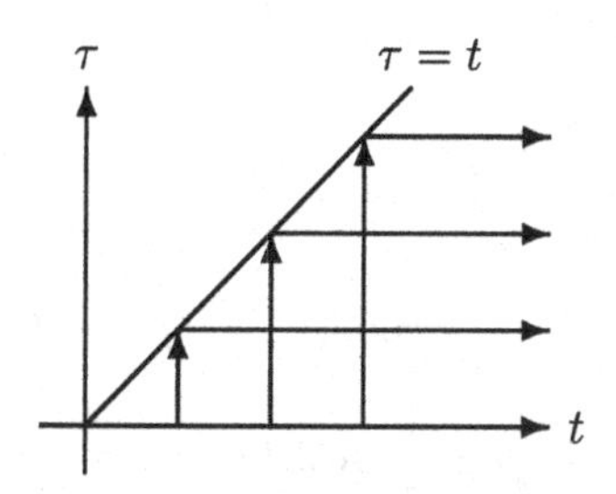

Fig. 10.1 Region of integration for convolution theorem.

Example 10.2.2. (a) Use the convolution theorem to find the Laplace transform of $h(t) = \int_0^t e^{-(t-\tau)} \cos \tau d\tau$.

(b) Verify (a) by computing $\mathcal{L}[h(t)] = \int_0^\infty e^{-st} \int_0^t e^{-(t-\tau)} \cos \tau d\tau dt$.

Solution: (a) In this problem we can set $f(t) = e^{-t}$ and $g(t) = \cos t$. Then, from Table 10.1 $F(s) = 1/(s+1)$ and $G(s) = s/(s^2+1)$. Therefore,

$$\mathcal{L}[h(t)] = F(s)G(s) = \frac{s}{(s+1)(s^2+1)}.$$

(b) As you see below, the trick is to change the order of integrations.

$$\begin{aligned}\mathcal{L}[h(t)] &= \int_0^\infty e^{-st}\int_0^t e^{-(t-\tau)}\cos\tau d\tau dt = \int_0^\infty\int_\tau^\infty e^{-(s+1)t}e^\tau\cos\tau dt d\tau\\ &= \int_0^\infty [\frac{e^{-(s+1)t}}{-(s+1)}]_\tau^\infty e^\tau\cos\tau dt d\tau\\ &= \int_0^\infty \frac{e^{-(s+1)\tau}}{(s+1)}e^\tau\cos\tau dt d\tau\\ &= \frac{1}{s+1}\int_0^\infty e^{-s\tau}\cos\tau d\tau = \frac{s}{(s+1)(s^2+1)}.\end{aligned}$$

Example 10.2.3. (a) Use the convolution theorem and Table 10.1 to find the inverse of $K(s) = 1/[s^2(s^2+4)]$.
(b) Use the partial fractions to find the inverse of $K(s)$.

Solution: (a) Since we use $\mathcal{L}^{-1}[F(s)G(s)] = \int_0^t f(\tau)g(t-\tau)d\tau$, where $\mathcal{L}^{-1}[F(s)] = f(t)$ and $\mathcal{L}^{-1}[G(s)] = g(t)$, we need to figure out what are $F(s)$ and $G(s)$ in this problem. Most reasonable choice is $F(s) = 1/s^2$ and $G(s) = 1/(s^2+4)$. Then, since

$$\mathcal{L}^{-1}[\frac{1}{s^2}] = t, \quad \mathcal{L}^{-1}[\frac{1}{s^2+4}] = \frac{1}{2}\sin 2t,$$

we obtain

$$\begin{aligned}\mathcal{L}^{-1}[\frac{1}{s^2(s^2+4)}] &= \frac{1}{2}\int_0^t \tau\sin 2(t-\tau)d\tau\\ &= \frac{1}{2}[\frac{1}{2}\tau\cos 2(t-\tau)]_0^t - \frac{1}{4}\int_0^t \cos 2(t-\tau)d\tau\\ &= \frac{1}{4}t + \frac{1}{8}[\sin 2(t-\tau)]_0^t = \frac{1}{4}t - \frac{1}{8}\sin 2t.\end{aligned}$$

(b) The partial fractions of $K(s)$ is

$$K(s) = \frac{1}{4}(\frac{1}{s^2} - \frac{1}{s^2+4}).$$

Therefore, from the table and adjusting the coefficients, we have

$$\mathcal{L}^{-1}[K(s)] = \frac{1}{4}\mathcal{L}^{-1}[\frac{1}{s^2}] - \frac{1}{8}\mathcal{L}^{-1}[\frac{2}{s^2+4}] = \frac{1}{4}t - \frac{1}{8}\sin 2t.$$

10.2.3 *Relation with the Fourier Transform*

In order to see the relation between the Laplace and the Fourier transforms, set $s = a + i\omega$ and rewrite the inverse transform (10.1.2). Then, since

$ds = id\omega$,

$$f(t) = \frac{1}{2\pi i}\int_{a-i\infty}^{a+i\infty} F(s)e^{st}ds = \frac{1}{2\pi}\int_{-\infty}^{\infty} F(a+i\omega)e^{(a+i\omega)t}d\omega, \quad (10.2.6)$$

where $F(a+i\omega)$ is given by

$$F(a+i\omega) = \int_0^{\infty} f(\tau)e^{-(a+i\omega)\tau}d\tau. \quad (10.2.7)$$

To avoid the confusion τ is used for t in (10.2.7). This can be thought of a Fourier transform of $f(\tau)e^{-a\tau}$ where the integration is performed in the positive real line instead of the whole real line. Substituting (10.2.7) to (10.2.6), we have after the change of the order of integration

$$\begin{aligned} f(t) &= \frac{1}{2\pi}\int_{-\infty}^{\infty}\int_0^{\infty} f(\tau)e^{-(a+i\omega)\tau}d\tau e^{(a+i\omega)t}d\omega \\ &= \frac{1}{2\pi}\int_0^{\infty} f(\tau)e^{a(t-\tau)}\int_{-\infty}^{\infty} e^{i(t-\tau)\omega}d\omega d\tau. \end{aligned}$$

Since from the Fourier transform we know that

$$\frac{1}{2\pi}\int_{-\infty}^{\infty} e^{i(t-\tau)\omega}d\omega = \delta(t-\tau),$$

we obtain

$$f(t) = \int_0^{\infty} f(\tau)e^{a(t-\tau)}\delta(t-\tau)d\tau = f(t).$$

This tells us that the inverse Laplace transform is essentially the Fourier inverse.

Exercises

1. Find the Laplace transform of the given functions.
   (a) $h(t) = \int_0^t (t-\tau)^2 \cos 2\tau d\tau$
   (b) $h(t) = \int_0^t (t-\tau)^3 e^{\tau} d\tau$

2. Find the Laplace transform of $h(t) = \int_0^t (t-\tau)e^{\tau}d\tau$ in the following ways.
   (a) Use the convolution theorem.
   (b) Apply the Laplace transform to $h(t)$.

3. Use the convolution theorem to find the inverse Laplace transform. Also use the partial fractions and compare the results.
   (a) $F(s) = \frac{s}{(s+1)(s^2+4)}$
   (b) $F(s) = \frac{1}{s^3(s^2+4)}$

10.3 Applications to Differential Equations

10.3.1 *Applications to ODE's*

For the sake of review we apply the Laplace transform to ODE problems before we go on to PDE's.

Example 10.3.1. Find the solution $u(t)$ to the initial value problem

$$u'' - 3u' + 2u = 0, \quad t > 0, \tag{10.3.1}$$

$$u(0) = 1, \quad u'(0) = 5. \tag{10.3.2}$$

Solution: Apply the Laplace transform to (10.3.1) using (10.3.2). Then,

$$s^2F(s) - su(0) - u'(0) - 3sF(s) + 3u(0) + 2F(s) = 0.$$

$$(s^2 - 3s + 2)F(s) = s + 2$$

$$F(s) = \frac{s+2}{(s-1)(s-2)}.$$

Applying the partial fraction and using the linearity, we have

$$\begin{aligned}\mathcal{L}^{-1}[F(s)] &= \mathcal{L}^{-1}[\frac{4}{s-2}] - \mathcal{L}^{-1}[\frac{3}{s-1}] \\ &= 4e^{2t} - 3e^t.\end{aligned}$$

The following example deals with the case where the source term is discontinuous.

Example 10.3.2. Find the solution $u(t)$ to the initial value problem

$$u'' - 3u' + 2u = f(t), \quad t > 0, \tag{10.3.3}$$

$$u(0) = 0, \quad u'(0) = 0, \tag{10.3.4}$$

where

$$f(t) = \begin{cases} 0 & 0 \le t < 1 \\ 1 & 1 \le t < 5 \\ 0 & 5 \le t. \end{cases} \tag{10.3.5}$$

Solution: This problem arises in the electric circuit where a unit voltage pulse is applied for $1 \le t < 5$. In the problem u is the charge in the circuit at t. Apply the Laplace transform to (10.3.3) using (10.3.4) and (10.3.5). Then,

$$s^2F(s) - 3sF(s) + 2F(s) = \int_0^\infty f(t)dt = \frac{e^{-s} - e^{-5s}}{s}.$$

$$(s^2 - 3s + 2)F(s) = \frac{e^{-s} - e^{-5s}}{s}$$

$$F(s) = \frac{e^{-s} - e^{-5s}}{s(s-1)(s-2)}.$$

Applying the partial fraction, we have

$$F(s) = (e^{-s} - e^{-5s})(\frac{\frac{1}{2}}{s} - \frac{1}{s-1} + \frac{\frac{1}{2}}{s-2}).$$

Therefore,

$$\mathcal{L}^{-1}[F(s)] = H(t-1)[\frac{1}{2} - e^{-(t-1)} + \frac{1}{2}e^{-2(t-1)}]$$
$$-H(t-5)[\frac{1}{2} - e^{-(t-5)} + \frac{1}{2}e^{-2(t-5)}].$$

10.3.2 *Applications to PDE's*

For time dependent equations such as heat or wave equation, we apply the Laplace transform to the time variable. The Laplace transform gives an alternative to the methods we already know and it is especially useful when the boundary conditions are non-homogeneous. We illustrate the points using several examples starting from the heat equation in one dimension.

Example 10.3.3. Solve the heat equation

$$u_t - a^2 u_{xx} = 0, \quad 0 < x < L, \quad 0 < t,$$

$$u(0,t) = 2t, \quad u(L,t) = 2t,$$

$$u(x,0) = f(x) = x^2 - Lx.$$

Solution: Denote

$$U(x,s) = \int_0^\infty u(x,t)e^{-st}dt.$$

Then,

$$\int_0^\infty [u_t - a^2 u_{xx}]e^{-st}dt = [ue^{-st}]_0^\infty + s\int_0^\infty u(x,t)e^{-st}dt - a^2 U_{xx}$$
$$= -f(x) + sU - a^2 U_{xx} = 0$$

or

$$a^2 U_{xx} - sU = -f(x). \tag{10.3.6}$$

We find a particular solution for (10.3.6) by the method of undetermined coefficients. Assume that the particular solution is given by

$$U_p = cx^2 + dx + e \tag{10.3.7}$$

and find the coefficients c, d, and e. Substituting (10.3.7) in (10.3.6) and collecting like terms, we have

$$a^2 2c - s(cx^2 + dx + e) + (x^2 - Lx) = 0,$$

$$(1 - sc)x^2 + (-L - sd)x + (2a^2 c - se) = 0.$$

This yields

$$c = \frac{1}{s},\ d = -\frac{L}{s},\ e = \frac{2a^2 c}{s} = \frac{2a^2}{s^2},$$

$$U_p = \frac{1}{s}(x^2 - Lx) + \frac{2a^2}{s^2}.$$

The boundary conditions are

$$U(0,s) = \int_0^\infty u(0,t)e^{-st}dt = \frac{2a^2}{s^2},$$

$$U(L,s) = \int_0^\infty u(L,t)e^{-st}dt = \frac{2a^2}{s^2}.$$

The variation of parameters formula leads to

$$U(x,s) = c_1 e^{\frac{\sqrt{s}}{a}x} + c_2 e^{-\frac{\sqrt{s}}{a}x} + \frac{1}{s}(x^2 - Lx) + \frac{2a^2}{s^2}.$$

At $x = 0$

$$U(0,s) = c_1 + c_2 + \frac{2a^2}{s^2} = \frac{2a^2}{s^2},$$

and at $x = L$

$$U(L,s) = c_1 e^{\frac{\sqrt{s}}{a}L} + c_2 e^{-\frac{\sqrt{s}}{a}L} + \frac{2a^2}{s^2} = \frac{2a^2}{s^2}.$$

Therefore, $c_1 = c_2 = 0$. So,

$$U(x,s) = \frac{1}{s}(x^2 - Lx) + \frac{2a^2}{s^2}.$$

From Table 10.1 we obtain

$$u(x,t) = (x^2 - Lx) + 2a^2 t.$$

The next example is an extension of Example 10.3.2 to the heat equation.

Example 10.3.4. Solve

$$u_t = a^2 u_{xx}, \quad 0 < x < \infty,\ 0 < t, \tag{10.3.8}$$

$$u(0,t) = h(t) = \begin{cases} 0 & 0 \le t < 1 \\ 1 & 1 \le t < 5 \\ 0 & 5 \le t. \end{cases}$$

$$u(x,0) = 1.$$

Solution: Denote

$$U(x,s) = \int_0^\infty u(x,t)e^{-st}dt.$$

Taking the Laplace transform of the equation (10.3.8), we obtain

$$\int_0^\infty u_t(x,t)e^{-st}dt = [u(x,t)e^{-st}]_0^\infty + s\int_0^\infty u(x,t)e^{-st}dt = -1 + sU,$$

$$\int_0^\infty u_{xx}(x,t)e^{-st}dt = U_{xx}.$$

Therefore, we have

$$a^2 U_{xx} - sU = -1.$$

The general solution is

$$U(x,s) = c_1 e^{\frac{\sqrt{s}}{a}x} + c_2 e^{-\frac{\sqrt{s}}{a}x} + \frac{1}{s}. \tag{10.3.9}$$

We look for a bounded solution in x. So, we set $c_1 = 0$ and find c_2 from the boundary condition. Since the boundary condition is

$$U(0,s) = \int_0^\infty u(0,t)e^{-st}dt = \frac{1}{s}(e^{-s} - e^{-5s}), \tag{10.3.10}$$

equating (10.3.9) and (10.3.10) at $x = 0$ with $c_1 = 0$, we have

$$c_2 + \frac{1}{s} = \frac{1}{s}(e^{-s} - e^{-5s}).$$

Therefore,

$$U(x,s) = \frac{1}{s}(e^{-s} - e^{-5s} - 1)e^{-\frac{\sqrt{s}}{a}x} + \frac{1}{s}.$$

To find the inverse we use the convolution theorem for the first and the second terms. Their inverses are denoted as u_1 and u_2, respectively. We use the first term $\frac{1}{s}e^{-s}e^{-\sqrt{s}x/a}$ to illustrate this. The choice of $F(s)$ and $G(s)$ in (10.2.4) and (10.2.5) is not unique. The simplest choice is

$$F(s) = e^{-s}, \quad G(s) = \frac{1}{s}e^{-\frac{\sqrt{s}}{a}x}.$$

Then, from Table 10.1 and the convolution theorem

$$u_1(x,t) = \mathcal{L}^{-1}\{\frac{1}{s}e^{-s}e^{-\frac{\sqrt{s}}{a}x}\} = \int_0^t \delta(t-\tau-1)[1-\mathrm{erf}(\frac{x}{a\sqrt{4\tau}})]d\tau$$
$$= [1-\mathrm{erf}(\frac{x}{a\sqrt{4(t-1)}})]H(t-1).$$

Similarly,

$$u_2(x,t) = \mathcal{L}^{-1}\{\frac{1}{s}e^{-5s}e^{-\frac{\sqrt{s}}{a}x}\} = \int_0^t \delta(t-\tau-5)[1-\mathrm{erf}(\frac{x}{a\sqrt{4\tau}})]d\tau$$
$$= [1-\mathrm{erf}(\frac{x}{a\sqrt{4(t-5)}})]H(t-5).$$

Therefore,

$$u(x,t) = 1 - [1-\mathrm{erf}(\frac{x}{a\sqrt{4t}})] + u_1(x,T) - u_2(x,t).$$

Next, we consider the Laplace transform for the wave equation.

Example 10.3.5. Consider a semi-infinite string

$$u_{tt} - c^2u_{xx} = 0, \quad 0 < x < \infty,\ 0 < t, \tag{10.3.11}$$

$$u(0,t) = f(t),$$

$$u(x,0) = 0, \quad u_t(x,0) = 0.$$

Solution: Denote

$$U(x,s) = \int_0^\infty u(x,t)e^{-st}dt.$$

Then,

$$\int_0^\infty u_{tt}(x,t)e^{-st}dt = [u_t(x,t)e^{-st}]_0^\infty + s\int_0^\infty u_t(x,t)e^{-st}dt$$
$$= s[u(x,t)e^{-st}]_0^\infty + s^2U.$$

Therefore, the Laplace transform of (10.3.11) is

$$s^2U = a^2U_{xx} \tag{10.3.12}$$

and the boundary conditions are

$$U(0,s) = \int_0^\infty u(0,t)e^{-st}dt = \int_0^\infty f(t)e^{-st}dt = F(s).$$

Solving (10.3.12) in x, we have

$$U(x,s) = c_1 e^{\frac{s}{a}x} + c_2 e^{-\frac{s}{a}x}.$$

We look for a solution which decays as $x \to \infty$. So, we set $c_1 = 0$. Then, from the boundary condition,

$$c_2 = \int_0^\infty f(t)e^{-st}dt = F(s).$$

Therefore,

$$U(x,s) = F(s)e^{-\frac{s}{a}x}.$$

The inverse is

$$u(x,t) = \int_0^t f(t-\tau)\delta(\tau - \frac{x}{a})d\tau.$$

So,

$$u(x,t) = f(t - \frac{x}{a})H(t - \frac{x}{a}).$$

Exercises

1. Use Examples 10.3.1 and 10.3.2 to find the solution to

$$u'' + 3u' + 2u = f(t), \quad t > 0,$$

$$u(0) = 1, \quad u'(0) = 2,$$

where

$$f(t) = \begin{cases} 0 & 0 \le t < 5 \\ 1 & 5 \le t < 10 \\ 0 & 10 \le t. \end{cases}$$

2. Express the solution of the given initial value problem in terms of a convolution integral.

(a) $u'' + 2u' + 2u = \sin at, \quad u(0) = 0, \quad u'(0) = 0$
(b) $u'' + 4u' + 4u = h(t), \quad u(0) = 2, \quad u'(0) = -3$
(c) $u^{(4)} - u = h(t), \quad u(0) = 0,\ u'(0) = 0,\ u''(0) = 0,\ u'''(0) = 0$
(d) $u^{(4)} + 5u''(0) + 4u = h(t),$
$u(0) = 0,\ u'(0) = 0,\ u''(0) = 0,\ u'''(0) = 0$

3. Use the Laplace transform to solve

$$u_{tt} = c^2 u_{xx}, \quad 0 < x < \infty,\ 0 < t,$$

$$u_x(0,t) = f(t),$$

$$u(x,0) = 0, \quad u_t(x,0) = 0.$$

4. Use the Laplace transform to solve

$$u_{tt} = c^2 u_{xx}, \quad 0 < x < L,\ 0 < t,$$

$$u(0,t) = 0, \quad u(L,t) = 0,$$

$$u(x,0) = \sin\frac{\pi x}{L}, \quad u_t(x,0) = -\sin\frac{\pi x}{L}.$$

5. Use the Laplace transform to solve

$$u_t = a^2 u_{xx}, \quad 0 < x < \infty,\ 0 < t,$$

$$u(0,t) = f(t),$$

$$u(x,0) = 0.$$

6. Use the Laplace transform to solve

$$u_t = a^2 u_{xx}, \quad 0 < x < L,\ 0 < t,$$

$$u(0,t) = 0, \quad u(L,t) = 0,$$

$$u(x,0) = \sin\frac{\pi x}{L}.$$

Chapter 11

Higher Dimensional Problems - Other Approaches

This chapter is a collection of methods which do not use the separation of variables and are applicable to multi-dimensional problems. In Section 11.1 we study the method of spherical means and the method of descent for the wave equation. In Section 11.2 we also discuss the non-homogeneous problems and in particular we learn the Duhamel's principle for the heat and wave equations.

11.1 Spherical Means and Method of Descent

In this section we study the wave equation in higher dimensions. We obtain the explicit solution formula for the initial value problems in the three and two dimensional wave equations. First we find the explicit solutions for the three dimensional problems by the method of spherical means and then we reduce the dimension to find the solution to the two dimensional case by the method of descent.

11.1.1 *Method of Spherical Means*

We use a method called the method of spherical means to derive a solution for a wave equation in R^n $(n = 3)$

$$u_{tt} - c^2 \Delta u = 0, \quad x \in R^n, \ 0 < t, \tag{11.1.1}$$

with initial data

$$u(\mathbf{x}, 0) = f(\mathbf{x}), \quad u_t(\mathbf{x}, 0) = g(\mathbf{x}). \tag{11.1.2}$$

We restrict our attention to $n = 3$ even though it is possible to discuss the problem with any positive integer n.

We need some preparations. First, the Laplacian of a spherically symmetric function $v = v(r,t)$, where $r = \sqrt{x_1^2 + x_2^2 + x_3^2}$, has a simple form given as follows.

$$v_{x_i} = v_r \frac{\partial r}{\partial x_i} = v_r \frac{x_i}{r}, \quad v_{x_i x_i} = v_{rr} (\frac{x_i}{r})^2 + v_r \frac{r^2 - x_i^2}{r^3}, \tag{11.1.3}$$

$$\Delta v = \sum_{i=1}^{3} v_{x_i x_i} = v_{rr} + \frac{2}{r} v_r = \frac{1}{r^2} \frac{\partial}{\partial r} (r^2 v_r). \tag{11.1.4}$$

Next, define the average of a function $h(\mathbf{x})$ on the sphere to be

$$\bar{h}(\mathbf{x}, r) = \frac{1}{4\pi r^2} \iint\limits_{|\mathbf{y}-\mathbf{x}|=r} h(\mathbf{y}) dS(\mathbf{y}), \tag{11.1.5}$$

where $dS(\mathbf{y})$ means the surface integral with respect to $\mathbf{y}$. We call this the spherical mean. Then, we can show that $\bar{h}(\mathbf{x}, r)$ satisfies

Lemma 11.1.1. *The spherical means $\bar{h}(\mathbf{x}, r)$ of any function h satisfy*

$$\Delta_{\mathbf{x}} \bar{h}(\mathbf{x}, r) = \sum_{i=1}^{3} \bar{h}_{x_i x_i} = \frac{\partial^2}{\partial r^2} \bar{h}(\mathbf{x}, r) + \frac{2}{r} \frac{\partial}{\partial r} \bar{h}(\mathbf{x}, r),$$

where $\Delta_{\mathbf{x}}$ means the Laplacian with respect to $\mathbf{x}$. Furthermore, the following holds.

$$\bar{h}(\mathbf{x}, 0) = h(\mathbf{x}), \quad \frac{\partial}{\partial r} \bar{h}(\mathbf{x}, r)|_{r=0} = 0.$$

Proof. To evaluate $\partial \bar{h}_h / \partial r$ we change the integration variables so that in (11.1.5) the limit of integration will not depend on r. Set $\mathbf{y} = \mathbf{x} + r\boldsymbol{\xi}$ with $|\boldsymbol{\xi}| = 1$. Then,

$$\bar{h}(\mathbf{x}, r) = \frac{1}{4\pi} \iint\limits_{|\boldsymbol{\xi}|=1} h(\mathbf{x} + r\boldsymbol{\xi}) dS(\boldsymbol{\xi}), \tag{11.1.6}$$

where $dS(\boldsymbol{\xi})$ means the surface integral with respect to $\boldsymbol{\xi}$.

$$\begin{aligned} \frac{\partial}{\partial r} \bar{h}_h(\mathbf{x}, r) &= \frac{1}{4\pi} \iint\limits_{|\boldsymbol{\xi}|=1} \sum_{i=1}^{n} h_{x_i}(\mathbf{x} + r\boldsymbol{\xi}) \frac{d(x_i + r\xi_i)}{dr} dS(\boldsymbol{\xi}) \\ &= \frac{1}{4\pi} \iint\limits_{|\boldsymbol{\xi}|=1} \sum_{i=1}^{n} h_{x_i}(\mathbf{x} + r\boldsymbol{\xi}) \xi_i dS(\boldsymbol{\xi}). \end{aligned} \tag{11.1.7}$$

Since $\boldsymbol{\xi}$ is a unit normal vector to the surface, $|\boldsymbol{\xi}| = 1$ and

$$\sum_{i=1}^{n} h_{x_i}(\mathbf{x} + r\boldsymbol{\xi})\xi_i = \mathbf{n} \cdot \nabla_{\mathbf{x}} h.$$

Here, $\nabla_{\mathbf{x}}$ means the gradient vector with respect to $\mathbf{x}$. Therefore, applying the divergence theorem to (11.1.7), we obtain

$$\frac{\partial}{\partial r}\bar{h}_h(\mathbf{x}, r) = \frac{1}{4\pi} \iiint\limits_{|\boldsymbol{\xi}|<1} \nabla_{\boldsymbol{\xi}} \cdot \nabla_{\mathbf{x}} h(\mathbf{x} + r\boldsymbol{\xi}) d\boldsymbol{\xi} = \frac{r}{4\pi} \iiint\limits_{|\boldsymbol{\xi}|<1} \Delta_{\mathbf{x}} h(\mathbf{x} + r\boldsymbol{\xi}) d\boldsymbol{\xi},$$

where we used the relation $\nabla_{\boldsymbol{\xi}} = r\nabla_{\mathbf{x}}$. Setting $\mathbf{y} = \mathbf{x} + r\boldsymbol{\xi}$ and changing the integration variables back to $\mathbf{y}$, we have

$$\begin{aligned}\frac{\partial}{\partial r}\bar{h}_h(\mathbf{x}, r) &= \frac{r}{4\pi} \Delta_{\mathbf{x}} \iiint\limits_{|\boldsymbol{\xi}|<1} h(\mathbf{x} + r\boldsymbol{\xi}) d\boldsymbol{\xi} = \frac{r^{1-n}}{4\pi} \Delta_{\mathbf{x}} \iiint\limits_{|\mathbf{y}-\mathbf{x}|<r} h(\mathbf{y}) d\mathbf{y} \\ &= \frac{r^{1-n}}{4\pi} \Delta_{\mathbf{x}} \int_0^r d\rho \iint\limits_{|\mathbf{y}-\mathbf{x}|=\rho} h(\mathbf{y}) dS(\mathbf{y}).\end{aligned}$$

Definition (11.1.5) yields that for $n = 3$

$$\frac{\partial}{\partial r}\bar{h}_h(\mathbf{x}, r) = r^{-2} \Delta_{\mathbf{x}} \int_0^r \rho^2 \bar{h}(\mathbf{x}, \rho) d\rho.$$

Multiplying by r^2 and differentiating with respect to r yields

$$\frac{\partial}{\partial r}(r^2 \frac{\partial}{\partial r}\bar{h}(\mathbf{x}, r)) = \Delta_{\mathbf{x}} r^2 \bar{h}(\mathbf{x}, r).$$

This shows that the spherical means of any function satisfy

$$\frac{\partial^2}{\partial r^2}\bar{h}(\mathbf{x}, r) + \frac{2}{r}\frac{\partial}{\partial r}\bar{h}(\mathbf{x}, r) = \frac{1}{r}\frac{\partial^2}{\partial r^2}[r\bar{h}(\mathbf{x}, r)] = \Delta_{\mathbf{x}} \bar{h}(\mathbf{x}, r). \tag{11.1.8}$$

This is known as Darboux's equation.

Since $\bar{h}(\mathbf{x}, r)$ is the average of h about $\mathbf{x}$, it is not difficult to see that $\bar{h}(\mathbf{x}, 0) = h(\mathbf{x})$. From (11.1.6), we see that $\bar{h}(\mathbf{x}, r)$ is even in r and therefore

$$\frac{\partial}{\partial r}\bar{h}(\mathbf{x}, r)|_{r=0} = 0.$$

□

Now, we derive the solution to (11.1.1) and (11.1.2). An advantage of spherical means is that we can transform the initial value problem for the higher-dimensional wave equation into that of one-dimensional wave equation. Here is how this is done. Let $u(\mathbf{x},t)$ be a solution of (11.1.1), (11.1.2). We define the spherical means of u to be

$$\bar{u}(\mathbf{x},r,t) = \frac{1}{4\pi} \iint_{|\boldsymbol{\xi}|=1} u(\mathbf{x}+r\boldsymbol{\xi},t)dS(\boldsymbol{\xi}).$$

Note that

$$u(\mathbf{x},t) = \bar{u}(\mathbf{x},0,t).$$

Theorem 11.1.2. *The solution $u(\mathbf{x},t)$ to (11.1.1) and (11.1.2) is given by*

$$\begin{aligned} u(\mathbf{x},t) &= \frac{d}{dt}\left[\frac{1}{4\pi c^2 t} \iint_{|\mathbf{y}-\mathbf{x}|=ct} f(\mathbf{y})dS(\mathbf{y})\right] + \frac{1}{4\pi c^2 t} \iint_{|\mathbf{y}-\mathbf{x}|=ct} g(\mathbf{y})dS(\mathbf{y}) \\ &= \frac{1}{4\pi c^2 t^2} \iint_{|\mathbf{y}-\mathbf{x}|=ct} [tg(\mathbf{y}) + f(\mathbf{y}) + \sum_i f_{y_i}(\mathbf{y})(y_i - x_i)]dS(\mathbf{y}). \end{aligned}$$

Proof. Applying the Laplacian to $\bar{u}(\mathbf{x},r,t)$, we see that

$$\begin{aligned} \Delta_{\mathbf{x}}\bar{u} &= \frac{1}{4\pi} \iint_{|\boldsymbol{\xi}|=1} \Delta_{\mathbf{x}} u(\mathbf{x}+r\boldsymbol{\xi},t)dS(\boldsymbol{\xi}) \\ &= \frac{1}{c^2}\frac{\partial^2}{\partial t^2}\frac{1}{4\pi} \iint_{|\boldsymbol{\xi}|=1} u(\mathbf{x}+r\boldsymbol{\xi},t)dS(\boldsymbol{\xi}) = \frac{1}{c^2}\frac{\partial^2}{\partial t^2}\bar{u}. \end{aligned}$$

When $n = 3$, from (11.1.8)

$$\frac{\partial^2}{\partial t^2}(r\bar{u}) = c^2(r\frac{\partial^2}{\partial r^2} + 2\frac{\partial}{\partial r})\bar{u} = c^2\frac{\partial^2}{\partial r^2}(r\bar{u}).$$

Thus, $r\bar{u}$ is a solution of the one-dimensional wave equation with the initial condition

$$r\bar{u}|_{t=0} = r\bar{f}(\mathbf{x},r), \quad (\frac{\partial}{\partial t}r\bar{u})|_{t=0} = r\bar{g}(\mathbf{x},r).$$

Since $r\bar{f}(\mathbf{x},r)$ and $r\bar{g}(\mathbf{x},r)$ are odd in r and we take the limit as r approaches 0, we use the second part of (3.3.7). Then,

$$\begin{aligned} \bar{u}(\mathbf{x},r,t) &= \frac{1}{2r}[(r+ct)\bar{f}(\mathbf{x},r+ct) - (ct-r)\bar{f}(\mathbf{x},ct-r)] \\ &\quad + \frac{1}{2rc}\int_{ct-r}^{r+ct} s\bar{g}(\mathbf{x},s)ds. \end{aligned}$$

Now let r approach 0. Set $ct = s$. Then,

$$u(\mathbf{x},t) = \frac{d}{ds}(s\bar{f}(\mathbf{x},s)) + t\bar{g}(\mathbf{x},ct) = \frac{d}{dt}(t\bar{f}(\mathbf{x},ct)) + t\bar{g}(\mathbf{x},ct)$$
$$= \frac{d}{dt}[\frac{1}{4\pi c^2 t}\iint\limits_{|\mathbf{y}-\mathbf{x}|=ct} f(\mathbf{y})dS(\mathbf{y})] + \frac{1}{4\pi c^2 t}\iint\limits_{|\mathbf{y}-\mathbf{x}|=ct} g(\mathbf{y})dS(\mathbf{y}). \qquad (11.1.9)$$

The above integrals are the surface integral over the sphere

$$\sum_{i=1}^{n}(y_i - x_i)^2 = (ct)^2.$$

We perform the differentiation in t as we did in Lemma 11.1.1. We introduce $\boldsymbol{\xi}$ as $y_i - x_i = ct\xi_i$ satisfying $\sum \xi_i^2 = 1$. Then, for $n = 3$, we have $dS(\mathbf{y}) = (ct)^2 dS(\boldsymbol{\xi})$. We use the same procedure as in Lemma 11.1.1, *i.e.*, change the variable from $\mathbf{y}$ to $\boldsymbol{\xi}$ to eliminate the dependence of the limit of integration on t, perform the differentiation in t, and go back from $\boldsymbol{\xi}$ to $\mathbf{y}$. Then,

$$\frac{d}{dt}[\frac{1}{4\pi c^2 t}\iint\limits_{|\mathbf{y}-\mathbf{x}|=ct} f(\mathbf{y})dS(\mathbf{y})]$$
$$= \frac{d}{dt}[\frac{1}{4\pi c^2 t}\iint\limits_{|\boldsymbol{\xi}|=1} f(x_i + ct\xi_i)c^2t^2 dS(\boldsymbol{\xi})]$$
$$= \frac{1}{4\pi}\iint\limits_{|\boldsymbol{\xi}|=1} f(x_i + ct\xi_i)dS(\boldsymbol{\xi}) + \frac{t}{4\pi}\iint\limits_{|\boldsymbol{\xi}|=1} f_{y_i}(x_i + ct\xi_i)c\xi_i dS(\boldsymbol{\xi})$$
$$= \frac{1}{4\pi c^2 t^2}\iint\limits_{|\mathbf{y}-\mathbf{x}|=ct} f(\mathbf{y})dS(\mathbf{y}) + \frac{1}{4\pi}\iint\limits_{|\mathbf{y}-\mathbf{x}|=ct} f_{y_i}(\mathbf{y})(y_i - x_i)\frac{1}{c^2t^2}dS(\mathbf{y}).$$

Therefore,

$$u(\mathbf{x},t) = \frac{1}{4\pi c^2 t^2}\iint\limits_{|\mathbf{y}-\mathbf{x}|=ct} [tg(\mathbf{y}) + f(\mathbf{y}) + \sum_i f_{y_i}(\mathbf{y})(y_i - x_i)]dS(\mathbf{y}).$$

It is interesting to note that there is a possible loss of order of differentiability. □

11.1.2 *The Method of Descent*

A solution $u(x_1, x_2, t)$ of (11.1.1), (11.1.2) for $n = 2$ can be treated as a solution of the same problem for $n = 3$ which does not depend on x_3. Then, $u(x_1, x_2, t)$ is given by formula (11.1.9) with $x_3 = 0$ and

$$f(\mathbf{y}) = f(y_1, y_2), \quad g(\mathbf{y}) = g(y_1, y_2),$$

where the surface integrals are extended over the sphere

$$|\mathbf{y}-\mathbf{x}| = \sqrt{(y_1-x_1)^2+(y_2-x_2)^2+y_3^2} = ct.$$

On the sphere

$$y_3^2 = (ct)^2-(y_1-x_1)^2-(y_2-x_2)^2$$

$$\frac{\partial y_3}{\partial y_1} = -\frac{y_1-x_1}{y_3}$$

$$dS(\mathbf{y}) = \sqrt{1+(\frac{\partial y_3}{\partial y_1})^2+(\frac{\partial y_3}{\partial y_2})^2 dy_1dy_2} = \frac{ct}{|y_3|}dy_1dy_2,$$

where $|y_3| = \sqrt{c^2t^2-r^2}$, $r = \sqrt{(y_1-x_1)^2+(y_2-x_2)^2}$. Substituting $dS(\mathbf{y})$ in (11.1.9) and considering that (y_1, y_2, y_3) and $(y_1, y_2, -y_3)$ makes the same contribution, we arrive at

$$u(x_1,x_2,t) = \frac{\partial}{\partial t}\frac{1}{2\pi c}\iint_{r<ct}\frac{f(y_1,y_2)}{\sqrt{c^2t^2-r^2}}dy_1dy_2$$

$$+\frac{1}{2\pi c}\iint_{r<ct}\frac{g(y_1,y_2)}{\sqrt{c^2t^2-r^2}}dy_1dy_2. \quad (11.1.10)$$

Exercises

1. Verify (11.1.3) and (11.1.4) for $n = 3$. For $n = n$ what do we get for (11.1.4)?
2. Write down the surface integral (11.1.5) using the rectangular coordinates.
3. Write down (11.1.5) using the spherical coordinates

$$\boldsymbol{\xi} = [\sin\phi\cos\theta,\ \sin\phi\sin\theta,\ \cos\phi].$$

4. Verify $dS(\mathbf{y}) = (ct)^2 dS(\boldsymbol{\xi})$ if
   (a) $\boldsymbol{\xi}$ is the spherical coordinates given in Exercise 3.
   (b) $\boldsymbol{\xi}$ is the rectangular coordinates.

5. Start from (11.1.10) and use the method of descent to find a solution $u(x_1,t)$ of (11.1.1), (11.1.2) for $n = 1$.
   (a) The solution $u(x_1,t)$ can be treated as a solution of the same problem for $n = 2$ which does not depend on x_2. Then, $u(x_1,t)$ is given by the formula (11.1.10) with $x_2 = 0$ and

$$f(y) = f(y_1), \quad g(y) = g(y_1),$$

where the area integrals are performed over the circle

$$|\mathbf{y}-\mathbf{x}| = \sqrt{(y_1-x_1)^2+y_2^2} = ct.$$

Show that

$$u(x_1,t) = \frac{\partial}{\partial t}[\frac{1}{2\pi c}\int_{x_1-ct}^{x_1+ct}\int_{-a}^{a}\frac{f(y_1)}{\sqrt{a^2-y_2^2}}dy_2dy_1] + \frac{1}{2\pi c}\int_{x_1-ct}^{x_1+ct}\int_{-a}^{a}\frac{g(y_1)}{\sqrt{a^2-y_2^2}}dy_2dy_1, \quad (11.1.11)$$

where $a=\sqrt{c^2t^2-(y_1-x_1)^2}$.

(b) Perform the change of variable $y_2 = a\sin\theta$ in (11.1.11) and obtain

$$u(x_1,t) = \frac{1}{2}[f(x_1+ct)+f(x_1-ct)] + \frac{1}{2c}\int_{x_1-ct}^{x_1+ct} g(s)ds.$$

11.2 Duhamel's Principle

We consider the non-homogeneous problems for the initial value problems of the heat and wave equations in the multidimensional cases. We derive the Duhamel's principle where the solutions to the non-homogeneous problems are obtained as the time integral of the solutions to the corresponding homogeneous problems where the initial condition is given by the non-homogeneous term.

11.2.1 *Heat Equation*

We derive the solution to the initial value problem

$$u_t - \Delta u = h(\mathbf{x},t), \quad \mathbf{x}\in R^n,\ 0<t, \quad (11.2.1)$$

$$u(\mathbf{x},0) = 0. \quad (11.2.2)$$

As you see the equation is non-homogeneous. To find the solution we consider the initial value problem where the initial value is specified at $t=s$

$$u_t - \Delta u = 0, \quad \mathbf{x}\in R^n,\ t>s, \quad (11.2.3)$$

$$u(\mathbf{x},s;s) = h(\mathbf{x},s). \quad (11.2.4)$$

Then, the solution to (11.2.3) and (11.2.4) is given by

$$u(\mathbf{x}, t; s) = \int_{R^n} h(\mathbf{y}, s)G(\mathbf{x} - \mathbf{y}, t - s)d\mathbf{y}, \tag{11.2.5}$$

where G is the heat kernel given in (2.6.34) and (9.4.8), and

$$\int_{R^n} f(\mathbf{y})d\mathbf{y} = \int_{-\infty}^{\infty} \cdots \int_{-\infty}^{\infty} f(\mathbf{y})dy_1 \cdots dy_n.$$

Using this expression, we have

Theorem 11.2.1. ***(Duhamel's Principle)*** *The solution to (11.2.1) and (11.2.2) is given by*

$$u(\mathbf{x}, t) = \int_0^t u(\mathbf{x}, t; s)ds. \tag{11.2.6}$$

Proof. Substituting (11.2.5) in (11.2.6) we have

$$u(\mathbf{x}, t) = \int_0^t \int_{R^n} h(\mathbf{y}, s)G(\mathbf{x} - \mathbf{y}, t - s)d\mathbf{y}ds. \tag{11.2.7}$$

After the change of variables we obtain

$$u(\mathbf{x}, t) = \int_0^t \int_{R^n} h(\mathbf{x} - \mathbf{y}, t - s)G(\mathbf{y}, s)d\mathbf{y}ds.$$

We show that the above $u(x, t)$ satisfies the equation (11.2.1). The problem is that there is a singularity in $G(\mathbf{x}, s)$ at $s = 0$. As we did in the Laplace equation, we isolate the singularity by decomposing the time integral into two parts and make different estimates.

$$\begin{aligned} u_t - \Delta u &= \int_0^t \int_{R^n} (\frac{\partial}{\partial t} - \Delta_{\mathbf{x}})h(\mathbf{x} - \mathbf{y}, t - s)G(\mathbf{y}, s)d\mathbf{y}ds \\ &\quad + \int_{R^n} f(\mathbf{x} - \mathbf{y}, 0)G(\mathbf{y}, t)d\mathbf{y} \\ &= \int_\varepsilon^t \int_{R^n} [(-\frac{\partial}{\partial s} - \Delta_{\mathbf{y}})h(\mathbf{x} - \mathbf{y}, t - s)]G(\mathbf{y}, s)d\mathbf{y}ds \\ &\quad + \int_0^\varepsilon \int_{R^n} [(-\frac{\partial}{\partial s} - \Delta_{\mathbf{y}})h(\mathbf{x} - \mathbf{y}, t - s)]G(\mathbf{y}, s)d\mathbf{y}ds \\ &\quad + \int_{R^n} h(\mathbf{x} - \mathbf{y}, 0)G(\mathbf{y}, t)d\mathbf{y} \\ &= I_\varepsilon + J_\varepsilon + F. \end{aligned}$$

The singularity is in J_ε. If h_t and Δh are bounded by a constant M, *i.e.*, $|h_t| \le M$ and $|\Delta h| \le M$, (9.4.9) implies

$$|J_\varepsilon| \le (|h_t| + |\Delta h|) \int_0^\varepsilon \int_{R^n} G(\mathbf{y}, s)d\mathbf{y}ds \le 2\varepsilon M,$$

For I_ε, integrating by parts in s, we have

$$\begin{aligned}
I_\varepsilon &= \int_\varepsilon^t \int_{R^n} [(\frac{\partial}{\partial s} - \Delta_{\mathbf{y}})G(\mathbf{y}, s)]h(\mathbf{x} - \mathbf{y}, t - s)d\mathbf{y}ds \\
&\quad + \int_{R^n} G(\mathbf{y}, \varepsilon)h\mathbf{x} - \mathbf{y}, t - \varepsilon)d\mathbf{y} \\
&\quad - \int_{R^n} G(\mathbf{y}, t)h(\mathbf{x} - \mathbf{y}, 0)d\mathbf{y} \\
&= \int_{R^n} G(\mathbf{y}, \varepsilon)h(\mathbf{x} - \mathbf{y}, t - \varepsilon)d\mathbf{y} - F.
\end{aligned}$$

Combining the estimates, canceling F, and observing that the heat kernel is a delta function in the limit as $\varepsilon \to 0$ we obtain

$$u_t - \Delta u = \lim_{\varepsilon \to 0} \int_{R^n} G(\mathbf{y}, \varepsilon)h(\mathbf{x} - \mathbf{y}, t - \varepsilon)d\mathbf{y} = h(\mathbf{x}, t).$$

□

11.2.2 *Wave Equation*

Consider

$$u_{tt}(\mathbf{x}, t) - c^2 \Delta u(\mathbf{x}, t) = h(\mathbf{x}, t), \quad \mathbf{x} \in R^n,\ 0 < t, \tag{11.2.8}$$

$$u(\mathbf{x}, 0) = 0, \quad u_t(\mathbf{x}, 0) = 0. \tag{11.2.9}$$

We define $u = u(\mathbf{x}, t; s)$ to be the solution of

$$u_{tt}(\mathbf{x}, t; s) - c^2 \Delta u(\mathbf{x}, t; s) = 0, \quad \mathbf{x} \in R^n,\ t > s, \tag{11.2.10}$$

$$u(\mathbf{x}, t; s) = 0, \quad u_t(\mathbf{x}, t; s) = h(\mathbf{x}, s), \quad \mathbf{x} \in R^n. \tag{11.2.11}$$

Then, we have

Theorem 11.2.2. ***(Duhamel's principle)*** *The solution to (11.2.8) and (11.2.9) is given by*

$$u(\mathbf{x}, t) = \int_0^t u(\mathbf{x}, t; s)ds. \tag{11.2.12}$$

Proof. The proof is just calculation and goes as follows.

$$u_t(\mathbf{x}, t) = u(\mathbf{x}, t; t) + \int_0^t u_t(\mathbf{x}, t; s)ds = \int_0^t u_t(\mathbf{x}, t; s)ds,$$

$$u_{tt}(\mathbf{x},t) = u_t(\mathbf{x},t;t) + \int_0^t u_{tt}(\mathbf{x},t;s)ds$$
$$= h(\mathbf{x},t) + \int_0^t u_{tt}(\mathbf{x},t;s)ds.$$

Differentiation of (11.2.12) in space yields

$$c^2\Delta u(\mathbf{x},t) = \int_0^t c^2\Delta u(\mathbf{x},t;s)ds = \int_0^t u_{tt}(\mathbf{x},t;s)ds.$$

Therefore, (11.2.8) and (11.2.9) hold. □

Example 11.2.3. (a) Show that for $n = 1$, the solution to (11.2.10) and (11.2.11) is

$$u(x,t;s) = \frac{1}{2c}\int_{x-c(t-s)}^{x+c(t-s)} h(y,s)dy.$$

(b) Show that the solution to (11.2.8) and (11.2.9) is

$$u(x,t) = \frac{1}{2c}\int_0^t \int_{x-c(t-s)}^{x+c(t-s)} h(y,s)dyds.$$

Exercises

1. Write down an explicit formula for a solution of
$$u_t(\mathbf{x},t) - a^2\Delta u(\mathbf{x},t) = h(\mathbf{x},t), \quad \mathbf{x} \in R^n,\ 0 < t,$$
$$u(\mathbf{x},0) = f(\mathbf{x}), \quad \mathbf{x} \in R^n.$$

2. Write down an explicit formula for a solution of
$$u_t(\mathbf{x},t) - a^2\Delta u(\mathbf{x},t) + cu(\mathbf{x},t) = h(\mathbf{x},t), \quad \mathbf{x} \in R^n,\ 0 < t,$$
$$u(\mathbf{x},0) = f(\mathbf{x}), \quad \mathbf{x} \in R^n.$$

3. Solve
$$u_{tt}(x,t) - c^2 u_{xx}(x,t) = 1, \quad -\infty < x < \infty,\ 0 < t,$$
$$u(x,0) = \sin x, \quad u_t(x,0) = 0.$$

4. Consider the solution to (11.2.8) and (11.2.9) for $n = 3$.

   (a) Show that the solution to (11.2.10) and (11.2.11) is
   $$u(\mathbf{x},t;s) = (t-s)\frac{1}{4\pi c^2(t-s)^2}\iint_{|\mathbf{y}-\mathbf{x}|=c(t-s)} h(\mathbf{y},s)dS(\mathbf{y}).$$

   (b) Show that the solution to (11.2.8) and (11.2.9) is
   $$u(\mathbf{x},t) = \int_0^t (t-s)\frac{1}{4\pi c^2(t-s)^2}\iint_{|\mathbf{y}-\mathbf{x}|=c(t-s)} h(\mathbf{y},s)dS(\mathbf{y})ds.$$

Chapter 12

Green's Functions

In Chapter 2 we studied the Green's function for the initial value problem of the heat equation and in Chapter 4 we did the Green's functions for the Laplace equation. In this chapter with the help of the delta functions we study the Green's functions in more detail. The solution to a system of linear equations $A\mathbf{x} = \mathbf{b}$ is given by $\mathbf{x} = A^{-1}\mathbf{b}$, where A^{-1} is the inverse matrix provided that it exists. With the aid of the delta function and the Green's identity we will see that the Green's function plays a similar role in PDE's to the inverse matrices in systems of linear equations. Therefore, if we find the Green's function, we have a solution representation for the problem. This is one of the reasons why we study the Green's functions. In Section 12.1 we revisit the Green's functions for the Laplace Equation and construct the Green's function using the eigenfunction expansion. In Sections 12.2 and 12.3 we study the Green's function for the heat and wave equations, respectively. We use the eigenfunction expansions for the initial boundary value problems and the Duhamel's principle or the Fourier transform for the initial value problems to construct the Green's functions.

12.1 Green's Functions for the Laplace Equation

12.1.1 *Eigenfunction Expansion*

In Section 4.4 we studied the Green's function $G(\mathbf{x};\mathbf{x}_0)$ for the Laplace equation. In particular in (4.4.2) we derived

$$u(\mathbf{x}_0) = \iiint\limits_{\Omega} G\triangle u d\mathbf{x} - \iint\limits_{\partial\Omega} (G\frac{du}{dn} - u\frac{dG}{dn})dS. \qquad (12.1.1)$$

This relation holds even if u is not a solution to the Laplace or Poisson equation. If we choose u to be a test function, the boundary integrals vanish and we have

$$\iiint_{\Omega} G\Delta\phi d\mathbf{x} = \phi(\mathbf{x}_0).$$

This means that the distributional derivative of the Green's function is a delta function, *i.e.*,

$$\Delta G(\mathbf{x};\mathbf{x}_0) = \delta(\mathbf{x}-\mathbf{x}_0), \quad \mathbf{x}\in\Omega, \tag{12.1.2}$$

where $\mathbf{x}$ is the variable for the Green's function and $\mathbf{x}_0$ is the observation point for u. So, we modify the definition in Section 4.4 and define the Green's function $G(\mathbf{x};\mathbf{x}_0)$ for the Laplace or Poisson equation to be a function satisfying (12.1.2) with the corresponding homogeneous boundary conditions. This means that if $u(\mathbf{x})$ is the solution to the boundary value problem given by

$$\Delta u = h(\mathbf{x}), \quad \mathbf{x}\in\Omega, \tag{12.1.3}$$

$$\alpha u + \beta\frac{\partial u}{\partial n} = g(\mathbf{x}), \quad \mathbf{x}\in\partial\Omega, \tag{12.1.4}$$

we define the Green's function to be the solution to

$$\Delta G(\mathbf{x};\mathbf{x}_0) = \delta(\mathbf{x}-\mathbf{x}_0), \quad \mathbf{x}\in\Omega, \tag{12.1.5}$$

$$\alpha G(\mathbf{x};\mathbf{x}_0) + \beta\frac{\partial G(\mathbf{x};\mathbf{x}_0)}{\partial n} = 0, \quad \mathbf{x}\in\partial\Omega. \tag{12.1.6}$$

To see if the definition makes sense, we multiply (12.1.5) by $u(\mathbf{x})$ and use the Green's identity. Then, we have

$$\iiint_{\Omega}(G\Delta u - u\Delta G)d\mathbf{x} = \iint_{\partial\Omega}(G\frac{du}{dn} - u\frac{dG}{dn})dS.$$

Note that we do not need to remove the region $B(\mathbf{x}_0,\varepsilon)$. The delta function replaces the process of taking the limit as $\varepsilon\to 0$. Using (12.1.5), we obtain

$$u(\mathbf{x}_0) = \iiint_{\Omega} Ghd\mathbf{x} - \iint_{\partial\Omega}(G\frac{du}{dn} - u\frac{dG}{dn})dS. \tag{12.1.7}$$

This is the same as (12.1.1) if Δu is replaced with h. For the Green's function for a ball (4.4.15) constructed in Section 4.4, we had $\beta = 0$ and $h = 0$. Therefore, in this case (12.1.7) reduces to

$$u(\mathbf{x}_0) = \iint_{\partial\Omega} g\frac{dG}{dn}dS, \tag{12.1.8}$$

which has the same form as (4.4.15).

In Section 4.4 we constructed the Green's functions for a half plane and a ball through the fundamental solutions. In this section we use the formulation (12.1.5) and (12.1.6) to find the Green's function. In other words, we solve a special non-homogeneous equation where the non-homogeneous term is a delta function. In the following example we use the eigenfunction expansion to construct the Green's function when Ω is a rectangle.

Example 12.1.1. (a) Find the Green's function for the boundary value problem on the rectangle $0 < x < L,\ 0 < y < H$

$$\Delta u = h(x, y), \quad 0 < x < L, \quad 0 < y < H,$$

$$u(0, y) = f_1(y), \quad u_x(L, y) = f_2(y),$$

$$u_y(x, 0) = g_1(x), \quad u_y(x, H) = g_2(x).$$

(b) Express the solution u using the Green's function.

Solution: (a) The Green's function $G(x, y; x_0, y_0)$ satisfies

$$\Delta G = \delta(x - x_0)\delta(y - y_0), \tag{12.1.9}$$

$$G(0, y; x_0, y_0) = 0, \quad G_x(L, y; x_0, y_0) = 0,$$

$$G_y(x, 0; x_0, y_0) = 0, \quad G_y(x, H; x_0, y_0) = 0.$$

We use the method of eigenfunction expansion to determine $G(\mathbf{x}; \mathbf{x}_0)$. From the boundary conditions for G, the eigenfunction expansion of G is given by

$$G = \sum_{m=0}^{\infty} \sum_{n=1}^{\infty} a_{mn} \cos \frac{m\pi y}{H} \sin \frac{(2n-1)\pi x}{2L}. \tag{12.1.10}$$

Substituting (12.1.10) in (12.1.9), multiplying the resulting equation by the eigenfunctions $\cos(m\pi y/H)\sin[(2n-1)\pi x/(2L)]$, and integrating it over the region, we see that the coefficients satisfy

$$\begin{aligned}
-\frac{LH}{2}\left(\frac{(2n-1)\pi}{2L}\right)^2 a_{0n} &= \int_0^L \int_0^H \delta(x - x_0)\delta(y - y_0) \sin \frac{(2n-1)\pi x}{2L} dy dx \\
&= \sin \frac{(2n-1)\pi x_0}{2L}, \quad m = 0,
\end{aligned}$$

$$-\frac{LH}{4}\{(\frac{m\pi}{H})^2+(\frac{(2n-1)\pi}{2L})^2\}a_{mn}$$
$$=\int_0^L\int_0^H \delta(x-x_0)\delta(y-y_0)\cos\frac{m\pi y}{H}\sin\frac{(2n-1)\pi x}{2L}dydx$$
$$=\cos\frac{m\pi y_0}{H}\sin\frac{(2n-1)\pi x_0}{2L},\quad m\neq 0.$$

Therefore,

$$G=-\sum_{n=1}^{\infty}\frac{1}{\frac{LH}{2}(\frac{(2n-1)\pi}{2L})^2}\sin\frac{(2n-1)\pi x_0}{2L}\sin\frac{(2n-1)\pi x}{2L}$$
$$-\sum_{m=1}^{\infty}\sum_{n=1}^{\infty}\frac{1}{\frac{LH}{4}\{(\frac{m\pi}{H})^2+(\frac{(2n-1)\pi}{2L})^2\}}$$
$$\cos\frac{m\pi y_0}{H}\sin\frac{(2n-1)\pi x_0}{2L}\cos\frac{m\pi y}{H}\sin\frac{(2n-1)\pi x}{2L}.$$

(b) Now we represent the solution using the Green's function and (12.1.7). We need to evaluate the boundary integrals. Note that in (12.1.7) $du/dn = \mathbf{n}\cdot\nabla u$ etc and the line integral is evaluated counter clockwise as in Equation (4.2.4) and Figure 4.1. This leads to

$$\int_{\partial\Omega}(G\frac{du}{dn}-u\frac{dG}{dn})ds$$
$$=\int_0^H[1,0]\cdot\{G[\frac{\partial u}{\partial x},\frac{\partial u}{\partial y}]-u[\frac{\partial G}{\partial x},\frac{\partial G}{\partial y}]\}|_{x=L}dy$$
$$+\int_L^0[0,1]\cdot\{G[\frac{\partial u}{\partial x},\frac{\partial u}{\partial y}]-u[\frac{\partial G}{\partial x},\frac{\partial G}{\partial y}]\}|_{y=H}(-dx)$$
$$+\int_H^0[-1,0]\cdot\{G[\frac{\partial u}{\partial x},\frac{\partial u}{\partial y}]-u[\frac{\partial G}{\partial x},\frac{\partial G}{\partial y}]\}|_{x=0}(-dy)$$
$$+\int_0^L[0,-1]\cdot\{G[\frac{\partial u}{\partial x},\frac{\partial u}{\partial y}]-u[\frac{\partial G}{\partial x},\frac{\partial G}{\partial y}]\}|_{y=0}dx.$$

Therefore, we obtain

$$u(x_0,y_0)=\int_0^H\int_0^L Ghdxdy$$
$$+\int_0^H G(L,y;x_0,y_0)f_2(y)dy+\int_0^L G(x,H;x_0,y_0)g_2(x)dx$$
$$+\int_0^H f_1(y)\frac{\partial G}{\partial x}(0,y;x_0,y_0)dy-\int_0^L G(x,0;x_0,y_0)g_1(x)dx.$$

12.1.2 *Modified Green's Function*

In the system of linear equations we learned that when a homogenous system $A\mathbf{x} = 0$ has a nontrivial solution, the non-homogeneous system $A\mathbf{x} = \mathbf{b}$ has infinitely many solutions or no solutions depending on the values of $\mathbf{b}$. See Appendix A.4. A similar situation happens in the differential equations. We actually learned this in Subsection 2.4.1 when we studied the steady state solutions to the heat equation with Neumann boundary conditions. (See Section 2.4.) What we learned is that the non-homogeneous term and the boundary conditions have to satisfy some restriction. To illustrate a point consider the Green's function for the steady state heat equation. It satisfies

$$G_{xx}(x; x_0) = \delta(x - x_0), \quad 0 < x < L, \tag{12.1.11}$$

$$G_x(0; x_0) = 0, \quad G_x(0; x_0) = 0, \tag{12.1.12}$$

where $0 < x_0 < L$. This is the same as the Green's function for the one dimensional Laplacian. Then, integrating (12.1.11) once, we have

$$G_x(x; x_0) = H(x - x_0) + c_1,$$

where c_1 is a constant of integration. We choose c_1 so that the boundary conditions are satisfied. At $x = 0$,

$$G_x(0; x_0) = c_1 = 0$$

and at $x = L$,

$$G_x(L; x_0) = 1 + c_1 = 0.$$

There is no c_1 satisfying both boundary conditions. The reason we cannot find the solution is the same as the system of linear equations. The homogeneous problem corresponding to (12.1.11) and (12.1.12) is

$$u_{xx} = 0, \quad 0 < x < L, \tag{12.1.13}$$

$$u_x(0) = 0, \quad u_x(L) = 0, \tag{12.1.14}$$

and $u = c_1$ is a solution. The homogeneous problem has a nontrivial solution. Then, the non-homogeneous problem

$$u_{xx} = h(x), \quad 0 < x < L, \tag{12.1.15}$$

with the boundary conditions (12.1.14) has infinitely many solutions or no solutions. For example, if $h(x)$ satisfies $\int_0^L h(x)dx = 0$, then for any constant c,

$$u(x) = c + \int_0^x \int_0^y h(s)dsdy$$

is a solution. When $h(x) = \delta(x - x_0)$, there is no solution since $\int_0^L \delta(x - x_0)dx = 1 \neq 0$. To remedy this problem, we define the modified (or generalized) Green's function as the solution to

$$G_{xx}(x; x_0) = \delta(x - x_0) + g(x), \quad 0 < x < L, \tag{12.1.16}$$

$$G_x(0; x_0) = 0, \quad G_x(L; x_0) = 0, \tag{12.1.17}$$

where we add g so that (12.1.16) and (12.1.17) have a solution. Integrating (12.1.16), we have

$$G_x(x; x_0) = H(x - x_0) + \int_0^x g(s)ds + c_1.$$

At $x = 0$,

$$G_x(0; x_0) = c_1 = 0, \tag{12.1.18}$$

At $x = L$,

$$G_x(L; x_0) = 1 + \int_0^L g(s)ds + c_1 = 0. \tag{12.1.19}$$

Therefore, we choose g so that $\int_0^L g(s)ds = -1$ is satisfied. A well-known choice of g is a nontrivial solution to the homogeneous problem (12.1.13) and (12.1.14), *i.e.*, $g(x) = c$, a constant. Then, (12.1.18) and (12.1.19) yield $c = -1/L$ and the Green's function is

$$G(x; x_0) = -\frac{1}{2L}x^2 + H(x - x_0)(x - x_0) + c_2,$$

where c_2 is an arbitrary constant. The above Green's function is rewritten as

$$G(x; x_0) = \begin{cases} -\frac{1}{2L}x^2 + c_2, & x < x_0, \\ -\frac{1}{2L}x^2 + (x - x_0) + c_2, & x_0 < x. \end{cases}$$

This is no longer symmetric in exchanging x and x_0 in general. However, we achieve the symmetry by choosing $c_2 = x_0 - x_0^2/(2L)$.

In the next example we examine the case where the nontrivial solution is not a constant.

Example 12.1.2. Find the Green's function for the boundary value problem

$$u_{xx} + u = h(x), \quad 0 < x < \pi,$$

$$u_x(0) = 0, \quad u_x(\pi) = 0.$$

Solution: The Green's function satisfies

$$G(x; x_0)_{xx} + G(x; x_0) = \delta(x - x_0), \quad 0 < x < \pi, \tag{12.1.20}$$

$$G_x(0; x_0) = 0, \quad G_x(\pi; x_0) = 0. \tag{12.1.21}$$

By the variation of parameters formula, the solution is

$$\begin{aligned} G &= c_1 \cos x + c_2 \sin x \\ &\quad - \cos x \int_0^x \delta(s - x_0) \sin s ds + \sin x \int_0^x \delta(s - x_0) \cos s ds. \end{aligned}$$

Therefore,

$$G(x; x_0) = \begin{cases} c_1 \cos x + c_2 \sin x, & x < x_0, \\ c_1 \cos x + c_2 \sin x - \cos x \sin x_0 + \sin x \cos x_0 & x > x_0. \end{cases}$$

Now we check the boundary conditions.

$$G_x(0; x_0) = -c_1 \sin 0 + c_2 \cos 0 = 0 \Rightarrow c_2 = 0,$$

$$G_x(\pi; x_0) = -c_1 \sin x + \cos(x - x_0) \Rightarrow \cos(\pi - x_0) = 0?$$

So, there is no solution to (12.1.20) and (12.1.21).

A remedy is to consider the modified Green's function.

$$G(x; x_0)_{xx} + G(x; x_0) = \delta(x - x_0) + c \cos x, \quad 0 < x < \pi,$$

$$G_x(0; x_0) = 0, \quad G_x(\pi; x_0) = 0,$$

where $c \cos x$ is a solution to the corresponding homogeneous equation. The variation of parameters formula leads to the general solution given by

$$\begin{aligned} G &= c_1 \cos x + c_2 \sin x \\ &\quad - \cos x \int_0^x \delta(s - x_0) \sin s ds + \sin x \int_0^x \delta(s - x_0) \cos s ds \\ &\quad - \cos x \int_0^x c \cos s \sin s ds + \sin x \int_0^x c \cos s \cos s ds \\ &= c_1 \cos x + c_2 \sin x - \cos x \int_0^x \delta(s - x_0) \sin s ds \\ &\quad + \sin x \int_0^x \delta(s - x_0) \cos s ds + \frac{c}{4} x \sin x, \end{aligned}$$

$$\begin{aligned} G &= c_1 \cos x + c_2 \sin x - \cos x \sin x_0 + \sin x \cos x_0 + \frac{c}{4} x \sin x \\ &= c_1 \cos x + c_2 \sin x + \sin(x - x_0) + \frac{c}{4} x \sin x. \end{aligned}$$

Now we check the boundary conditions.

$$G_x = -c_1 \sin x + c_2 \cos x + \frac{c}{4} \sin x + \frac{c}{4} x \cos x, \quad x < x_0,$$

$$G_x = -c_1 \sin x + c_2 \cos x + \cos(x - x_0) + \frac{c}{4} \sin x + \frac{c}{4} x \cos x, \quad x_0 < x.$$

Then,

$$G_x(0; x_0) = c_2 = 0,$$

$$G_x(\pi; x_0) = -c_2 + \cos(\pi - x_0) - \frac{c}{4}\pi = 0.$$

This implies

$$c = \frac{4}{\pi} \cos(\pi - x_0) = -\frac{4}{\pi} \cos x_0. \tag{12.1.22}$$

Therefore, the modified Green's function is

$$G(x; x_0) = \begin{cases} c_1 \cos x - \frac{1}{\pi}(\cos x_0) x \sin x, & x < x_0, \\ c_1 \cos x + \sin(x - x_0) - \frac{1}{\pi}(\cos x_0) x \sin x, & x_0 < x. \end{cases}$$

To make G symmetric, choose $c_1 = -(x_0 \sin x_0)/\pi + \sin x_0$. Then,

$$G(x; x_0) = \begin{cases} -\frac{1}{\pi} x_0 \sin x_0 \cos x - \frac{1}{\pi}(\cos x_0) x \sin x + \sin x_0 \cos x, & x < x_0, \\ -\frac{1}{\pi} x_0 \sin x_0 \cos x - \frac{1}{\pi}(\cos x_0) x \sin x + \sin x \cos x_0, & x_0 < x. \end{cases}$$

Based on the two examples, when there is a nontrivial solution $\phi(\mathbf{x})$ to the homogeneous problem

$$\Delta u(\mathbf{x}) = 0, \quad \mathbf{x} \in \Omega, \tag{12.1.23}$$

$$\frac{\partial u}{\partial n}(\mathbf{x}) = 0, \quad \mathbf{x} \in \partial\Omega, \tag{12.1.24}$$

we modify the Green's function to be the solution of

$$\Delta G(\mathbf{x}; \mathbf{x}_0) = \delta(\mathbf{x} - \mathbf{x}_0) + c\phi(\mathbf{x}_0)\phi(\mathbf{x})$$

$$\frac{\partial G(\mathbf{x}; \mathbf{x}_0)}{\partial n} = 0,$$

where c is given by

$$c = -\frac{\int_\Omega \delta(\mathbf{x} - \mathbf{x}_0)\phi(\mathbf{x}) d\mathbf{x}}{\phi(\mathbf{x}_0) \int_\Omega \phi^2(\mathbf{x}) d\mathbf{x}}.$$

In the Neumann problem (12.1.23) and (12.1.24) we take $\phi(\mathbf{x})$ to be the eigenfunction for zero eigenvalue.

Exercises

1. Find the Green's function for

$$u_{xx} = h(x), \quad 0 < x < L,$$

$$u(0) = f(x), \quad u(L) = g(x).$$

2. Consider

$$\Delta G = \delta(\mathbf{x} - \mathbf{x}_0), \quad \mathbf{x} \in \Omega.$$

Use the method of eigenfunction expansion to determine $G(\mathbf{x}; \mathbf{x}_0)$.

(a) The domain Ω is the rectangle $0 < x < L$, $0 < y < H$ with

$$G_x = 0 \text{ on } x = 0, \quad G = 0 \text{ on } x = L,$$

$$G = 0 \text{ on } y = 0, \quad G = 0 \text{ on } y = H.$$

(b) The domain Ω is the rectangular box $0 < x < L$, $0 < y < H$, $0 < z < W$ with $G = 0$ on the six sides.

3. Use the one-dimensional eigenfunction expansion to determine $G(\mathbf{x}; \mathbf{x}_0)$.

$$\Delta G = \delta(\mathbf{x} - \mathbf{x}_0)$$

on the rectangle $0 < x < L$, $0 < y < H$ with

$$G = 0 \text{ on } x = 0, \quad G_x = 0 \text{ on } x = L,$$

$$G_y = 0 \text{ on } y = 0, \quad G_y = 0 \text{ on } y = H.$$

4. A modified Green's function $G(\mathbf{x}; \mathbf{x}_0)$ satisfies

$$\Delta G = \delta(\mathbf{x} - \mathbf{x}_0) + c$$

with

$$\nabla G \cdot \mathbf{n} = 0$$

on the boundary of the rectangle $0 < x < L$, $0 < y < H$.

(a) Find the value(s) of c for which the method of one-dimensional eigenfunction expansion $G(\mathbf{x}; \mathbf{x}_0) = \sum_{m=0}^{\infty} a_m(x) \cos \frac{m\pi y}{H}$ works. For this c, determine $G(\mathbf{x}; \mathbf{x}_0)$ and if possible, make $G(\mathbf{x}; \mathbf{x}_0)$ symmetric.

(b) Consider (a) with the method of two-dimensional eigenfunction expansion $G(\mathbf{x}; \mathbf{x}_0) = \sum_{m=0}^{\infty} \sum_{n=0}^{\infty} a_{mn} \cos \frac{n\pi x}{L} \cos \frac{m\pi y}{H}$.

12.2 Green's Functions for the Heat Equation

12.2.1 *Initial Boundary Value Problems*

We consider an initial boundary value problem on a domain $\Omega \in R^n$. We discuss the case where $n = 3$ with the remark that $n = 1$ or 2 is treated similarly.

$$u_t - a^2 \Delta u = h(\mathbf{x}, t), \quad \mathbf{x} \in \Omega,\ 0 < t, \tag{12.2.1}$$

$$\alpha u + \beta \frac{\partial u}{\partial n} = g(\mathbf{x}, t), \quad \mathbf{x} \in \partial\Omega, \tag{12.2.2}$$

with the initial data

$$u(\mathbf{x}, 0) = f(\mathbf{x}). \tag{12.2.3}$$

Motivated by the previous section, we define the Green's function $G(\mathbf{x}, t; \mathbf{x}_0, t_0)$ for the above problem to be a function satisfying the non-homogeneous heat equation

$$-G_t - a^2 \Delta G = \delta(\mathbf{x} - \mathbf{x}_0)\delta(t - t_0), \quad \mathbf{x} \in \Omega,\ 0 < t, \tag{12.2.4}$$

with the homogeneous boundary conditions

$$\alpha G + \beta \frac{\partial G}{\partial n} = 0, \quad \mathbf{x} \in \partial\Omega, \tag{12.2.5}$$

and with the homogeneous initial condition

$$G(\mathbf{x}, T; \mathbf{x}_0, t_0) = 0, \tag{12.2.6}$$

where $(\mathbf{x}_0, t_0)$ is the value of $(\mathbf{x}, t)$ at which u will be evaluated and T is some constant satisfying $T > t_0$. This T is introduced for our convenience. It looks strange to see that there is a negative sign before G_t and the initial condition (maybe called the terminal condition) is specified at $t = T$. This will become clear once we multiply (12.2.4) by u and integrate it in $\mathbf{x}$ and t over the region Ω and $0 < t < T$, respectively. Using the Green's identity and the integration by parts in t, we obtain

$$-\iiint_{\Omega} [Gu]_0^T d\mathbf{x} - a^2 \int_0^T \iint_{\partial\Omega} \left(u\frac{\partial G}{\partial n} - G\frac{\partial u}{\partial n}\right) ds dt$$

$$+ \int_0^T \iiint_{\Omega} G(u_t - a^2 \Delta u) d\mathbf{x} dt$$

$$= \int_0^T \iiint_{\Omega} \delta(\mathbf{x} - \mathbf{x}_0)\delta(t - t_0) u d\mathbf{x} dt = u(\mathbf{x}_0, t_0).$$

Because of the negative sign before G_t, in the third term of the left hand side we have $u_t - a^2\Delta u$ so that (12.2.1) can use used. Here, $\iint_{\partial\Omega}(u\nabla G - G\nabla u)\cdot \mathbf{n}ds$ is a surface integral and it can be simplified by the boundary conditions. For example, if $\beta \neq 0$, we use

$$\frac{\partial u}{\partial n} = \nabla u \cdot \mathbf{n} = \frac{1}{\beta}(g - \alpha u), \quad \nabla G \cdot \mathbf{n} = -\frac{\alpha}{\beta}G$$

to obtain

$$u(\mathbf{x}_0, t_0) = \iiint\limits_{\Omega} G(\mathbf{x}, 0; \mathbf{x}_0, t_0) f(\mathbf{x}) d\mathbf{x} + \frac{a^2}{\beta}\int_0^T \iint\limits_{\partial\Omega} G g ds dt + \int_0^T \iiint\limits_{\Omega} G h(\mathbf{x}, t) d\mathbf{x} dt. \quad (12.2.7)$$

The case where $\alpha = 0$ or $\beta = 0$ is treated similarly.

To solve (12.2.4) to (12.2.6) we introduce the change of variable for t and set $\tau = T - t$. We also denote $G(\mathbf{x}, T - \tau; \mathbf{x}_0, t_0) = K(\mathbf{x}, \tau; \mathbf{x}_0, t_0)$. Then, since $G_\tau = K_\tau$ and $G_t = G_\tau \frac{\partial \tau}{\partial t} = -K_\tau$, in terms of K the problem is reduced to

$$K_\tau - a^2\Delta K = \delta(\mathbf{x} - x_0)\delta(T - \tau - t_0), \quad \mathbf{x} \in \Omega,\ 0 < \tau, \quad (12.2.8)$$

$$\alpha K + \beta\frac{\partial K}{\partial n} = 0, \quad \mathbf{x} \in \partial\Omega, \quad (12.2.9)$$

$$K(\mathbf{x}, 0; \mathbf{x}_0, t_0) = 0. \quad (12.2.10)$$

We now have the usual initial boundary value problem for the non-homogeneous heat equation. This suggests that to solve (12.2.4) to (12.2.6) we solve (12.2.8) to (12.2.10) and replace τ with $T - t$.

As a specific example we construct the Green's function for the following initial boundary value problem. As in the boundary value problem for the Laplace equation we can use the eigenfunction expansion to find the Green's function.

Example 12.2.1. Find the Green's function for

$$u_t - a^2 u_{xx} = h(x, t), \quad 0 < x < L,\ 0 < t,$$

$$u(0, t) = g_1(t), \quad u(L, t) = g_2(t),$$

$$u(x, 0) = f(x).$$

Solution: We find the Green's function $K(x, \tau; x_0, t_0)$ satisfying

$$K_\tau - a^2 K_{xx} = \delta(x - x_0)\delta(T - \tau - t_0), \qquad (12.2.11)$$

$$K(0, \tau; x_0, t_0) = 0, \quad K(L, \tau; x_0, t_0) = 0, \qquad (12.2.12)$$

$$K(x, 0; x_0, t_0) = 0. \qquad (12.2.13)$$

An appropriate eigenfunction expansion for the solution to (12.2.11) to (12.2.13) is

$$K(x, \tau) = \sum_{n=1}^{\infty} b(\tau) \sin \frac{n\pi x}{L}.$$

Substituting this in (12.2.11), we have

$$\sum_{n=1}^{\infty} [b_n'(\tau) + (\frac{an\pi}{L})^2 b_n(\tau)] \sin \frac{n\pi x}{L} = \delta(x - x_0)\delta(T - \tau - t_0). \qquad (12.2.14)$$

Multiplying (12.2.14) by $\sin \frac{m\pi x}{L}$ and integrating over $0 < x < L$, we obtain

$$\frac{L}{2}[b_m'(\tau) + (\frac{am\pi}{L})^2 b(\tau)] = \delta(T - \tau - t_0) \sin \frac{m\pi x_0}{L}. \qquad (12.2.15)$$

This is a first order linear ODE. The integrating factor is $e^{(\frac{am\pi}{L})^2 \tau}$. Then, multiplying (12.2.15) by the integrating factor, we have

$$[b_m(\tau) e^{(\frac{am\pi}{L})^2 \tau}]' = \frac{2}{L} e^{(\frac{am\pi}{L})^2 \tau} \delta(T - \tau - t_0) \sin \frac{m\pi x_0}{L}.$$

Since (12.2.13) implies $b_m(0) = 0$, the integration in τ leads to

$$b_m(\tau) = e^{-(\frac{am\pi}{L})^2 \tau} \int_0^\tau \frac{2}{L} e^{(\frac{am\pi}{L})^2 s} \delta(T - s - t_0) \sin \frac{m\pi x_0}{L} ds.$$

This can be rewritten as

$$b_m(\tau) = \begin{cases} 0, & 0 \le \tau < T - t_0, \\ \frac{2}{L} e^{(\frac{am\pi}{L})^2 (T - \tau - t_0)} \sin \frac{m\pi x_0}{L}, & T - t_0 \le \tau \le T. \end{cases}$$

Therefore, using the Heaviside function, we obtain

$$\begin{aligned} K(x, \tau; x_0, t_0) &= \sum_{n=1}^{\infty} b_n(\tau) \sin \frac{n\pi x}{L} \\ &= \sum_{n=1}^{\infty} \frac{2}{L} H(\tau - T + t_0) e^{(\frac{an\pi}{L})^2 (T - \tau - t_0)} \sin \frac{n\pi x_0}{L} \sin \frac{n\pi x}{L}. \end{aligned}$$

After changing τ to t and K to G, we arrive at

$$G(x,t;x_0,t_0) = \sum_{n=1}^{\infty} \frac{2}{L} H(t_0 - t) e^{-(\frac{an\pi}{L})^2 (t_0 - t)} \sin \frac{n\pi x_0}{L} \sin \frac{n\pi x}{L}.$$

As an example consider the case where $h = g_1 = g_2 = 0$. Then, from (12.2.7) we obtain

$$\begin{aligned} u(x_0,t_0) &= \int_0^L G(x,0;x_0,t_0) f(x) dx \\ &= \sum_{n=1}^{\infty} (\frac{2}{L} \int_0^L \sin \frac{n\pi x}{L} f(x) dx) e^{-(\frac{an\pi}{L})^2 t_0} \sin \frac{n\pi x_0}{L}. \end{aligned}$$

We notice that this is the same as (2.3.16) in Chapter 2 where (2.3.19) is substituted.

12.2.2 *Initial Value Problems*

The initial value problem for the heat equation in $\Omega = R^n$ is given by

$$u_t - a^2 \Delta u = h(\mathbf{x}, t), \quad \mathbf{x} \in R^n,\ 0 < t, \tag{12.2.16}$$

$$u(\mathbf{x}, 0) = f(\mathbf{x}). \tag{12.2.17}$$

Therefore, the Green's function satisfies

$$-G_t - a^2 \Delta G = \delta(\mathbf{x} - \mathbf{x}_0) \delta(t - t_0), \tag{12.2.18}$$

$$G(\mathbf{x}, T) = 0. \tag{12.2.19}$$

As in the previous subsection we introduce $\tau = T - t$ and denote $G(\mathbf{x}, T - \tau; \mathbf{x}_0, t_0) = K(\mathbf{x}, \tau; \mathbf{x}_0, t_0)$. Then, in terms of K we have

$$K_\tau - a^2 \Delta K = \delta(\mathbf{x} - \mathbf{x}_0) \delta(T - \tau - t_0), \tag{12.2.20}$$

$$K(\mathbf{x}, 0) = 0. \tag{12.2.21}$$

In the following example we use the Duhamel's principle to find the Green's function. We can off course use the Fourier transform to find the Green's function. This approach is discussed in Exercises.

Example 12.2.2. Find the Green's function for the initial value problem by the Duhamel's principle.

$$u_t - a^2 u_{xx} = h(x,t), \quad -\infty < x < \infty,\ 0 < t,$$

$$u(x, 0) = f(x).$$

Solution: After changing to τ and K, we see that the Green's function $K(x,\tau;x_0,t_0)$ satisfies

$$K_\tau - a^2K_{xx} = \delta(x-x_0)\delta(T-\tau-t_0),$$

$$K(x,0;x_0,t_0) = 0.$$

We use the Duhamel's principle. Then,

$$\begin{aligned}
&K(x,\tau;x_0,t_0)\\
&= \int_0^\tau \int_{-\infty}^{\infty} \frac{1}{\sqrt{4a^2\pi(\tau-s)}} e^{-\frac{(x-y)^2}{4a^2\pi(\tau-s)}} \delta(y-x_0)\delta(T-s-t_0)dyds\\
&= \int_0^\tau \frac{1}{\sqrt{4a^2\pi(\tau-s)}} e^{-\frac{(x-x_0)^2}{4a^2\pi(\tau-s)}} \delta(T-s-t_0)ds\\
&= \frac{H(\tau-T+t_0)}{\sqrt{4a^2\pi(\tau-T+t_0)}} e^{-\frac{(x-x_0)^2}{4a^2\pi(\tau-T+t_0)}}.
\end{aligned}$$

Therefore, after changing back to t and G, we have

$$G(x,t;x_0,t_0) = \frac{H(t_0-t)}{\sqrt{4a^2\pi(t_0-t)}} e^{-\frac{(x-x_0)^2}{4a^2\pi(t_0-t)}}.$$

Since there is no boundary condition, setting $g=0$ in (12.2.7), we have

$$\begin{aligned}
u(x_0,t_0) &= \int_{-\infty}^{\infty} G(x,0;x_0,t_0)f(x)dx + \int_0^T \int_{-\infty}^{\infty} Gh(x,t)dxdt\\
&= \int_{-\infty}^{\infty} \frac{1}{\sqrt{4a^2\pi t_0}} e^{-\frac{(x-x_0)^2}{4a^2\pi t_0}} f(x)dx\\
&+ \int_0^{t_0} \int_{-\infty}^{\infty} \frac{1}{\sqrt{4a^2\pi(t_0-t)}} e^{-\frac{(x-x_0)^2}{4a^2\pi(t_0-t)}} h(x,t)dxdt. \qquad (12.2.22)
\end{aligned}$$

For $h=0$ (12.2.22) reduces to (2.6.4) in Section 2.6 and for $f=0$ this is the same as the one dimensional case ($n=1$) of (11.2.7) in Section 11.2.

Exercises

1. Use the eigenfunction expansion to construct the Green's function for the initial boundary value problems

$$u_t - a^2u_{xx} = h(x,t), \quad 0<x<L,\ 0<t,$$

$$u(x,0) = f(x)$$

with the following boundary conditions.

(a) $u(0,t) = g_1(t), \quad u_x(L,t) = g_2(t)$
(b) $u_x(0,t) = g_1(t), \quad u(L,t) = g_2(t)$

2. Find the Green's function for the two-dimensional problems.

$$u_t - a^2(u_{xx} + u_{yy}) = h(x,y,t), \quad 0 < x < L,\ 0 < y < H,\ 0 < t,$$

$$u(x,y,0) = f(x,y), \quad 0 < x < L,\ 0 < y < H.$$

(a) $u(0,y,t) = g_1(y,t),\ u(L,y,t) = g_2(y,t),$
$u(x,0,t) = g_3(x,t),\ u(x,H,t) = g_4(x,t)$
(b) $u(0,y,t) = g_1(y,t),\ u(L,y,t) = g_2(y,t),$
$u_y(x,0,t) = g_3(x,t),\ u(x,H,t) = g_4(x,t)$

3. Use the Fourier transform to construct the Green's function for

$$u_t - a^2 u_{xx} = h(x,t), \quad -\infty < x < \infty,\ 0 < t,$$

$$u(x,0) = f(x).$$

12.3 Green's Functions for the Wave Equation

12.3.1 *Initial Boundary Value Problems*

As an example of the initial boundary value problem for the wave equation, we consider

$$u_{tt} - c^2\Delta u = Lu = h(\mathbf{x},t), \quad \mathbf{x} \in \Omega,\ 0 < t, \tag{12.3.1}$$

$$\alpha u + \beta\frac{\partial u}{\partial n} = b(\mathbf{x},t), \quad \mathbf{x} \in \partial\Omega,$$

$$u(\mathbf{x},0) = f(\mathbf{x}),$$

$$u_t(\mathbf{x},0) = g(\mathbf{x}).$$

The Green's function $G(\mathbf{x},t;\mathbf{x}_0,t_0)$ is the function satisfying

$$G_{tt} - c^2\Delta G = \delta(\mathbf{x}-\mathbf{x}_0)\delta(t-t_0), \quad \mathbf{x} \in \Omega,\ 0 < t, \tag{12.3.2}$$

$$\alpha G + \beta\frac{\partial G}{\partial n} = 0, \quad \mathbf{x} \in \partial\Omega,$$

$$G(\mathbf{x},T;\mathbf{x}_0,t_0) = 0,$$

$$G_t(\mathbf{x},T;\mathbf{x}_0,t_0) = 0.$$

Since we differentiate in t twice, unlike the heat equation the sign in front of G_{tt} is positive. We multiply (12.3.2) by u and integrate it in $\mathbf{x}$ over the region Ω and $0 < t < T$. Using the Green's identity and the integration by parts in t, we obtain

$$\iiint_{\Omega} [G_t u - G u_t]_0^T d\mathbf{x} - c^2 \int_0^T \iint_{\partial\Omega} (u\frac{\partial G}{\partial n} - G\frac{\partial u}{\partial n}) ds dt$$

$$+ \int_0^T \iiint_{\Omega} G(u_{tt} - c^2 \Delta u) d\mathbf{x} dt$$

$$= \int_0^T \iiint_{\Omega} \delta(\mathbf{x} - \mathbf{x}_0)\delta(t - t_0) u d\mathbf{x} dt = u(\mathbf{x}_0, t_0). \qquad (12.3.3)$$

Here, $\iint_{\partial\Omega}(u\nabla G - G\nabla u) \cdot \mathbf{n} ds$ is a surface or line integral and it can be simplified by the boundary conditions. For example, if $\beta \neq 0$, we use

$$\frac{\partial u}{\partial n} = \nabla u \cdot \mathbf{n} = \frac{1}{\beta}(q - \alpha u), \quad \nabla G \cdot \mathbf{n} = -\frac{\alpha}{\beta} G$$

to obtain

$$u(\mathbf{x}_0, t_0) = \iiint_{\Omega} [G(\mathbf{x}, 0; \mathbf{x}_0, t_0) g(\mathbf{x}) - G_t(\mathbf{x}, 0; \mathbf{x}_0, t_0) f(\mathbf{x})] d\mathbf{x}$$

$$+ \frac{c^2}{\beta} \int_0^T \iint_{\partial\Omega} G q ds dt + \int_0^T \iiint_{\Omega} G h(\mathbf{x}, t) d\mathbf{x} dt. \qquad (12.3.4)$$

As in the heat equation, for the initial boundary value problems, we use the eigenfunction expansion to find the Green's function. As a specific example we consider a one-dimensional problem.

Example 12.3.1. Find the Green's function $G(x, t; x_0, t_0)$ for

$$u_{tt} - c^2 u_{xx} = h(x, t), \quad 0 < x < L, \ 0 < t,$$

$$u(0, t) = b_l(t), \quad u(L, t) = b_r(t),$$

$$u(x, 0) = f(x), \quad u_t(x, 0) = g(x).$$

Solution: The Green's function satisfies

$$G_{tt} - c^2 G_{xx} = \delta(x - x_0)\delta(t - t_0), \quad 0 < x < L, \ 0 < t,$$

$$G(0, t) = 0, \quad G(L, t) = 0,$$

$$G(x,T)=0, \quad G_t(x,T)=0.$$

As before we set $\tau = T-t$ and $G(\mathbf{x}, T-\tau; \mathbf{x}_0, t_0) = K(\mathbf{x}, \tau; \mathbf{x}_0, t_0)$. Then,

$$K_{\tau\tau} - c^2 K_{xx} = \delta(x-x_0)\delta(T-\tau-t_0), \quad 0<x<L,\ 0<t, \qquad (12.3.5)$$

$$K(0,T-\tau)=0, \quad K(L,T-\tau)=0,$$

$$K(x,0)=0, \quad K_t(x,0)=0.$$

The eigenfunction expansion is

$$K = \sum_{n=1}^{\infty} a_n(\tau) \sin \frac{n\pi x}{L}.$$

Substitute this in (12.3.5). Then,

$$\sum_{n=1}^{\infty} [a_n''(\tau) + (\frac{cn\pi}{L})^2 a_n(\tau)] \sin \frac{n\pi x}{L} = \delta(x-x_0)\delta(T-\tau-t_0). \qquad (12.3.6)$$

Multiplying by $\sin(m\pi x/L)$ and integrating in x, we obtain

$$a_m''(\tau) + (\frac{cm\pi}{L})^2 a_m(\tau) = \frac{2}{L}\delta(T-\tau-t_0)\sin\frac{m\pi x_0}{L}, \quad m=1,2,\ldots.$$

We use the variation of parameters formula available in Appendix A.3.4. Then,

$$\begin{aligned} a_m(\tau) &= b_m \cos\frac{cm\pi\tau}{L} + c_m \sin\frac{cm\pi\tau}{L} \\ &+ \frac{2}{L}[-\cos\frac{cm\pi\tau}{L}\int_0^\tau \frac{L}{cm\pi}\sin\frac{cm\pi s}{L}\delta(T-s-t_0)ds \\ &+ \sin\frac{cm\pi\tau}{L}\int_0^\tau \frac{L}{cm\pi}\cos\frac{cm\pi s}{L}\delta(T-s-t_0)ds]\sin\frac{m\pi x_0}{L}. \end{aligned}$$

So, for $0 \le \tau < T - t_0$

$$a_m(\tau) = b_m \cos\frac{cm\pi\tau}{L} + c_m \sin\frac{cm\pi\tau}{L}.$$

At $\tau = 0$,

$$a_m(0) = b_m = 0,$$

$$a_m'(0) = \frac{cm\pi}{L}c_m = 0.$$

Therefore, for $0 \le \tau < T - t_0$, $a_m(\tau) = 0$.

For $\tau \geq T - t_0$, the effect of non-homogeneous terms kicks in and we have

$$
\begin{aligned}
a_m(\tau) &= \frac{2}{cm\pi}[-\cos\frac{cm\pi\tau}{L}\int_0^\tau \sin\frac{cm\pi s}{L}\delta(T-s-t_0)ds \\
&\quad + \sin\frac{cm\pi\tau}{L}\int_0^\tau \cos\frac{cm\pi s}{L}\delta(T-s-t_0)ds]\sin\frac{m\pi x_0}{L} \\
&= \frac{2}{cm\pi}[-\cos\frac{cm\pi\tau}{L}\sin\frac{cm\pi(T-t_0)}{L} \\
&\quad + \sin\frac{cm\pi\tau}{L}\cos\frac{cm\pi(T-t_0)}{L}]\sin\frac{m\pi x_0}{L} \\
&= \frac{2}{cm\pi}\sin\frac{cm\pi(t_0-T+\tau)}{L}\sin\frac{m\pi x_0}{L}.
\end{aligned}
$$

We combine two cases using the Heaviside function $H(\tau - T + t_0)$ and have

$$
a_m(\tau) = \frac{2}{cm\pi}H(\tau - T + t_0)\sin\frac{cm\pi(t_0-T+\tau)}{L}\sin\frac{m\pi x_0}{L}.
$$

Therefore, the Green's function is

$$
\begin{aligned}
&G(x,t;x_0,t_0) \\
&= \sum_{n=1}^{\infty}\frac{2}{cn\pi}H(\tau-T+t_0)\sin\frac{cn\pi(t_0-T+\tau)}{L}\sin\frac{n\pi x_0}{L}\sin\frac{n\pi x}{L} \\
&= \sum_{n=1}^{\infty}\frac{2}{cn\pi}H(t_0-t)\sin\frac{cn\pi(t_0-t)}{L}\sin\frac{n\pi x_0}{L}\sin\frac{n\pi x}{L}. \qquad (12.3.7)
\end{aligned}
$$

12.3.2 *Initial Value Problems*

As an example of the initial value problem for the wave equation, consider

$$
u_{tt} - c^2\Delta u = Lu = h(\mathbf{x},t), \quad \mathbf{x}\in R^n,\ 0<t, \tag{12.3.8}
$$

$$
u(\mathbf{x},0) = f(\mathbf{x}), \tag{12.3.9}
$$

$$
u_t(\mathbf{x},0) = g(\mathbf{x}). \tag{12.3.10}
$$

We define the Green's function $G(\mathbf{x},t;\mathbf{x}_0,t_0)$ to be the function satisfying

$$
G_{tt} - c^2\Delta G = \delta(\mathbf{x}-\mathbf{x}_0)\delta(t-t_0), \quad \mathbf{x}\in R^n,\ 0<t, \tag{12.3.11}
$$

$$
G(\mathbf{x},T;\mathbf{x}_0,t_0) = 0,
$$

$$
G_t(\mathbf{x},T;\mathbf{x}_0,t_0) = 0.
$$

Or changing to $\tau = T - t$ and $G(\mathbf{x}, T-\tau; \mathbf{x}_0, t_0) = K(\mathbf{x}, \tau; \mathbf{x}_0, t_0)$ we have

$$K_{\tau\tau} - c^2 \Delta K = \delta(\mathbf{x} - \mathbf{x}_0)\delta(T - \tau - t_0), \tag{12.3.12}$$

$$K(\mathbf{x}, 0; \mathbf{x}_0, t_0) = 0, \quad K_\tau(\mathbf{x}, 0; \mathbf{x}_0, t_0) = 0. \tag{12.3.13}$$

The solution for the initial value problem (12.3.8) to (12.3.10) is given by

$$\begin{aligned} u(\mathbf{x}_0, t_0) = & \iiint_\Omega [G(\mathbf{x}, 0; \mathbf{x}_0, t_0) g(\mathbf{x}) - G_t(\mathbf{x}, 0; \mathbf{x}_0, t_0) f(\mathbf{x})] d\mathbf{x} \\ & + \int_0^T \iiint_\Omega G h(\mathbf{x}, t) d\mathbf{x} dt. \end{aligned} \tag{12.3.14}$$

We can use the Duhamel's principle or the Fourier transform to find the Green's function. In the following example, we practice the Fourier transform.

Example 12.3.2. (a) Use the Fourier transform to find the Green's function for the initial value problem

$$u_{tt} - c^2 u_{xx} = h(x, t),$$

$$u(x, 0) = f(x), \quad u_t(x, 0) = g(x).$$

(b) Express the solution using the Green's function found in (a).

Solution: (a) Define $U(\omega, \tau; x_0, t_0)$ to be the Fourier transform of G

$$U(\omega, \tau; x_0, t_0) = \int_{-\infty}^{\infty} G e^{-i\omega x} dx.$$

Apply the Fourier transform to (12.3.12), we obtain

$$U_{\tau\tau} + c^2 \omega^2 U = \delta(T - \tau - t_0) e^{-i\omega x_0},$$

$$U(\omega, 0; x_0, T - t_0) = 0, \quad U_\tau(\omega, 0; x_0, T - t_0) = 0.$$

This is the second order linear equation. By the variation of parameters formula in Appendix A.3 the solution is given by

$$\begin{aligned} U = & \, c_1 \cos c\omega\tau + c_2 \sin c\omega\tau \\ & - \frac{e^{-i\omega x_0}}{c\omega} [\cos c\omega\tau \int_0^\tau \delta(T - s - t_0) \sin c\omega s ds \\ & - \sin c\omega\tau \int_0^\tau \delta(T - s - t_0) \cos c\omega s ds]. \end{aligned}$$

For $0 \leq \tau < T - t_0$

$$U(\omega, \tau; x_0, t_0) = c_1 \cos c\omega\tau + c_2 \sin c\omega\tau.$$

From the initial condition

$$U(\omega, 0; x_0, t_0) = c_1 = 0,$$

$$U_\tau(\omega, 0; x_0, t_0) = c\omega c_2 = 0.$$

For $\tau \geq T - t_0$, the effect of non-homogeneous terms kick in and we have

$$\begin{aligned} U &= -\frac{e^{-i\omega x_0}}{c\omega}[\cos c\omega\tau \int_0^\tau \delta(T-s-t_0) \sin c\omega s ds \\ &- \sin c\omega\tau \int_0^\tau \delta(T-s-t_0) \cos c\omega s ds] \\ &= -\frac{e^{-i\omega x_0}}{c\omega}[\cos c\omega\tau \sin c\omega(T-t_0) - \sin c\omega\tau \cos c\omega(T-t_0)] \\ &= \frac{e^{-i\omega x_0}}{c\omega} \sin c\omega(\tau - T + t_0). \end{aligned}$$

Combining two cases

$$U = H(\tau - T + t_0)\frac{e^{-i\omega x_0}}{c\omega} \sin c\omega(\tau - T + t_0).$$

Take the inverse. Then,

$$\begin{aligned} &G(x, t; x_0, t_0) \\ &= \frac{H(t_0 - t)}{2\pi} \int_{-\infty}^{\infty} [\frac{e^{-i\omega x_0}}{c\omega} \sin c\omega(\tau - T + t_0)]e^{i\omega x} d\omega \\ &= \frac{H(t_0 - t)}{2\pi} \int_{-\infty}^{\infty} \frac{1}{2ic\omega}[e^{i\omega(x-x_0+c(\tau-T+t_0)} - e^{i\omega(x-x_0-c(\tau-T+t_0))}\}]d\omega \\ &= \frac{H(t_0 - t)}{2c}[H(x - x_0 + c(\tau - T + t_0)) - H(x - x_0 - c(\tau - T + t_0))] \\ &= \frac{H(t_0 - t)}{2c}[H(x - x_0 + c(t_0 - t)) - H(x - x_0 - c(t_0 - t))]. \quad (12.3.15) \end{aligned}$$

(b) Since

$$\begin{aligned} &G_t(x, t; x_0, t_0) \\ &= -\frac{H(t_0 - t)}{2}[\delta(x - x_0 + c(t_0 - t)) + \delta(x - x_0 - c(t_0 - t))] \\ &- \frac{\delta(t_0 - t)}{2c}[H(x - x_0 + c(t_0 - t)) - H(x - x_0 - c(t_0 - t))], \end{aligned}$$

by (12.3.14), we have

$$u(x_0,t_0) = \int_{-\infty}^{\infty} [G(x,0;x_0,t_0)g(x) - G_t(x,0;x_0,t_0)f(x)]dx$$
$$= \frac{1}{2}\int_{-\infty}^{\infty} [\delta(x - x_0 + ct_0) + \delta(x - x_0 - ct_0)]f(x)dx$$
$$+\frac{1}{2c}\int_{-\infty}^{\infty} [H(x - x_0 + ct_0) - H(x - x_0 - ct_0)]g(x)dx$$
$$= \frac{1}{2}[f(x_0 - ct_0) + f(x_0 + ct_0)] + \frac{1}{2c}\int_{x_0-ct_0}^{x_0+ct_0} g(x)dx.$$

Exercises

1. Find the Green's function $G(x,t;x_0,t_0)$ for

$$u_{tt} - c^2u_{xx} = h(x,t),$$

$$u(x,0) = f(x), \quad u_t(x,0) = g(x)$$

with the following boundary conditions.

(a) $u(0,t) = b_l(t), \quad u_x(L,t) = b_r(t)$
(b) $u_x(0,t) = b_l(t), \quad u(L,t) = b_r(t)$

2. Consider the initial boundary value problem for the wave equation

$$u_{tt} - c^2u_{xx} = 0, \quad 0 < x < L,\ 0 < t,$$

$$u(0,t) = 0, \quad u(L,t) = 0,$$

$$u(x,0) = f(x), \quad u_t(x,0) = g(x).$$

(a) Use (12.3.3) and (12.3.7) to show that the solution $u(x_0,t_0)$ is given by

$$u(x_0,t_0) = \sum_{n=1}^{\infty}[\frac{2}{L}\int_0^L f(x)\sin\frac{n\pi x}{L}dx]\cos\frac{cn\pi t_0}{L}\sin\frac{n\pi x_0}{L}$$
$$+\sum_{n=1}^{\infty}[\frac{2}{cn\pi}\int_0^L g(x)\sin\frac{n\pi x}{L}dx]\sin\frac{cn\pi t_0}{L}\sin\frac{n\pi x_0}{L}.$$

(b) Find the solution by the separation of variables and compare it with the solution in (a).

3. Use (12.3.14) and (12.3.15) to show that the solution $u(x_0, t_0)$ to

$$u_{tt} - c^2 u_{xx} = 0, \quad -\infty < x < \infty,\ 0 < t,$$

$$u(x, 0) = f(x), \quad u_t(x, 0) = g(x)$$

is given by

$$u(x_0, t_0) = \frac{1}{2}[f(x_0 - ct_0) + f(x_0 + ct_0)] + \frac{1}{2c}\int_{x_0 - ct_0}^{x_0 + ct_0} g(x)dx.$$

4. Find the Green's function to the initial value problem (12.3.8) to (12.3.10) using the Duhamel's principle.

Appendices

A.1 Exchanging the Order of Integration and Differentiation

In the book we often exchange the order of integration and differentiation. The question is when and how we can do it. A sufficient condition is given as follows.

Theorem A.1.1. *If both $u(x,t)$ and $u_t(x,t)$ are continuous in the rectangle $(x,t) \in [a,b] \times [c,d]$, where a, b, c, and d are constants. Then,*

$$\frac{d}{dt}\int_a^b u(x,t)dx = \int_a^b \frac{\partial}{\partial t}u(x,t)dx, \quad t \in [c,d].$$

Proof. We use the definition of derivatives.

$$\begin{aligned}
\frac{d}{dt}\int_a^b u(x,t)dx &= \lim_{h\to 0}\frac{\int_a^b u(x,t+h)dx - \int_a^b u(x,t)dx}{h} \\
&= \lim_{h\to 0}\int_a^b \frac{u(x,t+h)-u(x,t)}{h}dx \\
&= \int_a^b \frac{\partial u(x,t)}{\partial t}dx.
\end{aligned}$$

□

The question is what happens if a and b depend on t. We have the following result called Leibniz rule.

Theorem A.1.2. *Suppose $a(t)$ and $b(t)$ are differentiable. If both $u(x,t)$ and $u_t(x,t)$ are continuous, we have*

$$\frac{d}{dt}\int_{a(t)}^{b(t)} u(x,t)dx = u(x,t)\frac{db(t)}{dt} - u(x,t)\frac{da(t)}{dt} + \int_{a(t)}^{b(t)} \frac{\partial}{\partial t}u(x,t)dx.$$

Proof. We use the definition of derivatives again. For simplicity first we prove the case where a is a constant.

$$\begin{aligned}\frac{d}{dt}\int_a^{b(t)} u(x,t)dx &= \lim_{h\to 0}\frac{\int_a^{b(t+h)} u(x,t+h)dx - \int_a^{b(t)} u(x,t)dx}{h}\\ &= \lim_{h\to 0}\frac{\int_a^{b(t+h)} u(x,t+h)dx - \int_a^{b(t)} u(x,t+h)dx}{h}\\ &+ \lim_{h\to 0}\frac{\int_a^{b(t)} u(x,t+h)dx - \int_a^{b(t)} u(x,t)dx}{h}\\ &= u(x,t)\frac{db(t)}{dt} + \int_a^{b(t)}\frac{\partial}{\partial t}u(x,t)dx.\end{aligned}$$

If a depends on t, we can decompose the integral

$$\frac{d}{dt}\int_{a(t)}^{b(t)} u(x,t)dx = \frac{d}{dt}[\int_c^{b(t)} u(x,t)dx - \int_c^{a(t)} u(x,t)dx]$$

and apply the above result. □

A.2 Infinite Series

We summarize a few useful convergence theorems. See [Rudin (1976)] for more details.

Theorem A.2.1. *Suppose $|a_n| \le b_n$ for all n and $\sum_{n=1}^{\infty} b_n$ converges. Then, $\sum_{n=1}^{\infty} a_n$ converges absolutely. This implies that $\sum_{n=1}^{\infty} a_n$ converges.*

Theorem A.2.2. ***(Weierstrass M-test)*** *Suppose that $|f_n(x)| \le M_n$ for all n and for all $x \in [a,b]$, where M_n is a constant, and that $\sum_{n=1}^{\infty} M_n < \infty$. Then, $\sum_{n=1}^{\infty} f_n(x)$ converges uniformly in the interval $[a,b]$.*

A.3 Useful Formulas in ODE's

Solution methods for ODE's appearing in this book are listed. For more complete treatment of ODE's see [Boyce and DiPrima (2009)].

A.3.1 *First Order Linear Equations*

Suppose $y(t)$ satisfies

$$y' + a(t)y = b(t), \quad y(t_0) = y_0.$$

Multiply the equation by $\exp(\int a(t)dt)$. Then,

$$e^{\int a(t)dt}y' + a(t)e^{\int a(t)dt}y = (e^{\int a(t)dt}y)' = b(t)e^{\int a(t)dt}.$$

Integrating this, we obtain

$$e^{\int a(t)dt}y = c + \int_{t_0}^{t} b(s)e^{\int a(s)ds} \Rightarrow y(t) = e^{-\int a(t)dt}[c + \int_{t_0}^{t} b(s)e^{\int a(s)ds}ds].$$

A.3.2 *Bernoulli Equations*

If $y(t)$ satisfies

$$y' + a(t)y = b(t)y^n, \quad y(t_0) = y_0,$$

where $n \neq 0, 1$, the equation is called the Bernoulli equation. To find the solution we change the dependent variable to $v = y^{1-n}$. Then, $v' = (1-n)y^{-n}y'$. So, we multiply the equation by $(1-n)y^{-n}$. Then,

$$(1-n)y^{-n}y' + (1-n)a(t)y^{1-n} = (1-n)b(t).$$

Use v. Then,

$$v' + (1-n)a(t)v = (1-n)b(t).$$

This is a linear equation and we can use the method in Appendix A.3.1.

A.3.3 *Second Order Linear Constant Coefficient Equations*

Consider $y(t)$ satisfying

$$ay'' + by' + cy = 0, \tag{A.3.1}$$

where a, b, and c are constants. We assume that y is given by $y = e^{rt}$, where r is a constant. Then, substituting this in (A.3.1), we have

$$ar^2 + br + c = 0 \Rightarrow r = \frac{1}{2a}[-b \pm \sqrt{b^2 - 4ac}] = r_{\pm}.$$

The general solution is given by

$$y = c_1 e^{r_+ t} + c_2 e^{r_- t}, \quad b^2 - 4ac > 0,$$

$$y = c_1 e^{r_+ t} + c_2 t e^{r_+ t}, \quad b^2 - 4ac = 0,$$

$$y = c_1 e^{-\frac{b}{2a}t} \cos\sqrt{4ac - b^2}t + c_2 e^{-\frac{b}{2a}t} \sin\sqrt{4ac - b^2}t, \quad b^2 - 4ac < 0.$$

A.3.4 *Variation of Parameters Formula*

Suppose y_1 and y_2 are two linearly independent solutions to

$$y'' + p(t)y' + q(t)y = 0.$$

Then, the general solution to the non-homogeneous equation

$$y'' + p(t)y' + q(t)y = g(t)$$

is given by

$$y = c_1y_1 + c_2y_2 - y_1 \int_0^t \frac{y_2(s)g(s)}{W(y_1, y_2)(s)} ds + y_2 \int_0^t \frac{y_1(s)g(s)}{W(y_1, y_2)(s)} ds,$$

where

$$W(y_1, y_2)(s) = y_1(s)y_2'(s) - y_2(s)y_1'(s).$$

A.4 Linear Algebra

Basic knowledge of Linear Algebra appearing in this book are listed. For more complete treatment of the subject consult [Strang (2009)].

A.4.1 *Solutions to Systems of Linear Equations*

We summarize a few basic facts about the linear algebra. Let A be an $n \times n$ matrix. Concerning the solutions of $A\mathbf{x} = \mathbf{0}$ and $A\mathbf{x} = \mathbf{b}$, we have

1. If $\det A \neq 0$, $A\mathbf{x} = \mathbf{0}$ has a unique solution $\mathbf{x} = \mathbf{0}$.
2. If $\det A = 0$, $A\mathbf{x} = \mathbf{0}$ has infinitely many nontrivial solutions.
3. If $\det A \neq 0$, the inverse matrix A^{-1} exists and $A\mathbf{x} = \mathbf{b}$ has a unique solution $\mathbf{x} = A^{-1}\mathbf{b}$.
4. If $\det A = 0$, $A\mathbf{x} = \mathbf{b}$ has no solution or many solutions. If there are many solutions, the solution can be written as

$$\mathbf{x} = c\mathbf{x}_n + \mathbf{x}_p,$$

where c is a scalar, $\mathbf{x}_n$ is a solution to $A\mathbf{x} = \mathbf{0}$, and $\mathbf{x}_p$ is a solution to $A\mathbf{x} = \mathbf{b}$.

A.4.2 *Eigenvalues, Eigenvectors, and Diagonalization*

The values of λ for which $A\mathbf{x} = \lambda\mathbf{x}$ has a nontrivial solution are called eigenvalues and the corresponding nontrivial solutions $\mathbf{x}$ are called the eigenvectors. To find such λ and $\mathbf{x}$ we rewrite $A\mathbf{x} = \lambda\mathbf{x}$ as $(A - \lambda I)\mathbf{x} = \mathbf{0}$ and compute the following.

1. Solve $\det(A - \lambda I) = 0$ for λ. These λ are the eigenvalues.
2. For each λ in Step 1, find nontrivial $\mathbf{x}$ satisfying $(A - \lambda I)\mathbf{x} = \mathbf{0}$. They are called the eigenvectors of λ.

If λ is not a repeated eigenvalue, there is only one linearly independent eigenvector $\mathbf{x}$ for λ. However, if λ is an eigenvalue repeated k times ($k > 1$), we may or may not have k linearly independent eigenvectors for that λ.

Suppose $\lambda_1, \lambda_2, \ldots, \lambda_n$ are the eigenvalues and $\mathbf{x}_1, \mathbf{x}_2, \ldots, \mathbf{x}_n$ are the corresponding eigenvectors. We assume that they are linearly independent. Then, we have

$$A\mathbf{x}_1 = \lambda_1\mathbf{x}_1,\ A\mathbf{x}_2 = \lambda_2\mathbf{x}_2, \ldots,\ A\mathbf{x}_n = \lambda_n\mathbf{x}_n.$$

They are the column vectors.

$$\begin{aligned}[A\mathbf{x}_1, A\mathbf{x}_2, \ldots, A\mathbf{x}_n] = A[\mathbf{x}_1, \mathbf{x}_2, \ldots, \mathbf{x}_n] &= [\lambda_1\mathbf{x}_1, \lambda_2\mathbf{x}_2, \ldots, \lambda_n\mathbf{x}_n] \\ &= [\mathbf{x}_1, \mathbf{x}_2, \ldots, \mathbf{x}_n]\begin{bmatrix} \lambda_1 & & & \\ & \lambda_2 & & \\ & & \ddots & \\ & & & \lambda_n \end{bmatrix}. \end{aligned} \qquad \text{(A.4.1)}$$

Set $P = [\mathbf{x}_1, \mathbf{x}_2, \ldots, \mathbf{x}_n]$ and

$$\Lambda = \begin{bmatrix} \lambda_1 & & & \\ & \lambda_2 & & \\ & & \ddots & \\ & & & \lambda_n \end{bmatrix}.$$

Then, (A.4.1) is written as

$$A = P\Lambda P^{-1} \text{ or } \Lambda = P^{-1}AP.$$

This process is called the diagonalization and it is possible when there are n linearly independent eigenvectors.

Hints and Solutions to Selected Exercises

Chapter 1

Section 1.1

1. (a) Second order. (b) Third order.

Section 1.2

1. (a) and (d) are linear. (c) Nonlinear. Set $\mathcal{L}u = u_t + uu_x - u_{xx}$. Then,

$$\mathcal{L}(cu) = cu_t + c^2uu_x - cu_{xx} \neq c\mathcal{L}u = c(u_t + uu_x - u_{xx}).$$

3. (a) Differentiation of (1.2.14) in t implies that v satisfies $v_{tt} - c^2v_{xx} = 0$, and $v(x,0) = p(x)$ by (1.2.15). Since $v_t(x,0) = u_{tt}(x,0) = c^2u_{xx}(x,0)$, by (1.2.15) $v_t(x,0) = 0$.

4. (a) Linearly dependent. $f_1 + f_2 - f_3 = 0$.

Section 1.3

1. (b) $u(x,t) = \cos 3(t - \frac{x}{2})$.

4. (c) $u(x,t) = (t^2 - 4x)e^{\frac{1}{2}(t-\sqrt{t^2-4x})}$.

6. $u(x,t) = \frac{x^2}{t}$.

7. (c) $u(x,t) = f(x-t)e^{-t} + e^{-t}\int_0^t g(\xi + x - t, \xi)d\xi$.

8. $u(x,t) = f(x-ct)e^{-rt}$.

9. (b) $u = u_0(xe^{-t+t_0}, t_0)e^{t-t_0} = e^t f(xe^{-t})$.

Section 1.4

1. (a) Set $\xi = x + y$ and $\eta = 3x - y$. Then, $u_{\xi\xi} + u_{\eta\eta} = 0$.

2. (b) $y(y-4x^2) > 0$ hyperbolic. $y = 0$ or $y = 4x^2$ parabolic. $y(y-4x^2) < 0$ elliptic.

Chapter 2

Section 2.1

2. (d) $-k\nabla v = k[\frac{x}{8a^4\pi t^2}\exp(-\frac{x^2+y^2}{4a^2t}), \frac{y}{8a^4\pi t^2}\exp(-\frac{x^2+y^2}{4a^2t})]$.

4. (a) Use $[dx, dy] = [ds\cos\theta, ds\sin\theta]$ and $\mathbf{n} = [\sin\theta, -\cos\theta]$.

Section 2.2

1. (b) The temperature is fixed at T_1 at $x = 0$ and the boundary $x = L$ is insulated.

2. (b) $-ku_x(L,t)$ is positive if $u(L,t) > h(t)$ and negative if $u(L,t) < h(t)$. So, $d > 0$.

Section 2.3

1. (b) Yes. $X' + \mu X = 0$, $Y'' - \lambda Y = 0$, $(1-\mu)T' - \lambda T = 0$.

(c) No. (d) Yes. $\frac{dT}{dt} + \lambda t T = 0$, $\frac{d}{dr}(r\frac{dR}{dr}) + \lambda r R = 0$.

3. (d) $u(x,t) = \sum_{n=1}^{\infty} \frac{2}{(n-\frac{1}{2})\pi}(-1)^{n-1}e^{-(\frac{a(n-\frac{1}{2})\pi}{2})^2 t}\cos\frac{(n-\frac{1}{2})\pi x}{2}$.

5. (a) The boundaries are insulated. Or there is no heat flux (or flow) from the boundaries.
(b) $u(x,t) = \frac{1}{2} + \sum_{n=1}^{\infty}\frac{2}{(n\pi)^2}[1-(-1)^n]e^{-(\frac{an\pi}{2})^2 t}\cos n\pi x$.
(c) Since $\lim_{t\to\infty} e^{-k(n\pi)^2 t} = 0$ for $n = 1, 2, 3, \ldots$, $\lim_{t\to\infty} u(x,t) = 1/2$.

8. (b) $u(x,t) = \frac{L}{2}e^{-bt} + \sum_{n=1}^{\infty}\frac{2}{L}\frac{2}{(\frac{n\pi}{L})^2}[1-(-1)^n]e^{-[b+(\frac{an\pi}{L})^2]t}\cos\frac{n\pi x}{L}$.

Section 2.4

1. (a) $u = -\frac{x^2}{2} + (\frac{T_2-T_1}{L} + \frac{L}{2})x + T_1$. (c) $u(x) = -\frac{T}{L+1}x + T$.

2. (b) $v = \frac{1}{2L}x^2 + x + \frac{1}{L}\int_0^L f(x)dx - \frac{1}{6L}L^2 - \frac{1}{2}L$.

3. (a) $v(x) = x - L$

(b) $w_t = a^2 w_{xx}$, $0 < x < L$, $0 < t$, BC: $w_x(0,t) = 0$, $w(L,t) = 0$, IC: $w(x,0) = L$.
(c) $u(x,t) = \sum_{n=1}^{\infty} \frac{4L}{(2n-1)\pi}(-1)^{n-1} e^{-(\frac{a(2n-1)\pi}{2L})^2 t} \cos\frac{(2n-1)\pi x}{2L} + x - L$.

Section 2.5

3. Multiply $X_m'' - \lambda_m X_m = 0$ by X_n and integrate over $[0, L]$ and vice verse. Then, we have

$$-a_L X_m(L) X_n(L) - \int_0^L (X_m' X_n' + \lambda_m X_m X_n) dx = 0. \quad (1)$$

$$-a_L X_m(L) X_n(L) - \int_0^L (X_m' X_n' + \lambda_n X_m X_n) dx = 0. \quad (2)$$

Subtracting (1) from (2), we obtain

$$(\lambda_m - \lambda_n) \int_0^L X_m X_n dx = 0.$$

5. The eigenvalues are positive and satisfy $\cot \sqrt{\lambda_n} L = \sqrt{\lambda_n}$, $n = 1, 2, \ldots$. The solution is

$$u(x,t) = \sum_{n=1}^{\infty} a_n e^{-a^2 \lambda_n t} (\sqrt{\lambda_n} \cos \sqrt{\lambda_n} x + \sin \sqrt{\lambda_n} x),$$

where

$$a_n = \frac{\int_0^L (\sqrt{\lambda_n} \cos \sqrt{\lambda_n} x + \sin \sqrt{\lambda_n} x) f(x) dx}{\int_0^L (\sqrt{\lambda_n} \cos \sqrt{\lambda_n} x + \sin \sqrt{\lambda_n} x)^2 dx}.$$

Section 2.6

3. $u(x,t) = -a\sqrt{\frac{t}{\pi}}[\exp(-\frac{(x-1)^2}{4a^2 t}) - \exp(-\frac{x^2}{4a^2 t})] + \frac{x}{2}[\text{erf}(\frac{1-x}{2a\sqrt{t}}) - \text{erf}(\frac{-x}{2\sqrt{kt}})]$.
6. Set $z = (y - x)/(2a\sqrt{t})$. Then,

$$u(x,t) = \frac{1}{\sqrt{\pi}} \int_{-\frac{1+x}{2a\sqrt{t}}}^{\frac{1-x}{2a\sqrt{t}}} e^{-z^2} dz = \frac{1}{2}(\text{erf}(\frac{1-x}{2a\sqrt{t}}) + \text{erf}(\frac{1+x}{2a\sqrt{t}})).$$

7. (a) $u(x,t) = \frac{1}{2a\sqrt{\pi t}} \int_0^{\infty} \{e^{-\frac{(x-y)^2}{4a^2 t}} + e^{-\frac{(x+y)^2}{4a^2 t}}\} f(y)) dy$.

Section 2.7

3. (a) Use the maximum principle. Then, $\min\{0, \min_{0\le x\le 1} u(x,0)\} \le u(x,t) \le \max_{0\le x\le 1} u(x,0)$.

5. $\frac{1}{2}\int_0^L u^2(x,T)dx + a^2\int_0^T\int_0^L u_x^2(x,t)dxdt + a_0a^2\int_0^T u^2(0,t)dt$
$+a_La^2\int_0^T u^2(L,t)dt = \frac{1}{2}\int_0^L f^2(x)dx$.

Chapter 3

Section 3.1

1. $u_{tt} = c^2u_{xx} + h(x,t)$.

Section 3.2

1. (a) Region (1): $x + \frac{1}{2}t < -1$, $u(x,t) = 0$.
Region (2) $-1 < x + \frac{1}{2}t < 1$ and $x - \frac{1}{2}t < -1$, $u(x,t) = \frac{1}{2}[1 - (x + \frac{1}{2}t)^2]$.
Region (3) $-1 < x - \frac{1}{2}t$ and $x + \frac{1}{2}t < 1$,
$u(x,t) = \frac{1}{2}[1 - (x - \frac{1}{2}t)^2 + 1 - (x + \frac{1}{2}t)^2]$.
Region (4) $1 < x + \frac{1}{2}t$ and $-1 < x - \frac{1}{2}t < 1$, $u(x,t) = \frac{1}{2}[1 - (x - \frac{1}{2}t)^2]$
Region (5) $1 < x - \frac{1}{2}t$, $u(x,t) = 0$.
Region (6) $1 < x + \frac{1}{2}t$ and $x - \frac{1}{2}t < -1$, $u(x,t) = 0$.

2. (a) Region (1) $x + \frac{1}{2}t < -1$, $u(x,t) = 0$.
Region (2) $-1 < x + \frac{1}{2}t < 1$ and $x - \frac{1}{2}t < -1$, $u(x,t) = \int_{-1}^{x+\frac{1}{2}t}(1 - s^2)ds$.
Region (3) $-1 < x - \frac{1}{2}t$ and $x + \frac{1}{2}t < 1$, $u(x,t) = \int_{x-\frac{1}{2}t}^{x+\frac{1}{2}t}(1 - s^2)ds$.
Region (4) $1 < x + \frac{1}{2}t$ and $-1 < x - \frac{1}{2}t < 1$, $u(x,t) = \int_{x-\frac{1}{2}t}^{1}(1 - s^2)ds$.
Region (5) $1 < x - \frac{1}{2}t$, $u(x,t) = 0$.
Region (6) $1 < x + \frac{1}{2}t$ and $x - \frac{1}{2}t < -1$, $u(x,t) = \int_{-1}^{1}(1 - s^2)ds = \frac{4}{3}$.

6. $u(x,t) = [\sin(x + ct) - \sin(x - ct)]/(2c) + t + e^{-t} - 1$.

Section 3.3

2. (a) Region (1): $0 < x < -\frac{1}{2}t + 1$, $u(x,t) = 0$.
Region (2): $-\frac{1}{2}t + 1 < x < -\frac{1}{2}t + 3$ and $\frac{1}{2}t - 1 < x < \frac{1}{2}t + 1$,
$u(x,t) = \frac{1}{2}[0 + 1 - (x + \frac{1}{2}t - 2)^2]$.

Region (3): $\frac{1}{2}t+1 < x < -\frac{1}{2}t+3$,
$u(x,t) = \frac{1}{2}[1-(x-\frac{1}{2}t-2)^2+1-(x+\frac{1}{2}t-2)^2]$

Region (4): $\frac{1}{2}t+2 < x < \frac{1}{2}t+3$ and $-\frac{1}{2}t+3 < x$,
$u(x,t) = \frac{1}{2}[1-(x-\frac{1}{2}t-2)^2+0]$.

Region (5): $\frac{1}{2}t+3 < x$, $u(x,t)=0$.

Region (6): $\frac{1}{2}t-1 < x < \frac{1}{2}t+1$ and $-\frac{1}{2}t+3 < x$, $u(x,t)=0$.

Region (7): $0 < x < \frac{1}{2}t-1$ and $0 < x < -\frac{1}{2}t+3$,
$u(x,t) = \frac{1}{2}[1-(x-\frac{1}{2}t-2)^2+1-(x+\frac{1}{2}t-2)^2]$.

Region (8): $\frac{1}{2}t-3 < x < \frac{1}{2}t-1$ and $-\frac{1}{2}t+3 < x$,
$u(x,t) = \frac{1}{2}[1-(x-\frac{1}{2}t-2)^2+0]$.

Region (9): $0 < x < \frac{1}{2}t-3$, $u(x,t)=0$.

5. For $ct \le x$, the solution is the same as the initial value problem and therefore,

$$u(x,t) = F(x+ct)+G(x-ct) = \frac{1}{2}[f(x-ct)+f(x+ct)]+\frac{1}{2c}\int_{x-ct}^{x+ct} g(s)ds. \tag{3}$$

For $x < ct$, we need to find the expression for $G(x-ct)$. Since $u_x(x,t) = F'(x+ct)+G'(x-ct)$, on the boundary $x=0$ we have

$$u_x(0,t) = F'(ct)+G'(-ct) = h(t).$$

Integrating in t, we obtain

$$\frac{1}{c}F(ct) - \frac{1}{c}G(-ct) = \int_0^t h(s)ds + C,$$

Then,

$$\frac{1}{c}F(ct-x) - \frac{1}{c}G(x-ct) = \int_0^{(ct-x)/c} h(s)ds + C.$$

Therefore, we have

$$G(x-ct) = \frac{1}{2}[f(ct-x)+\frac{1}{c}\int_0^{ct-x} g(s)ds] - c\int_0^{(ct-x)/c} h(s)ds - cC.$$

Hence, for $x-ct<0$,

$$u(x,t) = \frac{1}{2}[f(x+ct)+f(ct-x)] + \frac{1}{2c}[\int_0^{x+ct} g(s)ds + \int_0^{ct-x} g(s)ds]$$
$$-c\int_0^{(ct-x)/c} h(s)ds + \frac{1}{2}K - cC. \tag{4}$$

We require that $u(0,0)$ in (3) and (4) are equal. Then, $K/2 - cC = 0$.

Section 3.4

2. (b) $a_m = \frac{2}{L}\int_0^L f(x)\sin\frac{(m-\frac{1}{2})\pi x}{L}dx$,
$b_m = \frac{L}{(m-\frac{1}{2})\pi c}\frac{2}{L}\int_0^L g(x)\sin\frac{(m-\frac{1}{2})\pi x}{L}dx$.

3. (b) $u(x,t) = \sum_{n=1}^{\infty}\frac{2}{L}(\frac{L}{(n-\frac{1}{2})\pi c})^2[(-1)^{n-1}-1]\sin\frac{(n-\frac{1}{2})\pi ct}{L}\sin\frac{(n-\frac{1}{2})\pi x}{L}$.

5. (a) $r^2 < 4\pi^2c^2/L^2$.

$$u(x,t) = \sum_{n=1}^{\infty}[a_n e^{-\frac{r}{2}t}\cos(t\sqrt{4\lambda_n c^2 - r^2})$$
$$+b_n e^{-\frac{r}{2}t}\sin(t\sqrt{4\lambda_n c^2 - r^2})]\sin\frac{n\pi x}{L},$$

where

$$a_n = \frac{2}{L}\int_0^L f(x)\sin\frac{n\pi x}{L}dx,$$

$$b_n = [\frac{2}{L}\int_0^L g(x)\sin\frac{\pi x}{L}dx + \frac{r}{2}a_n]/\sqrt{4\lambda_n c^2 - r^2}, \quad n = 1, 2, \ldots.$$

(b) $4\pi^2c^2/L^2 < r^2 < 16\pi^2c^2/L^2$. Set $s_{n\pm} = \frac{1}{2}(-r \pm \sqrt{r^2 - 4\lambda_n c^2})$.

$$u(x,t) = [a_1 e^{s_{1+}t} + b_1 e^{s_{1-}t}]\sin\frac{\pi x}{L}$$
$$+\sum_{n=2}^{\infty}[a_n e^{-\frac{r}{2}t}\cos(t\sqrt{4\lambda_n c^2 - r^2})$$
$$+b_n e^{-\frac{r}{2}t}\sin(t\sqrt{4\lambda_n c^2 - r^2})]\sin\frac{n\pi x}{L},$$

where

$$\begin{bmatrix} 1 & 1 \\ s_{1+} & s_{1-} \end{bmatrix}\begin{bmatrix} a_1 \\ b_1 \end{bmatrix} = \begin{bmatrix} \frac{2}{L}\int_0^L f(x)\sin\frac{\pi x}{L}dx, \\ \frac{2}{L}\int_0^L g(x)\sin\frac{\pi x}{L}dx \end{bmatrix},$$

$$a_n = \frac{2}{L}\int_0^L f(x)\sin\frac{n\pi x}{L}dx,$$

$$b_n = [\frac{2}{L}\int_0^L g(x)\sin\frac{\pi x}{L}dx + \frac{r}{2}a_n]/\sqrt{4\lambda_n c^2 - r^2}, \quad n = 1, 2, \ldots..$$

Section 3.5

2. (b) Suppose u and v are the solutions to (3.5.7) to (3.5.9). Then the difference $w = u - v$ satisfies

$$\int_0^L [w_t^2(x,t) + c^2 w_x^2(x,t)]dx + c^2[w^2(L,t) + w^2(0,t)] = 0.$$

Chapter 4

Section 4.1

1. Substituting $\mathbf{B} = \nabla \times \mathbf{A}$ in (4.1.1), we have

$$\nabla \times (\frac{\partial \mathbf{A}}{\partial t} + c\mathbf{E}) = 0.$$

Therefore, $\mathbf{E} = -\frac{1}{c}\frac{\partial \mathbf{A}}{\partial t} +$ const. Substitute this in (4.1.2) and use $\nabla \times \mathbf{B} = \nabla \times \nabla \times \mathbf{A} = -\triangle \mathbf{A}$. Then, (4.1.2) is rewritten as

$$\frac{1}{c}\frac{\partial^2 \mathbf{A}}{\partial t^2} - c\triangle \mathbf{A} = 4\pi \mathbf{J}.$$

Section 4.2

3. (a) $u(x,y) = \sum_{n=1}^{\infty} a_n \sinh \frac{n\pi(x-L)}{H} \sin \frac{n\pi y}{H}$,
$a_n = -\frac{H}{2\sinh \frac{n\pi L}{H}} \int_0^H f_2(y) \sin \frac{n\pi y}{H} dy.$

5. Assume $u(r,\theta) = R(r)\Theta(\theta)$. Then, $\Theta(\theta)$ satisfies

$$\frac{d^2\Theta}{d\theta^2} + \lambda\Theta = 0, \quad \frac{d\Theta}{d\theta}(0) = \frac{d\Theta}{d\theta}(\frac{\pi}{2}) = 0.$$

So, the eigenvalues and eigenfunctions are

$$\lambda = 4n^2, \quad \phi = \cos 2n\theta, \quad n = 0, 1, 2, \ldots.$$

$R(r)$ satisfies

$$r^2\frac{d^2R}{dr^2} + r\frac{dR}{dr} - 4n^2R = 0, \quad R = c_1r^{2n}, \quad n = 0, 1, 2, \ldots.$$

The solution is

$$u(r,\theta) = \sum_{n=0}^{\infty} a_n r^{2n} \cos 2n\theta,$$

where

$$a_0 = \frac{2}{\pi}\int_0^{\pi/2} f(\theta)d\theta. \quad a_n = \frac{4}{\pi}\int_0^{\pi/2} f(\theta)\cos 2n\theta d\theta, \quad n = 1, 2, \ldots.$$

6. (a) Solvability condition: $\int_{-\pi}^{\pi}\{bg(\theta) - af(\theta)\}d\theta = 0$. One way to solve the problem is to decompose the problem into two sub problems $u = u_1 + u_2$, where

$$(A): \nabla^2 u_1 = 0, \quad \frac{\partial u_1}{\partial r}(a,\theta) = 0, \quad \frac{\partial u_1}{\partial r}(b,\theta) = g(\theta).$$

$$(B): \nabla^2 u_2 = 0, \quad \frac{\partial u_2}{\partial r}(a,\theta) = f(\theta), \quad \frac{\partial u_2}{\partial r}(b,\theta) = 0.$$

The solution to (A) is

$$u_1(r,\theta) = a_0 + \sum_{n=1}^{\infty}(r^n + a^{2n}r^{-n})(a_n \cos n\theta + a_n \sin n\theta),$$

where

$$a_n = \frac{1}{n(b^{n-1} - a^{2n}b^{-n-1})\pi}\int_{-\pi}^{\pi} g(\theta)\cos n\theta d\theta,$$

$$b_n = \frac{1}{n(b^{n-1} - a^{2n}b^{-n-1})\pi}\int_{-\pi}^{\pi} g(\theta)\sin n\theta d\theta, \quad n = 1, 2, \ldots.$$

The solution to (B) is given by

$$u_2(r,\theta) = c_0 + \sum_{n=1}^{\infty}(r^n + b^{2n}r^{-n})(c_n \cos n\theta + d_n \sin n\theta),$$

where

$$c_n = \frac{1}{n(a^{n-1} + b^{2n}a^{-n-1})\pi}\int_{-\pi}^{\pi} f(\theta)\cos n\theta d\theta,$$

$$d_n = \frac{1}{n(a^{n-1} + b^{2n}a^{-n-1})\pi}\int_{-\pi}^{\pi} f(\theta)\sin n\theta d\theta \quad n = 1, 2, \ldots.$$

Note that a_0 and c_0 are arbitrary.

Section 4.3

2. $\varepsilon = \sqrt{x^2+y^2+z^2}$. $\mathbf{n}\cdot\nabla K = -\frac{[x,y,z]}{\sqrt{x^2+y^2+z^2}} \cdot \frac{[x,y,z]}{\sqrt{x^2+y^2+z^2}} K'(\varepsilon) = -K'(\varepsilon)$.

Section 4.4

3. If $n = 2$,

$$G(\mathbf{x}, \mathbf{x}_0) = \frac{\ln|\mathbf{x}_0 - \mathbf{x}|}{2\pi} - \frac{\ln|\frac{\mathbf{x}_0}{a}||\mathbf{x}_0^* - \mathbf{x}|}{2\pi}.$$

$$\frac{\partial G}{\partial x_i} = \frac{1}{2\pi}\{\frac{x_i - x_{0i}}{|\mathbf{x}_0 - \mathbf{x}|^2} - (\frac{|\mathbf{x}_0|}{a})^2\frac{x_i - x_{0i}^*}{|\mathbf{x}_0^* - \mathbf{x}|^2}\}.$$

Therefore, by (4.4.11) and (4.4.12),

$$\frac{\partial G}{\partial x_i}|_{\mathbf{x}\in\partial B(\mathbf{0},a)} = \frac{1}{2\pi|\mathbf{x}_0 - \mathbf{x}|^2}\{1 - (\frac{|\mathbf{x}_0|}{a})^2\}x_i.$$

The rest is carried out in the same way as $n = 3$ and $H(y, x)$ has the same form.

4. From (4.4.15) and (4.4.16)

$$
\begin{aligned}
u(\mathbf{x}_0) &= u(r,\theta) \\
&= \int_{|\mathbf{x}|=a} \frac{1}{2\pi|\mathbf{x}_0-\mathbf{x}|^2}\frac{(a^2-|\mathbf{x}_0|^2)}{a} g(\mathbf{x})ds \\
&= \frac{1}{2\pi}\int_0^{2\pi} \frac{g(a,\theta)}{(r\cos\theta - a\cos\phi)^2 + (r\sin\theta - a\sin\phi)^2}\frac{(a^2-r^2)}{a} a d\phi.
\end{aligned}
$$

Section 4.5

1. Maximum: 2, Minimum: 0.

2. (a) The maximum is 1 and the minimum is -1.

(b) From the mean value theorem, it is zero.

Section 4.6

3. (a) $u(x,t) = \varepsilon \cosh \frac{n\pi ct}{L} \cos \frac{n\pi x}{L}$ (b) $\varepsilon \cosh \frac{n\pi ct}{L}$ (c) $n \geq \frac{L}{\pi T} \cosh^{-1}(\frac{\Delta}{\varepsilon})$

Chapter 5

Section 5.1

3. (a) $u = \frac{\alpha x}{\alpha[1-e^{-t}]+1} e^{-t}$

(b) $\alpha[1-e^{-t}] + 1 = 0 \Rightarrow t = \ln(\frac{\alpha}{\alpha+1})$

4. (b) $u(x,t) = \begin{cases} 0, & x \leq 0, \\ \frac{2x}{(1+\sqrt{1+4xt})}, & 0 < x \leq 1+t, \\ 1, & 1+t < x. \end{cases}$

Section 5.2

1. $u(x,t) = \begin{cases} 0, & x < 1/2, \\ 1 & x > 1/2. \end{cases}$

3. (c) $u(x,t) = \begin{cases} 1, & x \leq t, \\ \frac{x+1}{1+t}, & t < x \leq 1+2t, \\ 2, & 1+2t < x. \end{cases}$

(d) For $t < 1$ we have

$$u(x,t) = \begin{cases} 2, & x \le 2t, \\ \frac{2-x}{1-t}, & 2t < x \le 1+t, \\ 1, & 1+t < x. \end{cases}$$

For $t \ge 1$ we have a discontinuous solution. By the Rankine-Hugoniot condition, the speed of the shock is

$$\sigma = \frac{\frac{1}{2}(2^2 - 1)}{2 - 1} = \frac{3}{2}.$$

The shock originates at $(x, t) = (2, 1)$. So, the location of the shock is $x = \frac{3}{2}t + \frac{1}{2}$.

$$u(x,t) = \begin{cases} 2, & x \le \frac{3}{2}t + \frac{1}{2}, \\ 1, & \frac{3}{2}t + \frac{1}{2} < x. \end{cases}$$

4. (b) For $t < 1$

$$u(x,t) = \begin{cases} 0, & x \le 0, \\ \frac{2x}{1+\sqrt{1-4xt}}, & 0 < x \le 1-t, \\ 1, & 1-t < x. \end{cases}$$

For $t \ge 1$ a shock develops.

$$u(x,t) = \begin{cases} 0, & x < \frac{1}{3}(-t+1), \\ 1, & \frac{1}{3}(-t+1) < x. \end{cases}$$

Section 5.3

1. $u = \frac{1}{2}(1 - x^2) \pm y$.

2. (a) $u = e^{1-\sqrt{x^2+t^2}}$.

Section 5.4

1. (a)$u = x - t$.

2. $u = 2(1 - 27t)(\frac{x}{1-9t})^{3/2}$.

3. (a) $b = 1 - \gamma$. $a' - rba - \frac{\gamma}{1-\gamma} b^{1-\frac{1}{\gamma}} a^{1-\frac{1}{\gamma}} = 0$.

(b) $V(x,\eta) = a(\eta)x^b = \{\frac{\gamma}{r(1-\gamma)^{1+\gamma}}[e^{r(\frac{1}{\gamma}-1)\eta} - 1] + e^{r(\frac{1}{\gamma}-1)\eta}\}^\gamma x^{1-\gamma}$.

Section 5.5

1. $\begin{bmatrix} u \\ v \end{bmatrix} = \begin{bmatrix} 1 & 1 \\ -4 & 1 \end{bmatrix} \frac{1}{5} \begin{bmatrix} \cos(x+3t) - \sin(x+3t) \\ 4\cos(x-2t) + \sin(x-2t) \end{bmatrix}$.

Chapter 6

Section 6.1

1. (a) Since $f(-x) = \cos(-2x)\sin(-3x) = -\cos 2x \sin 3x$, it is odd.

2. (a) The integrand is odd. So, $\int_{-3}^{3} \cos^2 2x \sin 3x dx = 0$.

3. (a) The fundamental period of tangent is π. Set $\frac{n\pi(x+T)}{L} = \frac{n\pi x}{L} + \pi$. Then, $T = \frac{L}{n}$.

Section 6.2

1. (a) $L = 1$. Since $f(x)$ is even, $b_n = 0$ and

$$f(x) = \frac{1}{2} + \sum_{n=1}^{\infty} \frac{2}{(n\pi)^2}[1 - (-1)^n] \cos(n\pi x).$$

2. (a) The Fourier cosine series is the same as the previous problem. The Fourier sine series is

$$f(x) = \sum_{n=1}^{\infty} \frac{2}{n\pi} \sin(n\pi x).$$

Section 6.3

1. (b) $S_N(x)$ converges pointwise to $f(x) = 0$, but does not converge uniformly nor in the mean square sense to $f(x) = 0$.

4. (b) The solution is given in Figure 1.

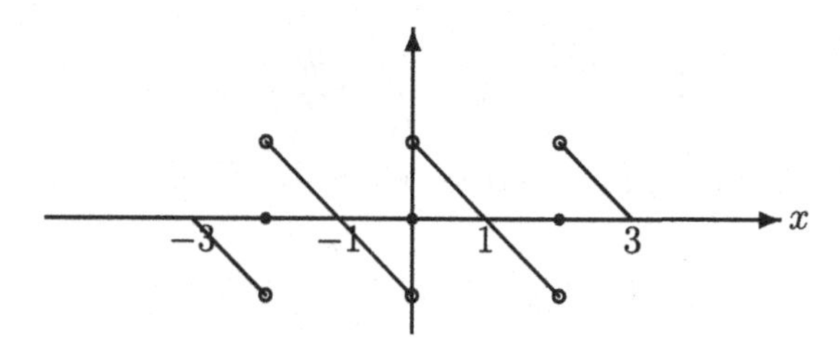

Fig. 1 The Fourier cosine series of $f(x) = \begin{cases} -1 - x, & -1 < x < 0, \\ 1 - x, & 0 < x < 1. \end{cases}$

5. (a) The solution is given in Figure 2.

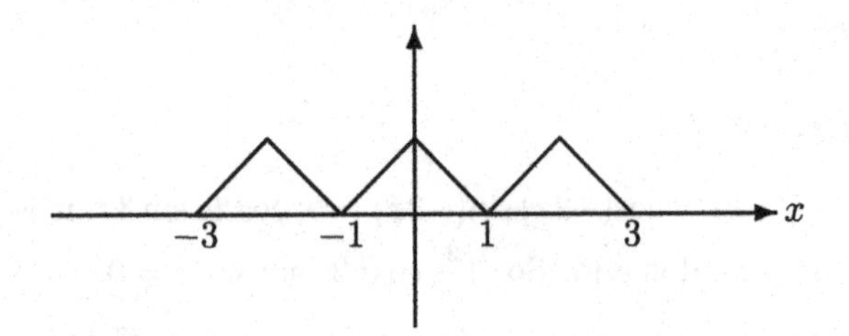

Fig. 2 The Fourier cosine series of $f(x) = 1 - x$ on $0 < x < 1$.

(b) The same as Figure 1.

6. (a) $b_n = \frac{2}{\pi}\int_0^\pi \sin(nx)dx = \frac{2}{\pi}[-\frac{1}{n}\cos(nx)]_0^\pi = \frac{2}{n\pi}[1-(-1)^n]$.

(b) $\sum_{n=1}^\infty b_n^2 \int_0^\pi \sin^2 nx dx = \sum \frac{16}{(2n-1)^2\pi^2}\frac{\pi}{2} = \int_0^\pi 1^2 dx = \pi$.

Section 6.4

2. (c) Yes. The function to which the Fourier cosine series converges is continuous.

(d) No. The function to which the Fourier sine series converges is not continuous.

Section 6.5

1. $\mu(x) = \exp(\int a(x)dx)$

3. (c) $u(x,t) = \exp[-(\frac{an\pi}{2L})^2 t + \frac{c(2x-ct)}{4a^2}]\sin\frac{n\pi x}{L}$

5. (a)

$$x\frac{d^2\phi}{dx^2} + \frac{d\phi}{dx} + \frac{\lambda}{x}\phi = \frac{d}{dx}(x\frac{d\phi}{dx}) + \frac{\lambda}{x}\phi = 0. \tag{5}$$

(b) Multiply (5) by ϕ and integrate. Then,

$$\int_1^b x(\frac{d\phi}{dx})^2 dx = \lambda\int_1^b \frac{1}{x}\phi^2 dx.$$

(c) Set $y = \ln x$ and change the independent variable. Then,

$$\frac{d^2\phi}{dy^2} + \lambda\phi = 0,$$

$$\frac{d\phi}{dy}(0) = \frac{d\phi}{dy}(\ln b) = 0,\ 0 < y < \ln b.$$

This is a usual eigenvalue problem.

$$\lambda_n = (\frac{n\pi}{\ln b})^2, \quad \phi_n = \cos\frac{n\pi \ln x}{\ln b}, \quad n = 0, 1, 2, \ldots.$$

(d) $\int_1^b \frac{1}{x} \cos\frac{m\pi \ln x}{\ln b} \cos\frac{n\pi \ln x}{\ln b} dx = \int_0^{\ln b} \cos\frac{m\pi y}{\ln b} \cos\frac{n\pi y}{\ln b} dy = \begin{cases} 0, & m \neq n, \\ 1, & m = n. \end{cases}$

(e) This should be clear from the form of $\phi_n = \cos(n\pi \ln x / \ln b)$.

Chapter 7

Section 7.1

1. The solution is given by

$$u(x,y,t) = \sum_{n=1}^{\infty} \frac{a_{0n}}{2} e^{-a^2\lambda_{0n}t} \sin\frac{n\pi y}{H} + \sum_{m=0}^{\infty}\sum_{n=1}^{\infty} a_{mn} e^{-a^2\lambda_{mn}t} \cos\frac{m\pi x}{L} \sin\frac{n\pi y}{H},$$

$$\lambda_{mn} = (\frac{m\pi}{L})^2 + (\frac{n\pi}{H})^2,$$

$$a_{mn} = \frac{4}{LH}\int_0^H \int_0^L f(x,y) \cos\frac{m\pi x}{L} \sin\frac{n\pi y}{H} dx dy, \quad m = 0, 1, 2, \ldots, \ n = 1, 2, \ldots.$$

3. (a) We solve two sub-problems.

$$\Delta u = u_{xx} + u_{yy} + u_{zz} = 0,$$

$$u_x(0,y,z) = 0, \quad u(x,0,z) = 0, \quad u(x,y,0) = f(x,y),$$
$$u_x(L,y,z) = 0, \quad u_y(x,W,z) = 0, \quad u(x,y,H) = 0$$

and

$$\Delta u = u_{xx} + u_{yy} + u_{zz} = 0,$$

$$u_x(0,y,z) = 0, \quad u(x,0,z) = 0, \quad u(x,y,0) = 0,$$
$$u_x(L,y,z) = 0, \quad u_y(x,W,z) = g(x,z), \quad u(x,y,H) = 0.$$

The solution is the sum of the solutions to the above sub-problems.

4. The solution is given by

$$
\begin{aligned}
u(x,y,z,t) \\
&= \sum_{n=1}^{\infty}\sum_{l=1}^{\infty}(\frac{a_{0nl}}{2}\cos c\sqrt{\lambda_{0nl}}t + \frac{b_{0nl}}{2}\sin c\sqrt{\lambda_{0nl}}t)\sin\frac{n\pi y}{W}\sin\frac{(l-\frac{1}{2})\pi z}{H} \\
&+ \sum_{m=1}^{\infty}\sum_{n=1}^{\infty}\sum_{l=1}^{\infty}(a_{mnl}\cos c\sqrt{\lambda_{mnl}}t + b_{mnl}\sin c\sqrt{\lambda_{mnl}}t) \\
&\cos\frac{m\pi x}{L}\sin\frac{n\pi y}{W}\sin\frac{(l-\frac{1}{2})\pi z}{H},
\end{aligned}
$$

where

$$
\lambda_{mnl} = (\frac{m\pi}{L})^2 + (\frac{n\pi}{W})^2 + (\frac{(l-\frac{1}{2})\pi z}{H})^2,
$$

$$
\begin{aligned}
a_{mnl} \\
&= \frac{8}{LWH}\int_0^H\int_0^W\int_0^L f(x,y,z)\cos\frac{m\pi x}{L}\sin\frac{n\pi y}{H}\sin\frac{(l-\frac{1}{2})\pi z}{H}dxdydz,
\end{aligned}
$$

$$
\begin{aligned}
b_{mnl} &= \frac{1}{c\sqrt{\lambda_{mn}}}\frac{8}{LWH}\int_0^H\int_0^W\int_0^L g(x,y,z) \\
&\cos\frac{m\pi x}{L}\sin\frac{n\pi y}{H}\sin\frac{(l-\frac{1}{2})\pi z}{H}dxdydz,
\end{aligned}
$$

$$
m = 0,1,\ldots,\ n = 1,2,\ldots,\ l = 1,2,\ldots.
$$

Section 7.2

1. $\phi_2 = \sin(x/2) - 2/\pi$.

4. The Rayleigh quotient is

$$
\lambda = \frac{\int_0^L[(1+x^2)(u')^2 + x^2u^2]dx}{\int_0^L u^2dx}.
$$

Choose $u = 1$. Then, $\lambda = L^2/3$.

5. (a) $\lambda\int_0^L u^2dx = \int_0^L[(1+x^2)(\frac{du}{dx})^2 + x^2u^2]dx$. (b) No.

Section 7.3

1. (a) Assume $u = v(x,t) + r(x,t)$, where

$$r(x,t) = b_l(t)(x - L) + b_r(t).$$

For v assume

$$v(x,t) = \sum_{n=0}^{\infty} a_n(t) \cos \frac{(n-\frac{1}{2})\pi x}{L}.$$

Then,

$$a_n(t) = a_n(0)e^{-k(\frac{(n-\frac{1}{2})\pi}{L})^2 t} + e^{-k(\frac{(n-\frac{1}{2})\pi}{L})^2 t} \int_0^t h_n(s) e^{k(\frac{(n-\frac{1}{2})\pi}{L})^2 s} ds,$$

$$a_n(0) = \frac{2}{L} \int_0^L \{f(x) - b_l(0)(x-L) - b_r(0)\} \cos \frac{(n-\frac{1}{2})\pi x}{L} dx.$$

(b) Assume

$$u = \sum_{n=1}^{\infty} a_n(t) \cos \frac{(n-\frac{1}{2})\pi x}{L}.$$

Then,

$$\begin{aligned} a_n(t) &= a_n(0)e^{-k(\frac{(n-\frac{1}{2})\pi}{L})^2 t} \\ &\quad + e^{-k(\frac{(n-\frac{1}{2})\pi}{L})^2 t} \int_0^t \left\{ \frac{2k}{L}[(-1)^{n-1} \frac{(n-\frac{1}{2})\pi}{L} b_r(t) - b_l(t)] \right. \\ &\qquad \left. + \frac{2}{L} h_n(t) \right\} e^{k(\frac{(n-\frac{1}{2})\pi}{L})^2 s} ds, \end{aligned}$$

$$a_n(0) = \frac{2}{L} \int_0^L f(x) \cos \frac{(n-\frac{1}{2})\pi x}{L} dx.$$

2. (a) Decompose into two problems $u = v + w$, where

$$\nabla^2 u = h, \quad u(0,y) = 0, u(L,y) = 0, u(x,0) = 0, u_y(x,H) = 0, \tag{6}$$

$$\nabla^2 w = 0, \quad w(0,y) = 0, w(L,y) = 0, w(x,0) = 0, w_y(x,H) = 1. \tag{7}$$

For (6), by the eigenfunction expansion, assume

$$v = \sum_{m=1}^{\infty} \sum_{n=1}^{\infty} a_{mn} \sin \frac{(m-\frac{1}{2})\pi y}{H} \sin \frac{n\pi x}{L}.$$

Then,

$$a_{mn} = -\frac{4}{LH\{(\frac{(m-\frac{1}{2})\pi}{H})^2 + (\frac{n\pi}{L})^2\}} \int_0^L \int_0^H h \sin\frac{(m-\frac{1}{2})\pi y}{H} \sin\frac{n\pi x}{L} dydx.$$

For (7), we use Section 4.2. Then, we have

$$w = \sum_{n=1}^{\infty} b_n \sinh\frac{n\pi y}{L} \sin\frac{n\pi x}{L},$$

$$b_n = \frac{2}{n\pi \cosh\frac{n\pi L}{H}} \int_0^L \sin\frac{n\pi x}{L} dx = \frac{2(1-\cos n\pi)L}{(n\pi)^2 \sinh\frac{n\pi L}{H}}.$$

(b) We use the eigenfunction expansion.

$$u = \sum_{m=1}^{\infty}\sum_{n=1}^{\infty} a_{mn}\phi_{mn} = \sum_{m=1}^{\infty}\sum_{n=1}^{\infty} a_{mn} \sin\frac{(m-\frac{1}{2})\pi y}{H} \sin\frac{n\pi x}{L},$$

where ϕ_{mn} are the eigenfunctions for $\nabla^2\phi = -\lambda\phi$ satisfying the corresponding homogeneous boundary condition. Then, by the Green's identity

$$\begin{aligned}
a_{mn} &= \frac{4}{LH}\int_0^L \int_0^H u\phi_{mn} dydx \\
&= -\frac{4}{LH\{(\frac{(m-\frac{1}{2})\pi}{H})^2 + (\frac{n\pi}{L})^2\}} \int_0^L \int_0^H u\nabla^2\phi_{mn} dydx \\
&= -\frac{4}{LH\{(\frac{(m-\frac{1}{2})\pi}{H})^2 + (\frac{n\pi}{L})^2\}} [\int_0^L \int_0^H h(x,t)\phi_{mn} dydx \\
&\quad + \int_{\partial\Omega} (u\nabla\phi_{mn} - \phi_{mn}\nabla u)\cdot \mathbf{n} ds] \\
&= -\frac{4}{LH\{(\frac{(m-\frac{1}{2})\pi}{H})^2 + (\frac{n\pi}{L})^2\}} [\int_0^L \int_0^H h(x,t)\phi_{mn} dydx \\
&\quad + \frac{L}{n\pi}\sin(m-\frac{1}{2})\pi(\cos n\pi - 1)].
\end{aligned}$$

Chapter 8

Section 8.1

3. (a) $u(r,\theta,t) = \sum_{n=0}^{\infty}\sum_{m=1}^{\infty} b_{nm} e^{-\alpha^2\lambda_{nm}t} J_n(\sqrt{\lambda_{nm}}r)\sin n\theta,$

where λ_{nm} is the mth root of $J_n(\sqrt{\lambda}a)=0$ and

$$b_{nm}=\frac{\int_0^\pi\int_0^a f(r,\theta)J_n(\sqrt{\lambda_{nm}}r)\sin n\theta drd\theta}{\int_0^\pi\int_0^a [J_n(\sqrt{\lambda_{nm}}r)\sin n\theta]^2 drd\theta}.$$

4. (b) The solution is given by

$$u(r,\theta,t)=\sum_{n=1}^{\infty}\sum_{m=1}^{\infty}(a_{nm}\cos c\sqrt{\lambda_{nm}}t+b_{nm}\sin c\sqrt{\lambda_{nm}}t)J_n(\sqrt{\lambda_{nm}}r)\sin n\theta,$$

where λ_{nm} is the mth root of $J_n'(\sqrt{\lambda}a)=0$ and

$$a_{nm}=\frac{\int_0^\pi\int_0^a f(r,\theta)J_n(\sqrt{\lambda_{nm}}r)\sin n\theta drd\theta}{\int_0^\pi\int_0^a [J_n(\sqrt{\lambda_{nm}}r)\sin n\theta]^2 drd\theta},$$

$$b_{nm}=\frac{\int_0^\pi\int_0^a g(r,\theta)J_n(\sqrt{\lambda_{nm}}r)\sin n\theta drd\theta}{c\sqrt{\lambda_{nm}}\int_0^\pi\int_0^a [J_n(\sqrt{\lambda_{nm}}r)\sin n\theta]^2 drd\theta}.$$

Section 8.2

2. (a) The solution is given by

$$u(r,\theta,z)=\sum_{n=0}^{\infty}\sum_{m=0}^{\infty}I_n(\sqrt{\lambda_{nm}}r)(a_{nm}\cos n\theta+b_{nm}\sin n\theta)\cos\frac{m\pi z}{H},$$

where $b_{0m}=0$ and

$$a_{nm}=\frac{\int_{-\pi}^{\pi}\int_0^H h(\theta,z)\cos n\theta\cos\frac{m\pi z}{H}dzd\theta}{I_n(\sqrt{\lambda_{nm}}a)\int_{-\pi}^{\pi}\int_0^H(\cos n\theta\cos\frac{m\pi z}{H})^2dzd\theta},$$

$$b_{nm}=\frac{\int_{-\pi}^{\pi}\int_0^H h(\theta,z)\sin n\theta\cos\frac{m\pi z}{H}dzd\theta}{I_n(\sqrt{\lambda_{nm}}a)\int_{-\pi}^{\pi}\int_0^H(\sin n\theta\cos\frac{m\pi z}{H})^2dzd\theta},\quad n=1,2,\ldots.$$

3. (a) With the eigenfunctions

$$v_{nml}^c=J_n(\sqrt{\lambda_{nml}}r)\cos n\theta\cos\frac{m\pi z}{H},\quad v_{nml}^s=J_n(\sqrt{\lambda_{nml}}r)\sin n\theta\cos\frac{m\pi z}{H},$$

the solution is given by

$$u(r,\theta,z,t)=\sum_{n=0}^{\infty}\sum_{m=0}^{\infty}\sum_{l=1}^{\infty}e^{-\lambda_{nml}\alpha^2t}(a_{nml}v_{nml}^c+b_{nml}v_{nml}^s),$$

where $b_{0ml}=0$, $\lambda_{nml}=(\frac{s_{nl}}{a})^2+(\frac{m\pi}{H})^2$, and

$$a_{nml}=\frac{\int_0^H\int_{-\pi}^{\pi}\int_0^a f(r,\theta,z)v_{nml}^c rdrd\theta dz}{\int_0^H\int_{-\pi}^{\pi}\int_0^a (v_{nml}^c)^2 rdrd\theta dz},$$

$$b_{nml}=\frac{\int_0^H\int_{-\pi}^{\pi}\int_0^a f(r,\theta,z)v_{nml}^s rdrd\theta dz}{\int_0^H\int_{-\pi}^{\pi}\int_0^a (v_{nml}^s)^2 rdrd\theta dz}.$$

Section 8.3

5. (a) The values of λ_{nk} are chosen so that $J_{l+\frac{1}{2}}(\sqrt{\lambda_{nk}}a) = 0$, $(k = 1, 2, \ldots)$. Then,

$$u = \sum_{n=0}^{\infty}\sum_{l=n}^{\infty}\sum_{k=0}^{\infty}[a_{nlk}e^{-\lambda_{nk}t}v_{nlk}^{c} + b_{nlk}e^{-\lambda_{nk}t}v_{nlk}^{s}],$$

where v_{nlk}^{c} and v_{nlk}^{s} are defined in (8.3.15) and

$$a_{nlk} = \frac{\int_{-\pi}^{\pi}\int_{0}^{\pi}\int_{0}^{a} f v_{nlk}^{c} r^2 \sin\phi dr d\phi d\theta}{\int_{-\pi}^{\pi}\int_{0}^{\pi}\int_{0}^{a}(v_{nlk}^{c})^2 r^2 \sin\phi dr d\phi d\theta},$$

$$b_{nlk} = \frac{\int_{-\pi}^{\pi}\int_{0}^{\pi}\int_{0}^{a} f v_{nlk}^{s} r^2 \sin\phi dr d\phi d\theta}{\int_{-\pi}^{\pi}\int_{0}^{\pi}\int_{0}^{a}(v_{nlk}^{s})^2 r^2 \sin\phi dr d\phi d\theta}.$$

7. Hint: Extend f as an odd function across the xy-plane.

Chapter 9

Section 9.1

1. Set $y = -x$ in $\int_{-\infty}^{\infty} f(x)\delta(-x)dx$ and change the variable. Then,

$$\begin{aligned}\int_{-\infty}^{\infty} f(x)\delta(-x)dx &= \int_{\infty}^{-\infty} f(-y)\delta(y)(-dy)\\ &= \int_{-\infty}^{\infty} f(-y)\delta(y)dy = f(-0) = f(0).\end{aligned}$$

Section 9.2

1. (a) $\int_{-\infty}^{\infty} f(x)e^{-i\omega x}dx = \int_{-a}^{a} e^{-i\omega x}dx = \frac{e^{-i\omega a} - e^{i\omega a}}{-i\omega} = \frac{\sin a\omega}{\omega}$.

2. $\frac{1}{2\pi}\int_{-\infty}^{\infty}\frac{\sin a\omega}{\omega}e^{i\omega x}d\omega = \frac{1}{2\pi}\int_{-\infty}^{\infty}\frac{e^{i\omega(x+a)} - e^{i\omega(x-a)}}{i\omega}d\omega$

$= H(x+a) - H(x-a) = \begin{cases} 0 & |x| > a \\ 1 & |x| < a \end{cases}.$

3. (a) We change the variable to $y = -x$ and use $f(-x) = -f(x)$. Then,

$$\begin{aligned}F(-\omega) &= \int_{\infty}^{-\infty} f(-y)e^{-i\omega y}(-dy) = \int_{-\infty}^{\infty} f(-y)e^{-i\omega y}dy\\ &= -\int_{-\infty}^{\infty} f(y)e^{-i\omega y}dy = -F(\omega).\end{aligned}$$

Section 9.3

2. $\int_{-\infty}^{\infty}|f(x)|^2dx=\int_{-\infty}^{\infty}e^{-2ax^2}dx=\sqrt{\frac{\pi}{2a}}$,
$F(\omega)=\int_{-\infty}^{\infty}e^{-ax^2}e^{-i\omega x}dx=\int_{-\infty}^{\infty}e^{-a(x+\frac{\omega}{2a})^2-\frac{\omega^2}{4a}}dx=\sqrt{\frac{\pi}{a}}e^{-\frac{\omega^2}{4a}}$,
$\frac{1}{2\pi}\int_{-\infty}^{\infty}|F(\omega)|^2d\omega=\frac{1}{2\pi}\frac{\pi}{a}\int_{-\infty}^{\infty}e^{-\frac{\omega^2}{2a}}d\omega=\frac{1}{2\pi}\frac{\pi}{a}\sqrt{2a\pi}=\sqrt{\frac{\pi}{2a}}$.
3. (b) $\mathcal{F}^{-1}(F)=\frac{1}{\pi}\int_{-\infty}^{\infty}g(y)\frac{a}{a^2+(x-y)^2}dy$.

Section 9.4

2. Multiply the equation by $e^{\gamma t}$ and then to set $v=e^{\gamma t}u$. Then, v satisfies

$$v_t=kv_{xx},\quad v(x,0)=f(x).$$

Therefore,

$$v=e^{\gamma t}u=\int_{-\infty}^{\infty}\frac{1}{\sqrt{4\pi kt}}f(y)e^{-\frac{(y-x)^2}{4kt}}dy,$$

$$u=e^{-\gamma t}\int_{-\infty}^{\infty}\frac{1}{\sqrt{4\pi kt}}f(y)e^{-\frac{(y-x)^2}{4kt}}dy.$$

3. (a) The Fourier transform is

$$\bar{U}(\omega,t)=\bar{U}(\omega,0)e^{-k\omega^2t}+e^{-k\omega^2t}\int_0^t e^{k\omega^2s}\bar{h}(\omega,s)ds,$$

where

$$\bar{U}=\int_{-\infty}^{\infty}u(x,t)e^{-i\omega x}dx,\quad \bar{h}=\int_{-\infty}^{\infty}h(x,t)e^{-i\omega x}dx.$$

(b) The inverse transform is

$$\begin{aligned}u(x,t)&=\frac{1}{2\pi}\int_{-\infty}^{\infty}\int_{-\infty}^{\infty}f(\bar{x})e^{-k\omega^2t+i\omega(x-\bar{x})}d\omega d\bar{x}\\&\quad+\frac{1}{2\pi}\int_0^t\int_{-\infty}^{\infty}\int_{-\infty}^{\infty}h(\bar{x},s)e^{-k\omega^2(t-s)+i\omega(x-\bar{x})}d\omega d\bar{x}ds\\&=\int_{-\infty}^{\infty}f(\bar{x})\frac{1}{\sqrt{4\pi kt}}e^{-\frac{(x-\bar{x})^2}{4kt}}d\bar{x}\\&\quad+\int_0^t\int_{-\infty}^{\infty}h(\bar{x},s)\frac{1}{\sqrt{4\pi k(t-s)}}e^{-\frac{(x-\bar{x})^2}{4k(t-s)}}d\bar{x}ds.\end{aligned}$$

4. The Fourier transform is

$$\bar{U}=\int_{-\infty}^{\infty}f(\mathbf{y})e^{-i\boldsymbol{\omega}\cdot\mathbf{y}}d\mathbf{y}e^{-i(\mathbf{v}_0\cdot\boldsymbol{\omega})t-k|\boldsymbol{\omega}|^2t},$$

where

$$\bar{U} = \int_{-\infty}^{\infty} u(\mathbf{y}, t) e^{-i\omega \cdot \mathbf{y}} d\mathbf{y}.$$

The inverse transform is

$$u(x,t) = (\frac{1}{2\pi})^n \int_{-\infty}^{\infty} \int_{-\infty}^{\infty} f(\mathbf{y}) e^{-i\omega \cdot \mathbf{y}} d\mathbf{y} e^{-i(\mathbf{v}_0 \cdot \omega)t - k|\omega|^2 t} e^{i\omega \cdot \mathbf{x}} d\omega$$
$$= \int_{-\infty}^{\infty} (\frac{1}{4\pi kt})^{n/2} f(\mathbf{y}) e^{-\sum_{i=1}^{n} \frac{(y_i - x_i - v_{i0} t)^2}{4kt}} d\mathbf{y}.$$

6. Use the Fourier series for y and the Fourier transform for x and t. Assume $u = v(x,t)Y(y)$. For Y we have

$$Y_{yy} + \lambda_m Y = 0, \quad Y(0) = 0, \quad Y(H) = 0,$$

$$\lambda_m = (\frac{m\pi}{H})^2, \quad Y_m = \sin\frac{m\pi y}{H}, \quad m = 1, 2, \ldots.$$

For v_m we have

$$(v_m)_t = k(v_m)_{xx} - k\lambda_m v_m, \quad v_m(x,0) = \frac{2}{H} \int_0^H f(x,y,0) \sin\frac{m\pi y}{H} dy.$$

By Exercise 2, we have

$$v_m(x,t) = e^{-\lambda_m t} \int_{-\infty}^{\infty} \frac{1}{\sqrt{4\pi kt}} v_m(\bar{x}, 0) e^{-\frac{(x-\bar{x})^2}{4kt}} d\bar{x}.$$

Then,

$$u(x,y,t) = \sum_{m=1}^{\infty} v_m(x,t) Y_m(y).$$

7. The Fourier transform is

$$U(\omega,t) = \cos c\omega t \int_{-\infty}^{\infty} f(y) e^{-i\omega y} dy + \frac{\sin c\omega t}{c\omega} \int_{-\infty}^{\infty} g(y) e^{-i\omega y} dy$$
$$-\frac{e^{-i\omega x_0}}{c\omega} [\cos c\omega t \int_0^t \delta(s - t_0) \sin c\omega s ds$$
$$- \sin c\omega t \int_0^t \delta(s - t_0) \cos c\omega s ds],$$

where

$$U(\omega,t) = \int_{-\infty}^{\infty} u(x,t) e^{-i\omega x} dx.$$

The inverse transform is

$$u(x,t) = \frac{1}{2\pi}\int_{-\infty}^{\infty} e^{i\omega x}[\cos c\omega t \int_{-\infty}^{\infty} f(y)e^{-i\omega y}dy$$
$$+\frac{\sin c\omega t}{c\omega}\int_{-\infty}^{\infty} g(y)e^{-i\omega y}dy]d\omega$$
$$-\frac{1}{2\pi}\int_{-\infty}^{\infty} e^{i\omega x}\frac{e^{-i\omega x_0}}{c\omega}[\cos c\omega t\int_0^t \delta(s-t_0)\sin c\omega s ds$$
$$-\sin c\omega t\int_0^t \delta(s-t_0)\cos c\omega s ds]d\omega$$
$$= \frac{1}{2}[f(x+ct)+f(x-ct)] + \frac{1}{2c}\int_{x-ct}^{x+ct} g(y)dy$$
$$+\frac{1}{2c}[H(x-x_0+c(t-t_0)) - H(x-x_0-c(t-t_0))]H(t-t_0).$$

Chapter 10

Section 10.1

2. (b) We do the substitution with $u = t - b$. Then,

$$\int_0^{\infty} H(t-b)f(t-b)e^{-st}dt = \int_b^{\infty} f(t-b)e^{-st}dt$$
$$= \int_0^{\infty} f(u)e^{-s(u+b)}du = e^{-bs}F(s).$$

3. (a) $\mathcal{L}^{-1}[\frac{s}{(s^2+1)(s^2+2)}] = \mathcal{L}^{-1}[\frac{s}{(s^2+1)} - \frac{s}{(s^2+2)}] = \cos t - \cos\sqrt{2}t.$

4. (a) $e^{a\sqrt{s}}[\frac{\sqrt{s}}{2}I + \frac{a}{4}J] = \sqrt{\pi}$ (b) $\frac{dI}{da} = -\frac{a}{2}J = \sqrt{s}I - 2\sqrt{\pi}e^{-a\sqrt{s}}$
(c) $\lim_{a\to\infty} I(a) = 0$

Section 10.2

1. (b) $\mathcal{L}[h(t)] = \frac{3!}{s^4(s-1)}$.

2. $\mathcal{L}[\int_0^t (t-\tau)e^{\tau}d\tau] = \frac{1}{s^2(s-1)}$.

3. (b) $\mathcal{L}^{-1}[F(s)] = \int_0^t \frac{1}{2!}(t-\tau)^2\frac{1}{2}\sin(2\tau)d\tau.$

Section 10.3

2. (b) $u(t) = 2e^{-2t} + te^{-2t} + \int_0^t (t-\tau)e^{-2(t-\tau)}h(\tau)d\tau.$
(c) $u(t) = -\frac{1}{2}\int_0^t \sin(t-\tau)h(\tau)d\tau + \int_0^t \sinh(t-\tau)h(\tau)d\tau.$

4. $u(x,t) = [\cos\frac{c\pi t}{L} - \frac{L}{c\pi}\sin\frac{c\pi t}{L}]\sin\frac{\pi x}{L}$.

Chapter 11

Section 11.1

3. $\bar{h}(\mathbf{x}, r) = \frac{1}{4\pi}\int_0^{2\pi}\int_0^{\pi} h(\mathbf{y})\sin\phi d\phi d\theta$, where

$$\mathbf{y} = [r\sin\phi\cos\theta + x_1, r\sin\phi\sin\theta + x_2, r\cos\phi + x_3].$$

5. (b) Performing the change of variable $y_2 = a\sin\theta$, we see

$$\begin{aligned} u(x_1,t) &= \frac{\partial}{\partial t}[\frac{1}{2\pi c}\int_{x_1-ct}^{x_1+ct} f(y_1)\int_{-\frac{\pi}{2}}^{\frac{\pi}{2}} \frac{1}{a\cos\theta}a\cos\theta d\theta dy_1] \\ &+\frac{1}{2\pi c}\int_{x_1-ct}^{x_1+ct} g(y_1)\int_{-\frac{\pi}{2}}^{\frac{\pi}{2}} \frac{1}{a\cos\theta}a\cos\theta d\theta dy_1 \\ &= \frac{1}{2}[f(x_1+ct)+f(x_1-ct)] + \frac{1}{2c}\int_{x_1-ct}^{x_1+ct} g(s)ds. \end{aligned}$$

Section 11.2

1. $u(\mathbf{x},t) = \int_{R^n} f(\mathbf{x}-\mathbf{y})G(\mathbf{y},t)d\mathbf{y} + \int_0^t\int_{R^n} h(\mathbf{x}-\mathbf{y}, t-s)G(\mathbf{y},s)d\mathbf{y}ds$.

3. $u(x,t) = [\sin(x+ct)+\sin(x-ct)]/2 + t^2/2$.

Chapter 12

Section 12.1

1. $G(x;x_0) = \begin{cases} \frac{x_0-L}{L}x, & x < x_0, \\ \frac{x-L}{L}x_0, & x_0 < x. \end{cases}$

2. (a) $G = -\sum_{m=1}^{\infty}\sum_{n=1}^{\infty}\frac{1}{\frac{LH}{4}\{(\frac{m\pi}{H})^2+(\frac{(2n-1)\pi}{2L})^2\}}$
$\sin\frac{m\pi y_0}{H}\cos\frac{(2n-1)\pi x_0}{2L}\sin\frac{m\pi y}{H}\cos\frac{(2n-1)\pi x}{2L}$.

4. Assume $G = \sum_{m=0}^{\infty} a_m(x)\cos\frac{m\pi y}{H}$. Then,

$$a_0(x) = \begin{cases} -\frac{1}{2LH}x^2 + a, & x < x_0, \\ -\frac{1}{2LH}x^2 + \frac{1}{H}(x-x_0) + a, & x > x_0, \end{cases}$$

where $a = -x_0^2/(2LH) + x_0/H$ makes $a_0(x)$ symmetric. For $m \neq 0$,

$$a_m(x) = \begin{cases} -2\dfrac{\cosh\frac{m\pi(L-x_0)}{H}\cosh\frac{m\pi x}{H}}{m\pi\sinh\frac{m\pi L}{H}}\cos\dfrac{m\pi y_0}{H}, & x < x_0, \\ -2\dfrac{\cosh\frac{m\pi(L-x)}{H}\cosh\frac{m\pi x_0}{H}}{m\pi\sinh\frac{m\pi L}{H}}\cos\dfrac{n\pi y_0}{H}, & x > x_0. \end{cases}$$

Section 12.2

1. (a)

$$G(x,t;x_0,t_0) = \sum_{n=1}^{\infty}\frac{2}{L}e^{-(\frac{a(n-\frac{1}{2})\pi x}{L})^2(t_0-t)}\sin\frac{(n-\frac{1}{2})\pi x_0}{L}\sin\frac{(n-\frac{1}{2})\pi x}{L}.$$

3. $G(x,t;x_0,t_0) = \frac{H(t_0-t)}{2a\sqrt{\pi(t_0-t)}}e^{-\frac{(x-x_0)^2}{4a^2(t_0-t)}}$

Section 12.3

1. (a)

$$G(x,t;x_0,t_0) = \sum_{n=1}^{\infty}\frac{2}{c(n-\frac{1}{2})\pi}H(t_0-t)$$

$$\sin\frac{c(n-\frac{1}{2})\pi(t_0-t)}{L}\sin\frac{(n-\frac{1}{2})\pi x_0}{L}\sin\frac{(n-\frac{1}{2})\pi x}{L}.$$

4. Use Example 11.2.3 with $h(y,s) = \delta(y-x_0)\delta(T-s-t_0)$.

$$K(x,\tau;x_0,t_0) = \frac{1}{2c}\int_0^{\tau}\int_{x-c(\tau-s)}^{x+c(\tau-s)}\delta(y-x_0)\delta(T-s-t_0)dyds,$$

$$G(x,t;x_0,t_0) = \frac{1}{2c}[H(x-x_0+c(t_0-t)) - H(x-x_0-c(t_0-t))].$$

Bibliography

Abramowitz, M. and Stegun, I.A. (1974). Handbook of Mathematical Functions. (Dover).

Berg, P.W. and McGregor, J.L. (1966). Elementary Partial Differential Equations. (Holden-Day).

Boyce, W.E. and DiPrima, R.C. (2009). Elementary Differential Equations and Boundary Value Problems, 10th ed. (Wiley).

Bray, W.O. (2011). A Journey into Partial Differential Equations. (Jones & Bartlett Learning).

Churchill, R.V. (1972). Operational Mathematics, 3rd ed. (McGraw-Hill).

Evans, L.C. (1998). Partial Differential Equations. (Amer. Math. Soc.).

Folland, G.B. (1992). Fourier Analysis and its Applications. (Brooks/Cole).

Haberman, R. (2012). Applied Partial Differential Equations with Fourier Series and Boundary Value Problems, 5th ed. (Pearson Prentice Hall).

Howell, K.B. (2001). Principles of Fourier Analysis. (Chapman & Hall/CRC).

John, F. (1995). Partial Differential Equations, 4th ed. (Springer-Verlag).

Rudin, W. (1976). Principles of Mathematical Analysis, 3rd ed. (McGraw-Hill).

Shreve, S.E. (2004). Stochastic Calculus for Finance Vol. II. (Springer-Verlag).

Smoller, J. (1983). Shock Waves and Reaction-Diffusion Equations. (Springer-Verlag).

Stakgold, I. (1997). Green's Functions and Boundary Value Problems, 2nd ed. (Wiley).

Strang, G. (2009). Introduction to Linear Algebra, 4th ed. (Wellesley-Cambridge Press).

Strauss, W.A. (2008). Partial Differential Equations. (Wiley).

Vretblad, A. (2003). Fourier Analysis and its Applications. (Springer-Verlag).

Watson, G.N. (1995). Treatise of the Theory of Bessel Functions, 2nd ed. (Cambridge University Press).

Whitham, G.B. (1974). Linear and Nonlinear Waves. (Wiley-Interscience).

Zachmanoglou, E.C. and Thope, D. (1989). Introduction to Partial Differential Equations with Applications. (Dover).

Zauderer, E. (1989). Partial Differential Equations of Applied Mathematics, 2nd ed. (Wiley).

Comments on Bibliography:

1. [Bray (2011); Haberman (2012); Strauss (2008); Zachmanoglou and Thope (1989)] are undergraduate level PDE textbooks which complement this book.
2. [Evans (1998); John (1995); Zauderer (1989)] are graduate level PDE textbooks.
3. [Boyce and DiPrima (2009)] has a comprehensive treatment of elementary ODE's.
4. [Abramowitz and Stegun (1974); Churchill (1972); Watson (1995)] are reference books on Bessel's and Legendre functions.
5. [Folland (1992); Howell (2001); Vretblad (2003)] discuss the Fourier analysis.
6. [Smoller (1983); Whitham (1974)] discuss the conservation laws.
7. [Strang (2009)] is an excellent undergraduate level textbook on linear algebra.
8. [Shreve (2004)] has applications of PDE to finance.
9. [Stakgold (1997); Zauderer (1989)] have detailed treatment of Green's functions.

Index